Surface and Interface Characterization by Electron Optical Methods

NATO ASI Series

Advanced Science Institutes Series

A series presenting the results of activities sponsored by the NATO Science Committee, which aims at the dissemination of advanced scientific and technological knowledge, with a view to strengthening links between scientific communities.

The series is published by an international board of publishers in conjunction with the NATO Scientific Affairs Division

A	**Life Sciences**	Plenum Publishing Corporation
B	**Physics**	New York and London
C	**Mathematical and Physical Sciences**	Kluwer Academic Publishers Dordrecht, Boston, and London
D	**Behavioral and Social Sciences**	
E	**Applied Sciences**	
F	**Computer and Systems Sciences**	Springer-Verlag
G	**Ecological Sciences**	Berlin, Heidelberg, New York, London,
H	**Cell Biology**	Paris, and Tokyo

Recent Volumes in this Series

Volume 184—Narrow-Band Phenomena—Influence of Electrons with Both Band and Localized Character
edited by J. C. Fuggle, G. A. Sawatzky, and J. W. Allen

Volume 185—Nonperturbative Quantum Field Theory
edited by G. 't Hooft, A. Jaffe, G. Mack, P. K. Mitter, and R. Stora

Volume 186—Simple Molecular Systems at Very High Density
edited by A. Polian, P. Loubeyre, and N. Boccara

Volume 187—X-Ray Spectroscopy in Atomic and Solid State Physics
edited by J. Gomes Ferreira and M. Teresa Ramos

Volume 188—Reflection High-Energy Electron Diffraction and Reflection Electron Imaging of Surfaces
edited by P. K. Larsen and P. J. Dobson

Volume 189—Band Structure Engineering in Semiconductor Microstructures
edited by R. A. Abram and M. Jaros

Volume 190—Squeezed and Nonclassical Light
edited by P. Tombesi and E. R. Pike

Volume 191—Surface and Interface Characterization by Electron Optical Methods
edited by A. Howie and U. Valdrè

Series B: Physics

Surface and Interface Characterization by Electron Optical Methods

Edited by

A. Howie

University of Cambridge
Cambridge, United Kingdom

and

U. Valdrè

University of Bologna
Bologna, Italy

Plenum Press
New York and London
Published in cooperation with NATO Scientific Affairs Division

Proceedings of a NATO Advanced Study Institute
on the Study of Surfaces and Interfaces
by Electron Optical Techniques
held April 4–15, 1987,
in Erice, Sicily, Italy

Library of Congress Cataloging in Publication Data

NATO Advanced Study Institute on the Study of Surfaces and Interfaces by Electron Optical Techniques (1987: Erice, Italy)
    Surface and interface characterization by electron optical methods / edited
by A. Howie and U. Valdrè.
        p.   cm.—(NATO ASI series. Series B, Physics; v. 191)
    "Proceedings of a NATO Advanced Study Institute on the Study of Surfaces
and Interfaces by Electron Optical Techniques, held April 4–15, 1987, in Erice,
Sicily, Italy"—CIP t.p. verso.
    Bibliography: p.
    Includes index.
    ISBN 0-306-43086-X
    1. Surfaces (Physics)—Technique—Congresses. 2. Electron microscopy—
Congresses. 3. Electron microscope, Transmission—Congresses. I. Howie, A. II.
Valdrè, U. (Ugo). III. Title. IV. Series.
QC173.4.S94N385   1987   88-28989
530.4'1—dc19   CIP

Printed in the United States of America

STM image of the 7 × 7 reconstruction at the Si (111) surface. The distance between the main 6-fold symmetry centers is 2.6 nm. Color enhancement of the intensity scale reveals fine details of atomic heights and bonding states at the surface. A surface step is visible as a discontinuity running vertically near the center of the figure. Atoms are blue. Colors from blue to red indicate details further away from the surface. (Courtesy Drs. J. E. Demuth, R. J. Hamers, R. M. Tromp, and M. E. Welland. Fig. 3 from chapter by Garcia.)

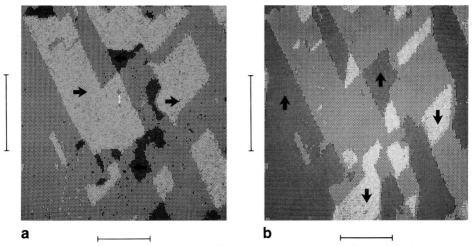

a   b

Magnetic domain images from an Fe (100) surface, measured simultaneously for the spin polarization vector component along the horizontal axis (a) and the vertical axis (b). Areas of the same color have the same magnetization (see arrows). Within the gray areas in (a) and (b) the horizontal (a) and vertical (b) polarization components, respectively, are zero, while the vertical and horizontal components are maximum. The scale markers each correspond to 60 μm (Fig. 2 from chapter by Kirschner.)

PREFACE

The importance of real space imaging and spatially-resolved spectroscopy in many of the most significant problems of surface and interface behaviour is almost self evident. To join the expertise of the traditional surface scientist with that of the electron microscopist has however been a slow and difficult process. In the past few years remarkable progress has been achieved, including the development of new techniques of scanning transmission and reflection imaging as well as low energy microscopy, all carried out in greatly improved vacuum conditions. Most astonishing of all has been the advent of the scanning tunneling electron microscope providing atomic resolution in a manner readily compatible with most surface science diagnostic procedures. The problem of beam damage, though often serious, is increasingly well understood so that we can assess the reliability and usefulness of the results which can now be obtained in catalysis studies and a wide range of surface science applications.

These new developments and many others in more established surface techniques are all described in this book, based on lectures given at a NATO Advanced Study Institute held in Erice, Sicily, at Easter 1987. It is regretted that a few lectures on low energy electron diffraction and channeling effects could not be included.

Fifteen lecturers from seven different Countries and 67 students from 23 Countries and a wide variety of backgrounds attended the school. Substantial problem-solving sessions and computer demonstrations were arranged to back up the lectures and a number of these problems have been reproduced in the book.

We wish to thank all the lecturers for their effort and patience in producing the material for the book. For permission to include various figures and quotations (specifically identified where appropriate in the text) we are grateful to the American Institute of Physics, the American Physical Society, the International Business Machines Corporation and the Nobel Foundation.

The two enthralling lectures given enthusiastically by Dr Wolfgang Telieps were a high point of the School. It was thus with great shock and sorrow that we learnt of his death so soon afterwards. We are grateful to his colleague, Professor Bauer, for providing the obituary notice included here and for completing the write up of the relevant lecture notes.

The success of the School was greatly assisted by a contribution from NATO and from the facilities of the Ettore Majorana Centre for Scientific Culture. For secretarial support we are grateful to Mrs Irene Salerno, Mrs Mary Waterworth and Miss Vanna Valdrè.

A. Howie and U. Valdrè

June 1988

v

OBITUARY

Wolfgang Telieps, who gave one of the most interesting lectures in this school, died in a tragic car accident on May 31, 1987 at the age of 36. He had studied physics at the Technical University Clausthal, where he received a M.S. degree in physics in 1977 and his Ph.D. in 1983. For his M.S. thesis work he studied the imaging column of the low energy electron microscope; in his Ph.D. project he took over the responsibility for the complete instrument. In spite of strong discouragement by referees, he succeeded in demonstrating the viability of LEEM. This achievement and the work which he did in the following years in the course of his preparation for the "habilitation" rapidly brought him international recognition.

Wolfgang Telieps was not only an outstanding physicist but also a great human being who was liked by all who knew him for his pleasant personality, his objectiveness, his tolerance, his willingness to help and many other positive characteristics which made him an ideal collaborator. His untimely death has thus robbed, not only the scientific community of a very talented young member, but also his colleagues of an unforgettable friend.

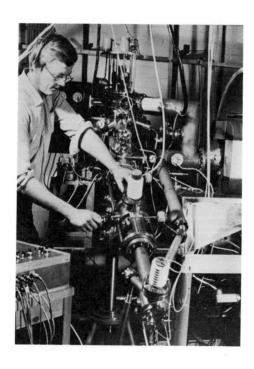

CONTENTS

Principles and techniques of transmission imaging of surfaces   1
    A. Howie

Catalyst studies by scanning transmission electron microscopy 11
    A. Howie

Localised surface imaging and spectroscopy in the scanning
    transmission electron microscope 19
    A. Howie

Fundamentals of high resolution transmission electron microscopy. . . . 31
    D.J. Smith

Profile imaging of small particles, extended surfaces and dynamic
    surface phenomena. 43
    D.J. Smith

Transmission electron microscopy and diffraction from semiconductor
    interfaces and surfaces. 55
    J.M. Gibson

The transmission electron microscopy of interfaces and multilayers. . . 77
    W.M. Stobbs

Surface microanalysis and microscopy by X-ray photoelectron
    spectroscopy (XPS), core-loss spectroscopy (CLS) and
    Auger electron spectroscopy (AES). 89
    J. Cazaux

Reflection electron microscopy. 127
    J.M. Cowley

An introduction to reflection high energy electron diffraction. 159
    P.J. Dobson

Intensity oscillations in reflection high energy electron
    diffraction during epitaxial growth. 185
    P.J. Dobson, B.A. Joyce, J.H. Neave, J. Zhang

Emission and low energy reflection electron microscopy. 195
    E. Bauer and W. Telieps

Scanning tunneling microscopy and spectroscopy. 235
    N. Garcia

Spin-polarized secondary electrons from ferromagnets. 267
    J. Kirschner

Electronically stimulated desorption: mechanisms, applications
    and implications 285
  D. Menzel

Structure and catalytic activity of surfaces. 301
  V. Ponec

Subject Index 315

PRINCIPLES AND TECHNIQUES OF TRANSMISSION IMAGING OF SURFACES

A. Howie

Cavendish Laboratory
Madingley Road
Cambridge CB3 OHE, U.K.

INTRODUCTION

Although many surface characterisation techniques are essentially broad beam methods yielding diffraction or other information averaged over large areas of surface, crucial properties of the surface may depend on steps or other inhomogeneities. There is indeed an instability, shown schematically in figure 1, whereby even a uniform single crystal overgrowth such as an expitaxial layer or adsorbed oxide tends to become inhomogeneous, forming island structures or localised defects. The overgrowth has an energy E proportional to thickness t at small t where it matches its lattice parameter with the substrate crystal at the cost of elastic strain energy (point A). At larger values of t, when elastic strain energy becomes prohibitive, misfit dislocations are introduced to relieve it. Eventually where the overgrowth has completely relaxed to its own lattice parameter, E becomes the characteristic interfacial energy of the boundary, independent of t (point C). Since the E versus t curve is concave downwards, a homogeneous film represented by point B can reduce its energy to the point D by breaking up into regions A and C.

The study of these important and almost inevitable inhomogeneous features of extended surfaces, as well as of more obviously heterogeneous surface structures like catalyst particles, clearly requires a high resolution, real space imaging technique. For such a purpose, electron

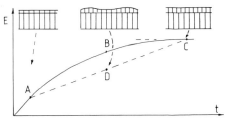

Fig. 1. Energy per unit area of an epitaxial overgrowth as a function of thickness t, showing instability at B.

1

microscopy has many natural advantages which have enabled it to produce some very striking results, despite the fact that it still has some severe disadvantages associated with rather poor vacuum conditions, electron beam damage and often relatively low sensitivity to surface structure. With the recent progress in instrumentation and technique, described in many places in this volume, it seems that many of these difficulties can be overcome and that a revolution in our knowledge of surface structure and properties may be at hand.   Here we discuss the application to surface problems of conventional transmission electron microscopy at near normal incidence.

DIFFRACTION CONTRAST IMAGING

Image contrast in transmission electron microscopy[1] is mainly generated by elastic scattering for which the dominant process, in the small-angle region accessible to conventional axial microscopy, is Bragg reflection or coherent addition of contributions from many atoms to the same scattering event.   In a crystalline material this is demonstrated most directly by the diffraction pattern of Bragg spots, corresponding for such high energy electrons to a nearly planar section of the reciprocal lattice, as shown in figure 2.   Different regions of the diffraction pattern can be selected with an aperture to form images in the microscope such as bright field (i.e. involving the centre spot) and various dark field images (using other spots in the pattern).   High resolution structure images are obtained when several Bragg spots are included in the aperture and interfere together to produce fringes with the spacing of the Bragg planes.

Some of these transmission diffraction contrast techniques are illustrated in figure 3 which shows an oriented oxide overgrowth on a (111) Ni crystal[2].   The oxide was grown under controlled conditions on a clean surface in a UHV chamber.   The specimen was then conveyed to the electron microscope using a UHV transfer-device.   Moiré fringes (a kind of vernier effect corresponding to the difference in lattice parameter between the nickel and the oxide) are visible in figure 3(a) and can be seen to arise from the interference of a 220 diffracted beam from the Ni with a 220 beam from the oxide, corresponding (see problem 1) to the two spots included in the aperture.   Figure 3(b) is formed using one of the inner ring of 200 oxide spots and reveals that, in addition to the continuous oxide overgrowth oriented parallel to the nickel, there are islands of oxide present with a systematic but quite different orientation relation.   The formation of three-dimensional oxide islands under a continuous oxide overgrowth or adsorbed layer may be an example of the surface instability

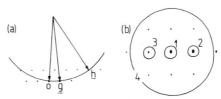

Fig. 2. Ewald sphere construction (a) for selecting reciprocal lattice vectors g associated with Bragg waves $\chi + g$ with the same energy as the incident wave χ. At high energies, the important g vectors (for small-angle Bragg reflections) lie in the zero order Laue zone but weak higher Laue zone reflections can be excited at larger angles. By selecting different parts of the diffraction pattern (b) a variety of images can be formed: (1) bright field, (2) dark field, (3) weak beam, (4) structure image.

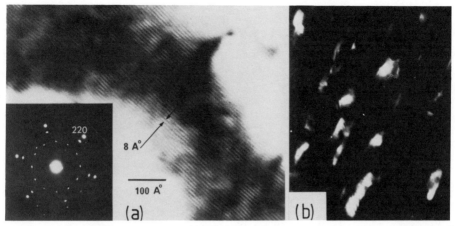

Fig. 3. Oxide overgrowth on a (111) Ni single crystal. The oxide consists of a misfitting epitaxial layer (a) with islands of differently oriented oxide (b).

described in the introduction. Very detailed EM studies have been made of the oxide island nucleation growth process after the formation of adsorbed oxygen layers on clean copper surfaces[3]. There is conclusive, although indirect, evidence that surface steps act as nucleation sites.

Classical diffraction contrast techniques[1] can image the strain fields in the vicinity of dislocations and other crystal defects. The basic mechanism is the local tilting of the Bragg planes or the shift in phase of the diffracted wave as a result of the displacement $R(r)$ of the crystal at position $r = r(x,y,z)$. These effects are expressed by the simple kinematical theory expression[1] for the diffracted amplitude

$$\phi_g(x,y,t) = (\pi i/\xi_g) . \int_0^t \exp[-2\pi i s_g z - 2\pi i \underline{g}.R(\underline{r})]\ dz \qquad (1)$$

The reciprocal lattice vector $\underline{g}$ and Bragg deviation parameter s_g are shown in figure 2; ξ_g is the extinction distance given by the expression

$$\chi/\xi_g = U_g = \frac{\pi}{\Omega}\ \frac{m}{m_0}\ \Sigma_j\ f_j{}^B(\underline{g})\ \exp(-2\pi i\underline{g}.\underline{r}_j)\ \exp(-M_g{}^j) \qquad (2)$$

Here $\chi = \lambda^{-1}$ is the electron wave vector, m the relativistic electron mass, m_0 the rest mass, Ω the volume of the unit cell, $M_g{}^j$ the Debye-Waller term and $f_j{}^B$ the Born electron scattering amplitude of the jth atom in the unit cell. The single scattering assumption of the kinematical theory is not really valid for most situations in electron microscopy, where the multiple scattering dynamical theory outlined below is needed. It does, however, give a useful quantitive account of many important effects in the transmission imaging of defects.

Dislocations can be generated by a number of surface reactions and are easily studied by diffraction contrast techniques. Figure 4(a) shows misfit dislocation trigons, nucleated by a climb process from the surface[3] and lying at the interface between a gold (111) film and a Pd overgrowth. Much sharper, though noiser, images can be obtained [figure 4(b)] by weak beam dark field imaging using a diffracted beam far from the Bragg position. It can be seen from equation (1) using Fourier transform ideas, that the large value of s_g in this situation picks out for contribution to ϕ_g those regions of the crystal very close to the centre of the defect where the displacement varies rapidly with depth z in the specimen.

3

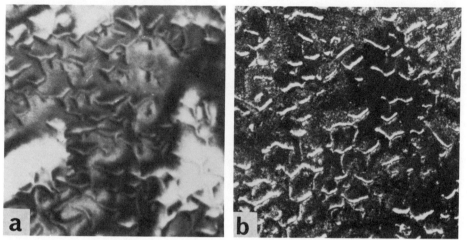

Fig. 4. Misfit dislocation trigons lying at the interface between an epitaxial Pd overgrowth and a (111) Au film[3]. The trigons are imaged in bright field (a) and, at higher resolution (b), in a 220 weak beam dark field image. The picture width is 200 nm.

IMAGING WITH FORBIDDEN REFLECTIONS

A technique first employed by Cherns[4] for imaging surface structures such as steps, even when no strain effects are present, can be regarded as an extreme example of the weak beam method. For the case of a perfect crystal with $\underline{R}$ = 0, equation (1) can be integrated directly to give the result

$$|\phi_g(x,y,t)|^2 = \sin^2(\pi t s_g)/(s_g \xi_g)^2 \qquad (3)$$

The image intensity thus shows thickness fringes, varying sinusoidally with t. For sufficiently large values of s_g, these can be sufficiently closely spaced to show detectable intensity changes across individual crystal steps of atomic height. Cherns[4] employed the so-called forbidden 1/3 (422) reflection from a (111) Au crystal to reveal these steps (see figure 5).

Reference to the reciprocal lattice of Au and figure 1 shows however that this reflection may be regarded as a (200) reflection in a higher order Laue zone of the diffraction pattern with s_g = 1/a where a is the cubic lattice parameter. Since the thin crystal can be formed by the stacking of (111) close packed layers at a spacing of a/3, t = na/3 and $t s_g$ = n/3. The intensity in the forbidden reflection is therefore zero when n is a multiple of 3 (and the crystal contains an integral number of unit cells) but becomes non-zero when a single close-packed layer is added or removed. This result suggests a simple way (see problem 2) of looking at the effect in terms of the contribution which a single hexagonal layer makes to the diffraction rod passing through the 1/3(422) position. When 3 layers are stacked up to make a unit cell of the f.c.c. structure, the diffracted amplitude in this position vanishes because the structure factor is zero. It is less clear on the basis of this simple argument however, that the intensity will only vanish for thicknesses corresponding to an integral number of unit cells if s_g has the value given above, i.e. if the Ewald sphere (figure 1) passes exactly through the 1/3(422) position.

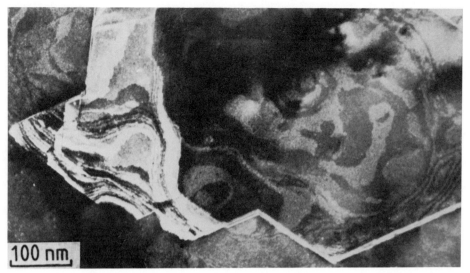

Fig. 5. A dark field image of a thin (111) Au thin film formed from the
forbidden 1/3(422) reflection. Bright incoherent twin boundaries
can be seen separating different regions within which the different
intensity levels indicate thickness changes of one atomic layer.

For reasons that are only partially clear, there have been rather few
subsequent applications of Cherns' imaging method. Apart from some further
observations in Au and Pt films[5,6], Iijima[7] and Ourmazd et al[8] were able to
resolve steps on (111) crystals of Si. Tu and Howie[9] showed that in cases
where the surface is parallel to a mirror plane [e.g. (100)] cancellation
at the expected forbidden reflection position can occur due to simultaneous
contributions from reciprocal lattice points at equal distances above and
below the zero order layer. The simple kinematical picture can also be
complicated by a variety of multiple scattering effects (see below) as well
as by overlapping interference from steps and surface reconstructions on
both surfaces. Since the forbidden reflection intensities are rather low
(typically 2×10^{-4} relative to the incident intensity) it is often assumed
that a small amount of surface contamination would be able to mask the
terrace contrast but this point has yet to be investigated quantitatively.

HIGH RESOLUTION IMAGING

High resolution imaging, using forbidden spots, has not so far been
very successful. Useful structure images have however been obtained from
thin oxide overgrowths on Ag[10] and Ba[11] crystals from which the crystal
structure and epitaxy as well as some details of the morphology of the
overgrowth can be deduced.

Surface reconstructions, usually giving structures with larger unit
cells than the bulk lattice, give extra diffraction spots which are readily
visible in a sufficiently thin crystal. Such reconstructions have also
been successfully imaged for the (100) Au surface[12] and compared with LEED
data. In the case of the 7 x 7 reconstruction on (111) Si, the images
obtained so far have not been of high quality, but analysis of the extra
diffraction spot intensities on a kinematical basis led[13] (see chapter by
Gibson) to the first solution for the currently accepted structure.

5

MULTIPLE SCATTERING (DYNAMICAL DIFFRACTION) EFFECTS

Multiple scattering effects are nearly always very significant in electron microscopy and diffraction so that the simple kinematical theory has only a very qualitative value. Furthermore, multiple Bragg reflection gives rise to a number of important new channelling effects in the propagation of electrons through crystals, whereby the probabilities of transmission, X-ray production, backscattering, etc. can all depend strongly on the Bragg conditions and hence of the direction of the incident beam in the crystal. These backscattering effects have been used for imaging of near surface defects in bulk crystals[14]. Here we simply outline very briefly the various ways in which the dynamical theory of electron diffraction can be developed to deal with channeling phenomena as well as with the contrast of transmission images.

In a perfect crystal the electron wave function satisfies the equation

$$\nabla^2 \psi + \chi^2 \psi + \sum_g U_g \exp(2\pi i \underline{g} \cdot \underline{r}) \; \psi = 0 \tag{4}$$

where $\chi = \lambda^{-1}$. The Fourier components of the potential U_g are related to the contents of the unit cell by equation (2) but are often given an additional imaginary part[1] to represent the effects of absorption due to inelastic scattering out of the imaging aperture. This equation can be solved in various ways.

One method[1] which is relatively efficient for thick crystals with small unit cells and is also useful for discussions of inelastic scattering effects, is similar to the almost free electron method in energy band theory. The wave function is expanded as a sum of Bloch waves $b(\underline{k}, \underline{r})$

$$\psi(\underline{r}) = \sum_j \psi^j \; b^j(\underline{k}^j, \underline{r}) \tag{5}$$

$$b^j(\underline{k}^j, \underline{r}) = \sum_g C_g^{\;j}(\underline{k}^j) \exp[2\pi i(\underline{k}^j + \underline{g}) \cdot \underline{r}] \tag{6}$$

Each Bloch wave consists of a number N of plane wave components, linked by the Bragg reflection process and having amplitudes $C_g^{\;j}$ which, because of equations (4) and (6), satisfy the equations

$$[\chi^2 + U_o -(\underline{k}^j + \underline{g})^2] \; C_g^{\;j} + \sum_{h \neq o} U_h \; C^j_{g-h} = 0 \tag{7}$$

Usually two components k_x, k_y of the wave vectors are fixed by the need to match the wave vector $\underline{\chi}$ of the wave incident on the entrance surface of the crystal $z = 0$. The remaining component k_z has N possible (positive) values $k_z^{\;1}$, $k_z^{\;2}, \ldots, k_z^{\;N}$ appearing like eigenvalues of eqns (7) and each corresponding to a Bloch wave eigenvector $C_o^{\;j}$, $C_g^{\;j}, \ldots, C_h^{\;j}$. In cases where the wave function can be well described for moderate values of $N \leq 100$, the problem can be conveniently solved numerically by matrix diagonalisation methods.

The results can be depicted in a useful way by means of dispersion surfaces in k space (figure 6) each of which shows the points $\underline{k}^j$

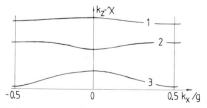

Fig. 6. High energy electron dispersion surfaces.

corresponding to a given energy E (which determines the magnitude of χ). The waves which are actually excited with $k_x = \chi_x$, $k_y = \chi_y$ are then readily found on each branch of the surface. The excitation amplitudes ψ^j of the Bloch waves also follow from the entrance surface boundary condition and in simple cases of transmission at near normal incidence is $\psi^j = C_o{}^j$.

In another approach to dynamical theory, the solution to equation (4) is sought in the form

$$\psi(\underline{r}) = \sum \phi_g(\underline{r}) \exp [2\pi i(\underline{\chi} + \underline{g} + \underline{s}_g)\cdot\underline{r}] \tag{8}$$

where $\underline{s}_g$ is parallel to z and is $\underline{g}$ defined in figure 1. Here the wave amplitudes $\phi_g(\underline{r})$ are assumed to be slowly varying and obey the coupled equations

$$(\underline{\chi} + \underline{g} + \underline{s}_g)\cdot\nabla\phi_g(\underline{r}) = \pi i \sum_h U_{g-h} \phi_h \exp[2\pi i(s_h - s_g)z] \tag{9}$$

In the case of a perfect crystal, when the wave amplitudes are purely functions of distance z from the entrance surface, these equations are ordinary first-order differential equations and can be integrated by a variety of simple numerical procedures.

The slice method[15] widely used for the simulation of high resolution images, extends this approach to imperfect crystals by treating them as periodic structures with a very large unit cell. The number of diffracted beams considered can then easily reach 1000.

In an alternative approach, used to treat diffraction contrast effects in imperfect crystals with relatively small unit cells[16], the potential coefficients U_g are regarded as slowly varying [e.g., $U_g \rightarrow U_g\exp(-2\pi i\underline{g}.\underline{R})$ for strain contrast cases]. It is often then still possible to ignore the dependence of ϕ_g on x and y (the column approximation[1,16]). The kinematical expression in equation (1) arises by direct integration of equations (9) in a very thin crystal when $\phi_0 \doteq 1$ and all the other wave amplitudes on the right hand side are negligibly small.

MORE COMPLEX EFFECTS IN TRANSMISSION IMAGING OF SURFACE STRUCTURES

The simple kinematical picture of the imaging of overgrowths, surface reconstructions and surface steps can be complicated (e.g. Tu and Howie[9]) by dynamical diffraction effects within the bulk and surface regions. In thin crystals, anomalies in forbidden reflection intensities due to atomic step changes in thickness were however first noted in many-beam slice calculations[17]. Individual surface steps can be detected even in bright field imaging[18] if the diffraction conditions are suitably adjusted. Steps on the entrance surface (see figure 7) may be more visible than those on the exit surface since the abrupt (though very small) phase shift that occurs on either side of the step is equivalent (via a Fourier transform) to an increased angular spread in the waves passing into the crystal below. If this crystal is oriented close to a Bragg reflection position, further phase shifts varying rapidly with θ are introduced which can lead to enhanced contrast[19]. Iijima[7] found some indications of these effects at steps in Si.

The phase shift effect across a step is a prominent source of contrast in reflection imaging (REM) and diffraction (RHEED), as shown in chapters by Cowley and by Dobson and as well as in LEED[20] because in these cases the scattering vector employed, $\underline{g} + \underline{s}_g$, has a large component normal to the surface. In principle such an effect could also be achieved in trans-

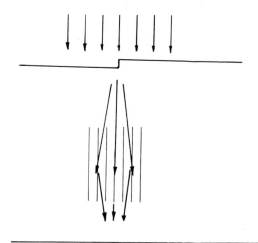

Fig. 7. Angular spread of waves generated by a step at the entrance surface of the crystal. Image contrast can then be created by angle-dependent Bragg reflection processes in the crystal.

mission microscopy by out-of-focus dark field imaging with the rather high-order spots where the Ewald sphere intersects a higher-order layer in the reciprocal lattice (see figure 2) but this experiment has not yet been attempted.

PROBLEMS

1. A (100) crystal of a face-centred-cubic metal of lattice parameter a_0 has a thin overgrowth of the same structure and orientation but with a slightly different lattice parameter a_1. Sketch the diffraction pattern. A dark field image is formed by using two of the lowest order diffraction spots (one from each material). Show that the interference fringes produced can be simply related to the places where the two lattices are in and out of registry. How might you detect in the image the formation of misfit dislocations in the out-of-registry positions because of interaction between the two materials? Note that the presence of misfit dislocations (see fig. 1) means that the registration between the two crystals, though still a periodic function of position, will not vary in a purely sinusoidal manner but will have Fourier components of higher order.

2. Show how forbidden reflections arise in the case of an (111) Au crystal film by adding the contributions of individual layers. You should find that the zero-order layer of the pattern has a hexagonal form but on taking account of the structure of the unit cell (containing 3 layers A,B,C) the innermost ring of spots (as stated in the text) vanishes if the film contains $3n$ layers. What forbidden reflections would you expect to arise in the case of a (100) film of (a) Au and (b) of Si?

REFERENCES

1. P.B. Hirsch, A. Howie, R.B. Nicholson, D.W. Pashley and M.J. Whelan, Electron Microscopy of Thin Crystals, Krieger, New York (1977).
2. B.K. Ambrose, J. Phys. E 9:382 (1976).
3. D. Cherns and M.J. Stowell, Thin Solid Films 29:127 (1975).

4. D. Cherns, Phil. Mag. 30:549 (1974).
5. R.L. Hines, Thin Solid Films 35:229 (1976).
6. J.P.F. Levitt and A. Howie, J. Microsc. 116:89 (1979).
7. S. Iijima, Ultramicrosc. 6:41 (1981).
8. A. Ourmazd, G.R. Anstis and P.B. Hirsch, Phil. Mag. A48:139 (1983).
9. K.N. Tu and A. Howie, Phil. Mag. 37:73 (1978).
10. W. Krakow, Surf. Sci. 140:137 (1984).
11. K. Abdelmoula, D. Renard and G. Nihoul, Ultramicrosc. 14:337 (1984).
12. K. Takayanagi, J. Microsc. 136:287 (1984).
13. K. Takayanagi, Y. Tanishiro, S. Takahashi and M. Takahashi, J. Vac. Sci. and Techn. A3:1502 (1985).
14. P. Morin, M. Pitaval, D. Besnard and G. Fontaine, Phil. Mag. 40:511 (1979).
15. P. Goodman and A. Moodie, Acta Cryst. A30:280 (1974).
16. A. Howie and Z.S. Basinski, Phil. Mag. 17:1039 (1968).
17. D.F. Lynch, Acta Cryst. A27: 399 (1971).
18. G. Lempfuhl and Y. Uchida. Ultramicrosc. 4:275 (1979).
19. R.L. Hines and A. Howie, Phil. Mag. 32:257 (1975).
20. P.O. Hahn and M. Henzler, J. Vac. Sci. and Tech. A2:574 (1984).

CATALYST STUDIES BY SCANNING TRANSMISSION ELECTRON MICROSCOPY

A. Howie

Cavendish Laboratory
Madingley Road
Cambridge CB3 0HE, U.K.

INTRODUCTION

Although the conventional transmission electron microscope must be the preferred choice in many situations where atomic resolution imaging is required, scanning electron microscopes have a number of advantages because of the much richer variety of imaging signals which they can employ. Furthermore, following the classic work of Crewe et al[1], which brought to practical use the high brightness field emission gun, the resolution of the scanning transmission electron microscope (STEM) can attain the 0.5nm level. For imaging in this machine using the electrons transmitted with zero or small energy loss close to the incident direction, it can be shown[2] from the principle of reciprocity or reversibility that the result is fully equivalent to the corresponding conventional bright field image. However, as will be seen, a far greater range of output signals, including microanalytical information, is readily available in the STEM. The field of application of this instrument is consequently by now very wide and includes reflection as well as transmission imaging. However, the study of industrial catalyst systems, frequently consisting of small heavy-atom clusters lying on, or embedded in, a highly porous light-atom support, offers an excellent and more specific example of the capabilities of scanning transmission methods in the surface science field.

CONVENTIONAL IMAGING OF CATALYST PARTICLES

In favourable cases, involving either comparatively large particles or very thin supports (so that confusion from overlapping detail in particle and support can be avoided), very useful information about catalyst particles can be obtained from conventional coherent imaging techniques. An obvious example is the study of particle size distributions and of various particle coarsening mechanisms[3]. The main point to bear in mind is that such images are often more strongly influenced by orientation differences than by local thickness in the particles. Only a fraction of the particles may, therefore, be visible in a given diffraction condition. Furthermore apparent differences in particle image intensities should not be attributed to particle shape effects such as the presence of both spheres and rafts[4] until controlled tilting experiments have been carried out.

11

Subject to this constraint, interesting effects such as the presence of multiple twinned particles (MTP) can readily be detected. MTP (see figure 1) consist of tetrahedral segments of fcc structure fitted together (with appreciable inhomogeneous elastic strain) across internal twin boundaries. In virtue of their close packed (111) external facets with low energy, they can be stable compared with single crystal structures[5]. Although MTP have been observed[6] in working catalysts and presumably have some special surface sites, there is as yet no direct evidence that they have catalytic significance. Particle shape differences, including the formation of surface facets, for example after exposure to sulphur poisoning[7], can also readily be detected in bright field images. Accurate measurements of the shape of single crystal facetted particles have been made[8] from measurements of thickness fringes in weak beam images.

High resolution structure images have equally been extremely useful in revealing details of complex, multiphase particle morphology in, for example, MoS_2 hydro-desulphurisation catalysts[9] [see fig.1(b)] and rare earth/metal oxide methanation catalysts[10]. Although most conventional electron microscopes are unable to carry out microanalysis of individual particles by the spectroscopic methods available in the STEM, identification of particles is often possible from direct measurement of lattice spacings. The application of high resolution and profile imaging methods to the observation of the surface atomic structure of catalyst and other particles is described elsewhere in this volume by Smith.

The difficulties of coherent imaging methods become rather serious when applied to small particles on the typical highly disordered and porous catalyst support. A great deal of background image noise is thus generated by structural and thickness variations in the support. For instance, the chance alignment of three carbon atoms along the beam direction in a charcoal support can be just as effective as a single platinum atom in contributing to a conventional axial bright field or dark field image. An example of the fine scale speckle generated in a coherent dark field image by such purely statistical effects in a disordered carbon support is shown in figure 2. The small bright spots should definitely not be confused with small catalyst particles. These statistical fluctuations in image intensity can to some extent be washed out by the use of hollow cone illumination[4] [figure 2(b)] where incoherent imaging over a range of illuminating directions is employed in a single exposure.

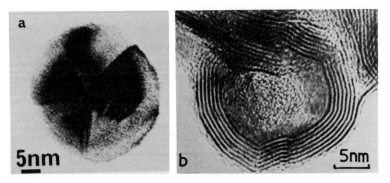

Fig. 1. Diffraction contrast image (a) of an Ag decahedron and structure image (b) of a hydro-desulphurisation MoS_2 particle[19].

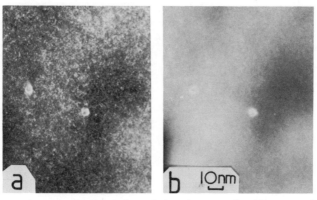

Fig. 2. Coherent dark field (a) and incoherent (hollow cone) dark field (b)
images of Pt particles on an amorphous C support.

DETECTION AND IMAGING OF CATALYST PARTICLES IN THE STEM

In the scanning transmission electron microscope (STEM) a beam of
typically 100 keV electrons is focussed to a small spot of diameter
$\simeq$ 0.5 nm which can be scanned over the specimen as shown schematically in
figure 3. As indicated there, images can then be formed from a variety of
different signals, using for example part of the transmitted energy loss
spectrum (EL), scattered electrons collected in an annular detector (AD),
characteristic X-ray signals (X) or secondary electron signals (S). In
addition, with the spot fixed on some image feature of interest, a complete
X-ray or energy loss spectrum can be obtained as well as a microdiffraction
pattern (M) on a TV camera or photographic film[11] inserted into the system.
In all this work, the ability to collect in a short time adequate signal
counts, to provide good statistics and noise free images, depends crucially
on the use of a high brightness, field emission electron source F which can
typically yield about 10^{10} electrons per second in the focussed spot.

Early experiments[1] with the high brightness STEM demonstrated the
ability of the annular detector to image single atoms of high Z supported
on very thin carbon films. The annular detector image is also very
effective[12] in imaging small heavy-atom clusters on the thicker, more
disordered, low Z supports found in typical catalysts. Initially the
images were formed by displaying a ratio of the annular detector signal to
the axial signal EL collected at about 20 eV energy loss (see figure 3).
By this method, known as Z contrast imaging, intensity changes due to local
variations in support thickness are normalised out leaving a variation due
to the high angle Rutherford scattering from heavy atoms. In many
situations however, the annular detector image alone provides adequate
imaging of small catalyst particles[13]. The improved visibility of small
catalyst clusters in this kind of image compared with the situation in a
STEM bright field image is shown in figure 4.

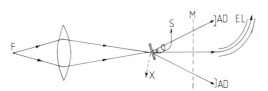

Fig. 3. Imaging modes in the STEM.

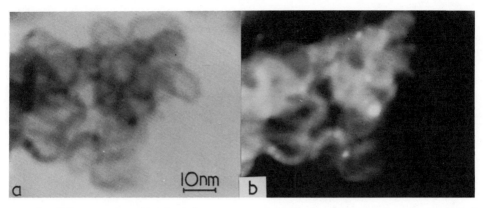

Fig. 4. STEM bright field (a) and annular dark field (b) images of Pt clusters on a disordered carbon support[19].

The annular detector signal in the STEM is related by the reciprocity principle[2] to hollow cone imaging in the conventional instrument. To some extent therefore, the ability to covercome the problems previously discussed for coherent axial imaging modes is due to the use of a complete annulus in both these techniques. In practice however, the STEM detector operates at scattering angles well beyond the range of any current hollow cone illumination system and this is a much more crucial factor in understanding its operation.

With increasing scattering angle θ, the intensity of the coherent (Bragg) scattering is reduced as a result of thermal vibration by the Debye-Waller factor $\exp(-M_g)$ where M_g increases with temperature and is proportional to the quantity $g^2 = 4\sin^2(\theta/2)/\lambda^2$. This drop in intensity goes into thermal diffuse scattering and the associated mean free path Λ_{TDS} can be computed by integrating the thermal diffuse scattering over all scattering angles

$$1/\Lambda_{TDS} = (\frac{8\pi^3\hbar^2}{\Omega m_0^2 v^2 \lambda^2}) \int_0^\pi |f^B(K)|^2 [1 - \exp(-M_g)] \sin\theta \, d\theta \qquad (1)$$

where $K = 2\sin(\theta/2)/\lambda$, Ω is the atomic volume, v the velocity and f^B the Born amplitude. Typical graphs for Λ_{TDS} as a function of electron energy are included in fig. 2 of the next chapter.

At relatively large scattering angles where $\exp(-M_g) \ll 1$, the (incoherent) thermal diffuse scattering dominates over the coherent scattering. Furthermore, introducing the X-ray scattering factor f_X via the relation

$$f^B(K) = [m_0 e^2/4\pi\epsilon_0 \hbar^2 K^2) (Z - f_X(K)],$$

we see that at such scattering angles where $f_X \to Z$, the intensity of the scattering is proportional to $Z^2/\sin^4(\theta/2)$ from each atom in the illuminated region. The annular detector signal is thus sensitive to atomic number but insensitive to crystal orientation.

NANOANALYSIS IN THE STEM

Chemical analysis can be carried out on a scale of better than 2 nm in the STEM using characteristic X-ray signals or characteristic edges in the 100 eV to 2 keV range of the electron energy loss spectrum. Either of these results can be related to inner shell binding energies. An example where

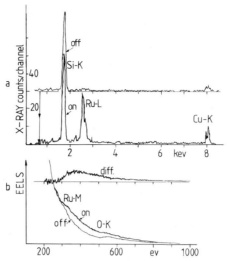

Fig. 5. X-ray spectra (a) and electron energy loss spectra (b) obtained
with the focussed spot on and just off a small Ru particle.

both techniques have been applied[13] to the same 7 nm diameter Ru catalyst
particles is shown in figure 5. Quantification of the results is usually
rather easier in the X-ray case because the signal to background ratio is
much better. In the energy loss case, some awkward extrapolation of the
spectrum is often needed to achieve background subtraction[14]. Generally
speaking, the K and L shell edges are sharper and easier to use than higher
shell edges in the energy loss spectrum and since the range of losses
accessible does not usually go beyond 2 keV, this method is often at its
best with light elements. The X-ray methods can only be extended to deal
with some of the important light elements if windowless detector systems
are used. Furthermore (see chapter by Cazaux) the detection efficiency of
the X-ray method is limited, first by the solid angle of the detector and
secondly by the fluorescence yield, whereas with the electron energy loss
method we can detect a fairly large fraction of the inner shell
ionisations. Figure 5 also indicates an additional difficulty of the X-ray
method where a Cu peak is significantly enhanced when the beam strikes the
Ru particle. This signal is generated in the specimum grid by electrons
scattered through large angles in the particle. Such spurious effects do
not arise in electron energy loss spectroscopy.

An important question for nanoanalysis is the spatial resolution
achievable. Since the inner shell ionisation occurs by Coulomb interaction
with the fast electrons rather than by a contact interation, it may not be
limited to the area of the 5 nm focussed spot. A simple classical picture,
originally due to Bohr and shown in figure 6(a), provides a useful estimate
of the degree of localisation achievable in the interaction. When an

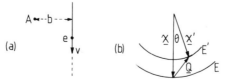

Fig. 6. Inelastic interactions of a fast electron showing (a) the real
space picture of localisation and (b) the momentum transfer
$hQ = \hbar q(\theta)$ in reciprocal space.

electron of velocity v passes an atom A at impact parameter b, the atom experiences an electromagnetic pulse whose duration is about b/v and has Fourier components of typical frequency $\omega = v/b$. Energy transfers $\Delta E = \hbar \omega = \hbar v/b$ can therefore occur. This rough argument suggests that an ionisation event of excitation energy ΔE will take place for impact parameters up to a value $b = \hbar v/\Delta E$.

From momentum transfer considerations [see figure 6(b)] it can be seen that the Bohr impact parameter $\hbar v/\Delta E = [2\pi q(0)]^{-1}$ where $\hbar q(0)$ is the minimum momentum transfer for an energy transfer ΔE and occurs when the scattering angle θ is zero. For finite scattering angles θ, where the momentum transfer increases [see fig.6(b)] with consequent improvement in the localisation we can use the approximate relation

$$b(\theta) = \frac{1}{2\pi q(\theta)} = \frac{\hbar v}{\Delta E} \frac{1}{\sqrt{(1 + \theta^2/\theta_E^2)}} \tag{2}$$

At a typical excitation energy of $\Delta E = 500$ eV, the Bohr impact parameter $\hbar v/\Delta E$ for 100 keV electrons is about 0.2 nm. This value is small enough to provide spatial resolution only slightly larger than the spot size but not so small (in relation to typical interplanar distances in crystals) to require serious spectroscopic corrections due to channelling effects which generally only become significant if higher-scattering angles are used[16].

So far electron energy loss spectroscopy in the STEM has mainly been used for compositional analysis of small particles, precipitates or internal defects in thin specimens with results which can be very impressive in terms of both sensitivity and resolution[17]. There are however a number of potential applications which are more specific to surface science. We ignore here the study of surface plasmons and other valence excitations which is discussed in the next chapter as well as the use of characteristic core losses excited in the glancing angle reflection mode which is treated in Cowley's chapter. Even in transmission energy loss spectra, significant shifts of the characteristic core edge position or changes in its shape can arise from modifications to the final empty state wave function into which the excited electron is transferred. The effects which are most relevant here are local changes in chemical binding, ionicity or valence as well as structural changes. Well above the edge, the results can in principle be analysed in terms of near-neighbour distances as in EXAFS but the subject is much less well developed in the electron case[14]. Figure 7 shows an example[18] of the effect of the surface on the shape of the K shell edge in a thin diamond crystal. The step indicated by the arrow and lying about 5 eV below the main edge has a height which is independent of crystal thickness and can thus be attributed to the

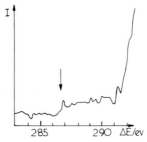

Fig. 7. K-shell edge loss spectrum in a thin crystal of diamond[18] showing a surface precursor (arrowed).

surface. It could arise from a genuine surface state but is more probably due to a thin contamination layer of amorphous carbon for which the K shell edge appears at a somewhat lower energy than in diamond.

Auger spectroscopy offers the possibility of extremely selective surface spectroscopy and chemical analysis (see chapter by Cazaux). Preliminary results have been obtained in the STEM using a conventional hemisperical Auger analyser with the specimen in a field-free position outside the objective lens. So far, partly because of the consequent increase in the size of the illuminating spot but more seriously because of the very low Auger count rates, the spatial resolution achieved has been limited to about 10 nm at best[19]. Nevertheless, if some improvements can be made in the collection efficiency and energy analysis of secondary electrons (see next chapter), it should be possible to get very useful information about such questions as the surface distribution of promoters and poisons on typical catalysts.

MICRODIFFRACTION PATTERNS

Microdiffraction patterns can be collected from the area illuminated by the stationary spot using post specimen scanning, T.V. cameras, optical transfer systems[20] or photographic procedures modified for ultra high vacuum operation[11]. The diffraction spots are considerably broadened because of the very small size of the specimen scattering volume but provide enough information to recognise simple crystal structures and orientations. Microdiffraction patterns have thus been very useful in verifying that very small catalyst particles produced by organo-metallic chemistry or other procedures were in fact metallic single crystals[13,21]. Figure 8 is a microdiffraction pattern[22] from a small gold particle in [111] orientation showing quite clearly the forbidden reflections (see previous chapter). The intensity of these reflections relative to the main reflections suggests that the particle must be only a very few atomic layers thick and may even have a flattened or raft-like structure.

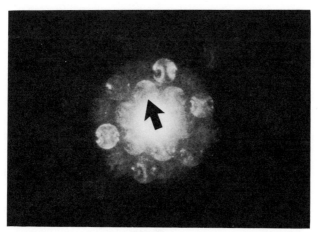

Fig. 8. Microdiffraction pattern[22] of a 5 nm Au particle in the [111] orientation. An arrow points to one of the inner ring of "forbidden" spots.

CONCLUSIONS

The STEM does not yet offer atomic resolution on the same almost routine basis (given suitable specimens) as the conventional electron microscope. However, the great variety of imaging procedures and other spatially localised signals which it can produce make it a very attractive instrument for the characterisation of complex specimens like heterogeneous catalysts. In many practical situations it is the combination of several such signals from a single specimen that is particularly useful.

REFERENCES

1. A.V. Crewe, J.P. Langmore and M.S. Isaacson, in: "Physical Aspects of Electron Microscopy and Microbeam Analysis" eds B.M. Siegel and D.R. Beaman. Wiley, New York (1975).
2. J.M. Cowley, Appl. Phys. Letts 15:58 (1969).
3. P.J.F. Harris, J. Catal. 97:527 (1986).
4. M.M.J. Treacy and A. Howie, J. Catal. 63:265 (1980).
5. A. Howie and L.D. Marks, Phil. Mag. A49:95 (1984).
6. L.D. Marks and A. Howie, Nature 282:196 (1979).
7. P.J.F. Harris, Nature, 323:792 (1986).
8. M.J. Yacaman in: "Chemistry and Physics of Solid Surfaces V", eds R. Vanselow and R. Howe. Springer, Berlin (1984); M.J. Yacaman and T. Ocana, Phys. Stat. Sol.(a) 42:571 (1977).
9. J.V. Sanders, J. Electron Microsc. Technique, 3:67 (1986).
10. Z.L. Wang, C. Colliex, V. Paul-Boncour, A. Percheron Guegan, J.C. Achard and J. Barrault, J. Catal. 105:120 (1987).
11. J.M. Rodenburg and D. McMullan, J. Phys. E 18:949 (1985).
12. M.M.J. Treacy, A. Howie and C.J. Wilson, Phil. Mag. A38:569 (1978).
13. A. Howie, L.D. Marks and S.J. Pennycook, Ultramicrosc. 8:163 (1982).
14. R.F. Egerton, "Electron Energy Loss Spectroscopy in the Electron Microscope" Plenum Press, New York (1986).
15. S.J. Pennycook and A. Howie, Phil. Mag. A41:809 (1980).
16. J. Taftø, O.L. Krivanek, J.C.H. Spence and J.M. Honig, Phys. Rev. Letts. 48:560 (1982).
17. S.D. Berger and S.J. Pennycook, Nature 298:635 (1982).
18. S.J. Pennycook, Ultramicrosc. 7:99 (1981).
19. D. Imeson, J. Microsc. 147:65 (1987).
20. J.M. Cowley, Chem. Scripta 14:33 (1979).
21. S.J. Pennycook, A. Howie, M.D. Shannon and R. Whyman, J. Mol. Catal. 20:345 (1983).
22. J.M. Rodenburg, Ph.D. Thesis, University of Cambridge (1986).

LOCALISED SURFACE IMAGING AND SPECTROSCOPY IN THE

SCANNING TRANSMISSION ELECTRON MICROSCOPE

A. Howie

Cavendish Laboratory
Madingley Road
Cambridge CB3 0HE, U.K.

INTRODUCTION

As indicated in the previous chapter, the scanning transmission electron microscope (STEM) can provide, on an almost routine basis, structural and compositional information on a near atomic level from suitably thinned, but possibly still very irregular, specimens. This is a very useful facility in the study of heterogeneous catalysts as well as in materials science more generally, irrespective of whether surfaces and interfaces are involved. It would clearly be even more valuable if the STEM could also supply localised chemical data for the surface scientist. Ideally this should include not only information such as the location of poison or promoter atoms on a catalyst, but also local data about the valence and excited state structure of the various surfaces or interfaces present.

Some of the surface chemical or electronic information required may be available from analysis of the shifts and shape changes in the structure of the characteristic inner shell excitations observed in the electron energy loss spectrum (see previous chapter). Relatively stronger energy loss signals are available in the valence loss region of the spectrum where the presence of a surface can induce profound changes, including the formation of surface collective modes such as surface plasmons. The interpretation of this part of the energy loss spectrum is, however, complicated by many-electron screening effects and depends on the dielectric excitation theory which is outlined in the next section. Later sections deal with recent advances in surface imaging using low energy loss electrons, as well as secondary or Auger electrons in the STEM. Finally we discuss some aspects of electron beam ionisation damage where combined imaging and spectroscopy studies have elucidated the primary processes responsible for loss of crystallinity or atomic desorption.

CLASSICAL DIELECTRIC EXCITATION THEORY

In this theory, originally due to Fermi[1], classical electrodynamics is employed[2] to calculate the electric field induced by a charged particle passing through a polarisable medium whose properties are specified by a complex dielectric function $\varepsilon(q,\omega)$. The dielectric function depends on the momentum $\hbar q$ and energy $\hbar\omega$ of the excitations generated; although, for simplicity here we will usually ignore the q-dependence. Neglect of these dispersion or non-local effects means that the polarisation induced at each point is directly related to the electric field at that point. The real and imaginary parts of ε are constrained[3] by various sum rules and by Kramers Kronig relations between them.

A simple example of the theory is the case (shown in figure 1) of an electron travelling with velocity v parallel to and at distance b from a planar interface between two different media specified by $\varepsilon(\omega)$ and $\varepsilon'(\omega)$. For this charge distribution, defined as a moving point by the equation

$$\rho(\underline{r}, b) = -e\, \delta(x - b)\, \delta(y)\, \delta(z - vt) \tag{1}$$

we can solve for the potential ϕ, working for convenience with Fourier components $\rho(\underline{q}, \omega, x)$ and $\phi(\underline{q}, \omega, x)$, where $\underline{q}$ is a two dimensional vector parallel to the surface. We then find, for $x \geq 0$ and in the non-relativistic limit,

$$\varepsilon(\omega)\,\left(\frac{d^2}{dx^2} - q^2\right)\,\phi(\underline{q}, \omega, x) = \frac{2\pi e}{\varepsilon_0}\,\delta(x - b)\,\delta(q_z v - \omega) \tag{2}$$

with solutions

$$\phi(\underline{q}, \omega, x) = \frac{\pi e\, \delta(q_z v - \omega)}{\varepsilon(\omega)\,\varepsilon_0\, q}\,\exp(-q|x - b|) + A\,\exp(-qx) \tag{3}$$

For $x \leq 0$, ε is replaced by ε' in equation (2) and the right hand side vanishes so that

$$\phi(\underline{q}, \omega, x) = B\,\exp(qx) \tag{4}$$

Having determined the constants A and B from the usual electrostatic boundary conditions at the interface $x = 0$, we can evaluate the work done against the electric field component E_z at the position of the particle and hence find the rate of energy loss. This quantity can then be expressed[4,5] in terms of a differential probability of energy loss $\hbar\omega$ per unit distance travelled.

$$\frac{dP(b,\omega)}{d\hbar\omega} = \frac{e^2}{2\pi^2 \varepsilon_0 \hbar^2 v^2}\,\left|\, \text{Im}\!\left(-\frac{1}{\varepsilon(\omega)}\right)\ln\left(\frac{2vq_c}{\omega}\right) + \right.$$
$$\left.\left\{ \text{Im}\!\left(-\frac{2}{\varepsilon(\omega) + \varepsilon'(\omega)}\right) - \text{Im}\!\left(-\frac{1}{\varepsilon(\omega)}\right)\right\} K_0\left(\frac{2\omega b}{v}\right)\,\right| \tag{5}$$

Here q_c is an upper limit cut-off on q_y and K_0 is the modified Bessel function defined by the relation:

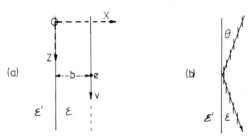

Fig. 1. Electron trajectories and coordinates near the interface between two dielectric media for (a) parallel incidence and (b) glancing angle incidence (parallel segment approximation).

$$K_0(x) = \int_0^\infty \exp[-x \sqrt{(1 + \xi^2)}] / \sqrt{(1 + \xi^2)} \; d\xi \qquad (6)$$

Equation (5) illustrates a number of aspects of valence loss spectroscopy of interfaces. At large values of the impact parameter b, the influence of the boundary (expressed by the K_0 function) falls off like $\exp(-2\omega b/v)/\sqrt{(2\omega b/v)}$ in agreement with the qualitative arguments given in the previous chapter about localisation effects in energy loss spectroscopy at very small scattering angles. The results are then eventually dominated by the first term in equation (5) corresponding to bulk losses originally identified by Fermi[1]. For free electron metals, where the complex dielectric constant is given by the expression

$$\varepsilon(\omega) = \varepsilon_1(\omega) + i\varepsilon_2(\omega) = 1 - \omega_p^2/\omega(\omega + i\gamma) \qquad (7)$$

the bulk loss is (in the limit of small damping γ) concentrated near the value $\Delta E = \hbar\omega_p$ where $\omega_p = \sqrt{(ne^2/\varepsilon_0 m)}$ is the bulk plasmon frequency for an electron gas of number density n. In other simple cases (see problem), peaks in the bulk loss function $\text{Im}[-1/\varepsilon(\omega)]$ can result either from zeros of $\varepsilon_1(\omega)$ (i.e. collective plasmon modes) or from maxima in $\varepsilon_2(\omega)$ (single electron excitations).

The bulk loss function, including improvements to cover dispersion effects, can be employed[6], as shown in figure 2, to calculate the mean free path Λ_{val} for valence excitations using the equation

$$\frac{1}{\Lambda_{val}} = \frac{e^2}{2\pi^2\varepsilon_0\hbar v^2} \int_0^\infty d\omega \int_{\omega/v}^{q_c} \text{Im}(- \frac{1}{\varepsilon(q,\omega)}) \frac{dq}{q} \qquad (8)$$

Near the boundary between two media, the second term in equation (5) depending on the difference between $\varepsilon(\omega)$ and $\varepsilon'(\omega)$ becomes important. The K_0 function varies rapidly when b is small and indeed diverges logarithmically unless some damping or aperture-dependent cut off is imposed in equation (6) to suppress the contribution from large values of the variable $\xi = vq_y/\omega$. Since the K_0 function and the logarithmic function tend to cancel in this region, it can be seen that the bulk loss function is effectively removed (the so-called Begrenzungseffekt) and replaced by an interface loss function $\text{Im}[-2 (\varepsilon + \varepsilon')]$. This function can again exhibit collective as well as single electron excitation losses. For instance at the interface between a vacuum ($\varepsilon' = 1$) and a free electron metal specified by equation (7), the collective surface plasmon mode excited has a frequency which is readily shown to be $\omega_p/\sqrt{2}$. In general however the relation between the bulk loss spectrum and the surface loss spectrum is more complex as will be shown in the next section where experimental loss spectra are discussed.

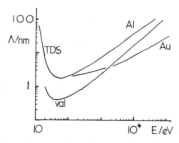

Fig. 2. Mean free paths for valence excitation Λ_{val} and thermal diffuse scattering Λ_{TDS} as a function of electron energy E.

The simple interface theory just outlined has been extended to include relativistic effects which can be important in parts of the loss spectrum of semiconductors or insulators when ε_1 is large. If the electron velocity v exceeds $c/\sqrt{\varepsilon_1}$, the velocity of light in the medium, Cerenkov radiation losses can occur. It is also possible, as discussed later, to extend the theory to more elaborate, planar sandwich structures as well as to spheres and cylinders. Another refinement of the theory concerns the deviation of the electron trajectory from a straight line as a result of the so called image force acting in the x direction in figure 1(a). This force can be calculated by using equation (3) to evaluate $\partial\phi/\partial x$ at the position of the incident electron but the result obtained[7] seems to have an almost negligible effect for the fast electrons used in the STEM.

The theory as developed here can be applied directly to compute energy loss spectra obtained in the STEM with the focussed probe placed at different distances from a planar interface aligned parallel to the incident beam. A more sophisticated quantum mechanical calculation[8] shows that the classical theory is usually valid provided the dependence on impact parameter is convoluted over the current distribution in the probe. For 100 keV electrons, a typical mean free path for plasmon excitation is, as can be seen from figure 2, about 100 nm so, depending on the specimen thickness, it may be necessary to take account of multiple losses.

The theory can also be modified very simply to deal with glancing angle reflection situations, by approximating the track by a series of parallel segments as indicated in figure 1(b). In the case of reflection at a small angle θ from an infinite free surface, the "probability" $Q(\theta)$ for energy loss (or more strictly the ratio of the first loss to the zero loss intensity) is then given[5] by the expression

$$Q(\theta) = \frac{e^2}{4\pi\varepsilon_0 \hbar v \theta} \int_0^\infty Im(\frac{-2}{\varepsilon + 1}) \frac{d\omega}{\omega} \qquad (9)$$

The use of this expression to analyse the loss spectra observed when valence excitations are generated by 100 keV electrons in glancing angle situations is described in the next section. Furthermore, given a dielectric function which includes the contribution of optical phonons, equation (9) also describes the long range dipole losses experienced in reflection experiments with very low energy electrons (see chapter by Kirschner).

VALENCE LOSS SPECTRA AT PLANAR SURFACES AND INTERFACES

The classical dielectric theory just described is capable of explaining in quantitative detail many of the features so far observed in electron energy loss spectra taken from various surfaces in the scanning transmission electron microscope. For parallel incidence conditions [fig 1(a)], fairly exhaustive studies have been made[9,10] of MgO smoke cubes. It has been shown (M.G. Walls private communication) that the bulk loss spectrum (obtained when the beam passes through the cube) and the surface loss spectrum (obtained with so-called ALOOF beam outside the cube but travelling parallel to a face) can both be fitted by the same complex dielectric function $\varepsilon(\omega)$ using equation (5).

Planar interfaces between different dielectric media can be conveniently studied as a function of impact parameter in transmission through thin samples which have been ion-milled and oriented so that the interface is normal to the free surfaces and parallel to the electron beam. Fig 3 shows some experimental and theoretical loss spectra[5] for the Si-SiO$_2$ interface. Note the appearance of an interface plasmon at 8 eV for impact parameters within approximately 2 nm of the interface.

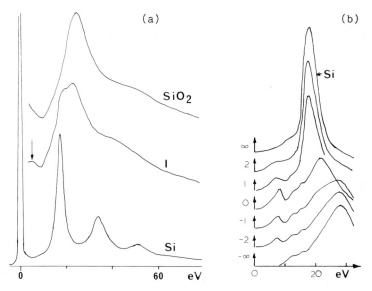

Fig. 3. Si-SiO$_2$ interface loss spectrum compared in (a) with spectra far
from the interface. Computed spectra for various distances b (in
nm) from the interface are shown in (b).

Plane surfaces (possibly with thin oxide or other surface overgrowths)
can be most easily studied in glancing angle mode. To explain the spectra
observed from a clean surface, we can employ equation (9) or refinements of
it to take account of relativistic effects and some penetration of the beam
into the medium before reflection. Fig.4 shows an experimental spectrum
obtained[12] for a GaAs (110) surface compared with various theoretical
computations. These comparisons suggest an effective depth of penetration
of about 2.5 nm at this angle of incidence corresponding to a total beam
path length in the GaAs of about 150 nm. This figure was roughly confirmed
by measuring the Ga L shell loss intensity at 1116 eV since, following the
analysis of the previous chapter, this can only be generated at rather
small impact parameters when the beam is either inside the surface or very
close outside it.

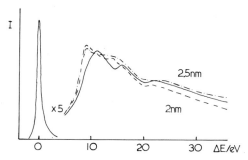

Fig. 4. Experimental glancing angle loss spectrum at the 660 Bragg position
on a GaAs surface compared with computed spectra for the
penetration depths indicated[12].

From observations of this kind (see also the results described in the chapter by Cowley) it is clear that energy loss spectroscopy with fast electrons at glancing angles probably does offer some significant opportunities for surface studies in highly localised regions of relatively easily prepared bulk samples. The main difficulty for work in the valence region is that the signal, though strong, is often dominated by contributions from excitations at relatively low values of q of order ω/v which are not very closely confined to the surface or interface and thus not very sensitive to the presence of the adsorbed layers or surface reconstructions likely to be of interest. As discussed by Howie and Milne[5], this is probably the reason why the strong dependence of surface plasmon energies on surface reconstruction in Si which was observed with low energy electrons is not detectable with higher energy electrons. It may however be possible to get improved surface sensitivity (at the expense of signal strength) with fast electrons by collecting the spectra at non-zero scattering angles (i.e. for larger values of q).

These questions can be answered more precisely by comparing experiments and computations from dielectric theory applied to more complex interfaces including, for example, a thin layer of a third dielectric material at the interface. Interface plasmons set up at the two surfaces of this thin layer are then coupled and have energies which may depend sensitively on the layer thickness. Figure 5 shows some results for the case of an oxide film on Al.

VALENCE EXCITATIONS ON SPHERES AND CYLINDERS

Classical dielectric excitation theory has also been applied to a number of non-planar geometries, particularly to spheres in connection with studies e.g.[13] of catalysts or other small particle sytems. Even for the simplest case when the electron beam passes outside an isolated sphere, many surface modes are excited on the sphere[14] and the calculations must extend to quite high angular momentum values ℓ to obtain good agreement with experiment[15]. The ℓ = 1 dipole approximation, which is frequently made, is only valid for spheres of radius much less than v/ω. Modes with high ℓ values are particularly important in the description of the effect of any surface modification, such as an oxide coating on the sphere[15]. Preliminary calculations have recently been made[16] for the case when the fast electron travels through the sphere. The case of a truly isolated sphere is rarely encountered in practice and, although it becomes rather complicated, the dielectric theory has been successfully extended to deal with interactions between pairs of spheres[17] and with a sphere semi-embedded in another medium[18].

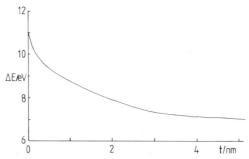

Fig. 5. Computed values of the interface plasmon energy for an Al surface as a function of the oxide overgrowth thickness.

The interpretation of energy loss spectra from cylindrical holes lying parallel to the beam is of considerable interest since (as discussed later) these can be formed in many materials by beam damage in the STEM. Once again it is usually necessary to include a rather large number of angular momentum components[19] (unless the beam passes exactly down the axis of the hole). Preliminary calculations (C. Walsh, unpublished) give quite good agreement with experimental spectra.

ENERGY-SELECTED SURFACE IMAGES

Images formed from electrons of a given energy loss can readily be obtained in the STEM and, in the right circumstances, these can have some surface specificity. Figures 6(a) and 6(b) show images of an MgO cube aligned with the electron beam and taken[11] with electrons of energy loss value $\Delta E = 14$ eV and 22 eV respectively. Reference to the dielectric data for MgO shows[11] that at 14 eV the surface loss function predominates over the bulk loss function thus explaining the bright region visible just inside and outside the edges of the particle in figure 6(a). Similar results have been obtained by imaging Ag catalyst particles in surface plasma loss electrons (D.J. Wheatley, unpublished). Such axial energy loss images have a large contribution from low q excitations and consequently a spatial resolution of order v/ω. It may be possible to improve this by placing the spectrometer off-axis to form images exclusively from the weaker excitations at higher q.

These images are an extreme high resolution elaboration of the low-loss imaging method of Wells[20] used to give greater surface specificity in reflection microscopy. Although the low-loss method does not really depend on the distinction between surface and bulk losses it has been applied at moderate resolution in the STEM[21]. The success of such comparatively crude energy selection procedures in low-loss reflection microscopy is related to the success of high resolution, glancing angle reflection microscopy without any need for energy filtering at all. In the latter case the images formed in zero loss and in surface plasma loss electrons are virtually identical[5,11].

Surprisingly high resolution images of surface structures can be formed in the STEM[22] by collecting the signal from low energy secondary electrons. These electrons can be collected over a larger range of emergent angles by allowing them to spiral up the magnetic field lines in the objective lens before applying an electric field to deflect them to an Everhart-Thornley detector. Figure 7 shows images obtained by D. Imeson[23] of a Pt on C catalyst where the small Pt clusters show up clearly in the annular detector image (a) but the complex structure of the support is much more clearly visible in (b).

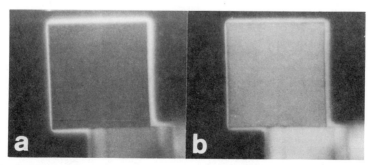

Fig. 6. STEM images of an MgO cube taken[11] with the 14 eV (a) and 22 eV (b) energy loss electrons. The cube edge is 120 nm.

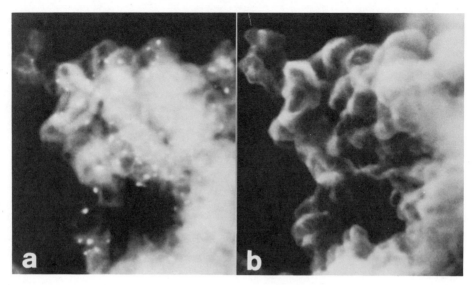

Fig. 7. Annular dark field (a) and secondary electron image (b) of Pt particles on C catalyst specimen[23]. Picture width: 160 nm.

The sensitivity of the secondary electron signal to surface topography is well established in scanning electron microscopy and is due to the short escape distance Λ_{esc} of such electrons from the solid. For low energy ~50 eV electrons Λ_{esc} is mainly controlled by Λ_{val} and Λ_{TDS} already shown in figure 2 and is usually appreciably below 1 nm. A figure showing Λ_{esc} as a function of energy is given in the chapter by Dobson.

A complete explanation of the spatial resolution achieved in secondary electron images also requires consideration of localisation effects in the generation of plasmons, their propagation and decay into secondaries which eventually escape. Precise measurements of the secondary signal from an MgO cube showed[5] that it could arise when the beam was as much as 2 nm outside the cube as a result of the generation of surface valence excitations in this situation. However, the secondary electron signal falls rapidly as the cube becomes positively charged so that they can no longer escape. Observations of shifts in Auger peaks indicated that the cubes could typically become charged to +40 V. It may be therefore that high resolution secondary electron images of the type shown in figure 7 are actually formed from electrons of somewhat higher energy than conventional secondaries. Efforts are being made in several laboratories to design a secondary electron analyser to operate on the electrons emerging from the objective lens in the STEM. It may even be possible to acquire adequate Auger signals with good spatial resolution if such analysis can be developed.

IONISATION DAMAGE PROCESSES

The charging effects noted in the previous section are by no means the only sign of the strong interaction between the electron beam and the specimen which makes localised electron energy loss spectroscopy possible. Under the intense irradiation conditions of the high resolution electron microscope, and even more the STEM, ionisation damage processes can occur leading to desorption effects or permanent structural damage. Only in good conductors such as metals and semiconductors, where enough free carriers are available to fill very quickly the holes generated, can ionisation damage be ignored. In most other materials it is a much more serious effect than the alternative knock-on damage process by which very high energy electrons can displace atomic nuclei from their crystal sites in high angle collisions.

In most situations the amount of damage produced is simply proportional to the electron dose received and is independent of the dose rate so that the damage process is not related to any beam-heating effect. The dose required to cause a given amount of damage can be measured as a function of electron energy in different crystalline materials by observing the fading of the diffraction pattern or alternatively the changes in the energy loss spectrum or mass loss from the specimen. Such measurements[24] indicate enormous differences between the sensitivity of different materials to ionisation damage. Some at least of these differences are determined by the nature of the primary damage event which in some cases can be a valence excitation leading to electron-hole pair formation and bond breaking but in other cases follows from an inner shell ionisation process. This distinction seems to explain the relative stability of π-bonded organic molecules compared with saturated organic molecules. Damage experiments[25] carried out with aromatic organic crystals using low energy electrons show a damage threshold consistent (after allowing for the high rate of energy loss of these electrons) with carbon K-shell ionisation. Infra-red spectroscopy of irradiated coronene specimens suggests[26] that the damage is mainly to the C-H bonds at the periphery of the molecules rather than to the C-C bonds and is consistent with the observation that no ionisation damage at all occurs in graphite with a structure very similar to the internal part of the coronene molecule.

It appears that the covalent C-H bond in a molecule like coronene is not damaged by the removal of one bonding electron which can easily occur by direct valence excitation. The Auger process which usually follows a carbon K-shell ionisation can however leave such a bond <u>doubly</u> ionised and lead to its failure. Such double ionisation events involving C-C bonds do not apparently lead to damage.

This Auger mechanism of damage has already been identified in electron and photon-induced desorption phenomena observed in oxides and other materials (see chapter by Menzel for more details). It is interesting to note that it also seems to be the main mechanism in the rather dramatic hole drilling and other damage processes observed[27-29] in the STEM. In contradiction to most other electron beam damage events these are however very strongly dependent on the electron beam current density so that more or less simultaneous ionisation events by different electrons must play some role. The non-linear dependence on beam current enables holes to be drilled, slots to be cut and in some cases metallic lines to be drawn on a scale of about 2 nm and close to the size of the focussed beam. Many aspects of the damage process remain unclear, but it appears possible that, following the ejection of a number of electrons from the irradiated region (perhaps by a cascade of Auger processes like those discussed above), cations are repelled from the resulting positively charged region. In some cases, such as amorphous Al_2O_3, there is evidence from EELS spectra[29] of a high density of oxygen just prior to appearance of a hole. Small droplets of Al form in the surrounding region. Measurements of the impact parameter dependence of the damage process suggest that the initial event is in the Al L-shell rather than the Al or O K-shell. Apart from the intrinsic interest of the damage process however, the phenomenon has obvious potential in nanolithography and the fabrication of complex structures on the near atomic scale.

Attention should also be drawn to the rather dramatic damage events recently observed in atomic resolution electron microscopy of metal or oxide small particles and their surfaces (see chapter by Smith). At beam current densities of about 10^6 Am^{-2}, video sequences show abrupt changes of structure on a time scale of about 0.1 s. Although the degree of thermal contact between the particle and the support is important, the effect is certainly not simply a thermal one due to direct valence excitation. Inner shell ionisation processes may again be the initial event responsible.

Although these damage processes may constitute a profitable field of study for the electron microscopist, they inevitably suggest that there may be severe limitations on the application of both the conventional instrument and the STEM to sensitive surfaces particularly those with adsorbed layers. In this respect the scanning tunneling microscope enjoys an enormous advantage.

AKNOWLEDGEMENTS

I am indebted to a number of Cambridge colleagues notably D. Imeson, R. Milne, M.N. Mohd Muhid, M. Walls and C. Walsh for making available their recent results as well as to Professors R.H. Ritchie and P. Echenique for valuable discussions on dielectric theory. The work using the STEM at Cambridge has been supported by S.E.R.C. and B.P. plc.

PROBLEM

In a simple picture of dielectric response, where an insulator is modelled by a density N of valence electrons in simple harmonic oscillator potential wells with resonant frequency ω_0 and damping constant γ, the complex dielectric constant has the form

$$\varepsilon(\omega) = 1 - \frac{Ne^2/m\varepsilon_0}{\omega(\omega + i\gamma) - \omega_0^2}$$

Assuming γ to be small, investigate the usefulness of this expression for Ge where $a_0 = 0.566$ nm, $\varepsilon(0) = 16$, band gap $E_g = 0.65$ eV and the bulk energy loss spectrum shows a sharp peak at 16 eV. What structure would you expect in the surface loss spectrum?

The solution depends on noting that, for small γ, the maximum of $Im(-1/\varepsilon)$ occurs when $\varepsilon_1 = 0$, i.e. for $\omega = \sqrt{(\omega_0^2 + \omega_e^2)}$. For Ge $h\omega_e = 15.6$ eV and all the dielectric data can be fitted by taking $h\omega_0 = 4$ eV which can be taken to represent an average value of the band gap appreciably larger than the quoted minimum value E_g. Using similar arguments, the surface loss spectrum can be shown to have a sharp peak at $\omega = \sqrt{(\omega_0^2 + 0.5\omega_e^2)}$.

REFERENCES

1. E. Fermi, Phys. Rev. 57:485 (1940).
2. L.D. Landau and E.M. Lifshitz, "Electrodynamics of Continuous Media", Pergamon Press, (1960).
3. H. Raether, "Excitation of Plasmons and Interband Transitions by Electrons", Springer, Berlin (1980).
4. P.M. Echenique and J.B. Pendry, J. Phys. C 8:2936 (1975).
5. A. Howie and R.H. Milne, Ultramicrosc. 18:427 (1985).
6. R.H. Ritchie and A. Howie, Phil. Mag. 36:463 (1977).
7. P.M. Echenique and A. Howie, Ultramicrosc. 16:269 (1985).
8. R.H. Ritchie and A. Howie, Phil. Mag. (in press).

9. L.D. Marks, Solid. State Comm. 43:727 (1982).
10. J.M. Cowley, Surf. Sci. 114:587 (1982).
11. A. Howie and R.H. Milne, J. Microsc. 136:279 (1984).
12. A. Howie, R.H. Milne and M.G. Walls, Proc. EMAG 85, Inst. of Phys., London, 117 (1985).
13. A. Acheche, C. Colliex, H. Kohl, A. Nourtrier and P. Trebbia, Ultramicrosc. 20:99 (1986).
14. T.L. Ferrell and P.M. Echenique, Phys. Rev. 55:1526 (1985).
15. P.M. Echenique, A. Howie and D.J. Wheatley, Phil. Mag. B56:335 (1987).
16. P.M. Echenique, J. Bausells and A. Rivacoba, Phys.Rev. B35:1521 (1987).
17. P.E. Batson, Surf. Sci. 156:720 (1985).
18. Z.L. Wang and J.M. Cowley, Ultramicrosc. 21:77 (1987).
19. Y.T. Chu, R.J. Warmack, R.H. Ritchie, J.W. Little, R.S. Becker and T.L. Ferrell, Particle Accelerators 16:13 (1981).
20. O.C. Wells, Appl. Phys. Letts. 19:232 (1971).
21. M.M.J. Treacy, W. Krakow, D.A. Smith and G. Trafas, Appl. Phys. Letts 38:341 (1981).
22. D. Imeson, R.H. Milne, S.D. Berger and D. McMullan, Ultramicrosc. 17:243 (1985).
23. D. Imeson, J. Microsc., 147:65 (1987).
24. L. Reimer and J. Spruth, J. Microsc. Spectrosc. Electron. 3:579 (1987).
25. A. Howie, F.J. Rocca and U. Valdrè, Phil. Mag. B52:751 (1985).
26. A. Howie, M.N. Mohd Muhid, F.J. Rocca and U. Valdrè, Proc. EMAG 87, Inst. of Phys., London (1987).
27. M.E. Mochel, C.J. Humphreys, J.A. Eades, J.M. Mochel and A.K. Petford, Appl. Phys. Letts, 42:392 (1983).
28. A. Muray, M.S. Isaacson and I. Adesida, Appl. Phys. Letts, 45:589 (1984).
29. S.D. Berger, I.G. Salisbury, R.H. Milne, D. Imeson and C.J. Humphreys, Phil. Mag. B55:341 (1987).

FUNDAMENTALS OF HIGH RESOLUTION TRANSMISSION ELECTRON MICROSCOPY

David J. Smith

Center for Solid State Science and Department of Physics
Arizona State University, Tempe, AZ 85287, USA

INTRODUCTION

The recent emergence of high-resolution electron microscopes (HREMs) capable of resolving sub-2-Ångstrom detail on a routine basis has led to an enormous increase in the range of materials which can be usefully studied. Not only is it possible to resolve individual atomic columns in low index zones of most common metals but observations of semiconductors, for example, are no longer restricted to the traditional [110] zone, thereby making it feasible at last to obtain two-dimensional information about surfaces, interfaces and other planar defects. There is a worldwide upsurge of interest in the capabilities of these machines and the so-called intermediate voltage (300-400kV) HREMs are selling rapidly despite their considerable expense. Novel applications in such diverse fields as solid state chemistry, mineralogy and materials science are appearing almost daily.

The primary purpose of this short review is to introduce the basic concepts required to understand the technique of high-resolution electron microscopy. The essential aspects of image formation are outlined, including the use of contrast transfer theory as a guide to correct imaging conditions. Approaches to image simulations are described, in particular the multi-slice method, and the usefulness of Weak-Phase-Object (WPO) images for preliminary assessment of image detail is stressed. The likely occurrence of imaging artefacts, such as atom-pair dumbbells, due to second-order non-linear interference effects needs to be realized. General applications of the high-resolution technique to the study of surface and interfaces are then briefly described, in part to demonstrate how the method can provide useful qualitative and quantitative information but also to develop some awareness of its limitations. Further specific details related to these types of applications can be found in chapters of this book by Gibson, Stobbs and Smith. Reference should be made elsewhere, for example to the articles by van Dyck (1978) and Hawkes (1978) and the monographs by Saxton (1978) and Cowley (1981) for more comprehensive treatments of the imaging theory sketched out below. General information about applications of high-resolution electron microscopy can be found, in the first instance, by reference to the proceedings of any recent international electron microscopy meeting.

IMAGING THEORY

The imaging process within the electron microscope, represented schematically in Fig. 1, can be thought of as occurring in two sequential steps: scattering of the electron beam by the sample, followed by the process of image formation by the objective lens with successive enlargements by the projector lens system (not shown).

Image formation

As a first approximation, the electron beam is treated as a monochromatic plane wave incident onto the sample, taken as having a two-dimensional transmission function q(r). The diffracted wavefront generated by the object then forms an electron (Fraunhofer) diffraction pattern in the back focal plane (BFP) of the objective lens, a process represented formally by a Fourier transform of the object transmission function:

$$Q(u) = \mathcal{F}[q(r)] \tag{1}$$

The image amplitude is then expressed formally by a second Fourier transform as:

$$\Psi(r) = \mathcal{F}[Q(u)] = q(-r/M) \tag{2}$$

where M is the lens magnification.

In practice, the diffraction pattern is normally restricted by the presence of an objective aperture, represented as A(u), and modified by a phase factor $\exp\{i\chi(u)\}$ given by

$$\chi(u) = \pi\Delta f\lambda u^2 + \pi C_s \lambda^3 u^4/2 \tag{3}$$

This latter expression reflects the fact that the diffracted waves suffer phase changes which depend both on the objective lens defocus, Δf, and its spherical aberration coefficient, C_s. Typical curves for $\chi(u)$ are drawn in Fig. 2. The diffracted amplitude in the BFP is represented by

$$\Psi(u) = Q(u) \, A(u)\exp\{i\chi(u)\} \tag{4}$$

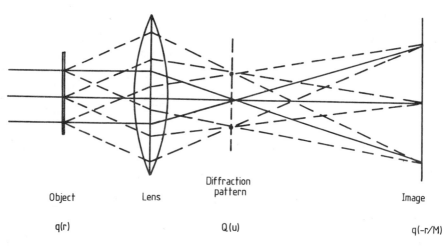

Object
q(r)

Lens

Diffraction
pattern
Q(u)

Image
q(-r/M)

Fig. 1. Schematic representation of the image formation process by the objective lens of an electron microscope.

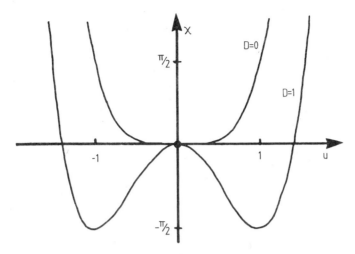

Fig. 2. Phase shift $\chi(u)$ introduced by the defocus and spherical aberration of the objective lens.

The function $T(u)=A(u)\exp\{i\chi(u)\}$ is usually called the transfer function (TF) of the objective lens. The image amplitude, referred to the object plane, then becomes

$$\Psi'(r) = \Psi(r)* t(r) \tag{5}$$

where * represents the convolution integral, and the image intensity is given by

$$I(r) = |\Psi(r)* t(r)|^2 \tag{6}$$

The functions $\Psi(r)$ and $t(r)$ are both complex and, clearly, the effect of $t(r)$ is to smear out $\Psi(r)$, equivalent to a loss of resolution.

Electron scattering

For imaging with electrons, the scattering angles are small (typically ~10^{-2} radians), and scattering by the object, represented as $q(r)$, can be treated in terms of successively more realistic approximations. In the kinematical (weak scattering) approximation, basically only appropriate for electrons in a qualitative sense, the scattering amplitudes are considered as very small in comparison with the incident amplitude, and scattering from each atom of the object can be added separately. In the phase object approximation (POA), any (angular) spread of the beam due to the object scattering is neglected, and the object transmission function due to the potential ϕ is $\exp\{-i\sigma\phi(r)\}$, where $\phi(r)=\int\phi)xyz)dz$ and the interaction constant $\sigma=\pi/\lambda E$. If the object is regarded as a weak-phase-object (WPO), with $\sigma\phi(r)\ll1$, then $q(r)\sim1+i\sigma\phi(r)$ so that the image amplitude is given by

$$\Psi(r) = 1 + \sigma\phi(r)*s(r) - i\sigma\phi(r)*c(r) \tag{7}$$

where $c(r)= \mathcal{F}[A \cos \chi(u)]$ and $s(r)= \mathcal{F}[A \sin \chi(u)\}$, and the image intensity becomes

$$I(r) = 1 + 2\sigma\phi(r)*s(r) \tag{8}$$

which is equivalent to an amplitude in the BFP of

$$Q(u) = \delta(u) + \sigma\Phi(u)A(u)\sin\chi(u) \tag{9}$$

33

It is therefore very important, for the purposes of image interpretation, to appreciate the behaviour of $\chi(u)$, in particular to find the value(s) of Δf which maximize $\sin\chi(u)$ over the widest possible range of u (see below for further discussion).

In reality, electron scattering should be regarded as a dynamical process for anything but a very thin object, involving large phase changes and multiple scattering, and it is therefore important that the relative height distribution of atoms should be properly taken into account. The immediate effect of the dynamical scattering is that there is no longer any simple relationship between the projection of the crystal structure and the electron wavefunction emerging from the exit surface of the specimen. Image simulations for the purposes of image interpretation, particularly in the region of crystal defects become an unavoidable necessity. In the multislice approach (Cowley and Moodie, 1957a), it is assumed that all atoms in the object are projected onto N planes (slices), normal to the beam direction and separated by height ΔZ. For each successive slice, the POA is applied, before propagation of the electron waves by Fresnel diffraction to the next slice. This process, which is represented schematically in Fig. 3, lends itself very easily to computation (Goodman and Moodie, 1974). The greatest accuracy accrues from letting $\Delta Z \rightarrow 0$ and $N \rightarrow \infty$ but slice thicknesses of 2-3Å prove sufficient for most practical purposes.

TRANSFER FUNCTION THEORY

For a phase object, the transfer of image contrast by the objective lens can be represented by $\sin\chi(u)$ rather than the exponential form. The basic <u>shape</u> of $\sin\chi(u)$ is actually microscope- and specimen-independent. It is therefore very convenient to express it in an alternative form using generalized units:

$$\sin\chi(k) = \sin\pi k^2(D-k^2/2) \tag{10}$$

where $k[=(C_s\lambda^3)^{1/4}]$ is the generalized spatial frequency and $D[=(C_s\lambda)^{1/2}]$ is the generalized defocus. Universal curves can be generated for $\sin\chi$ which, by using an appropriate scaling factor, are then applicable for any objective lens or operating voltage. The representative curves shown in Fig. 4 are for a 500kV HREM, with a C_s-value of 2.0mm, operating at the optimum objective lens defocus where there is the broadest possible

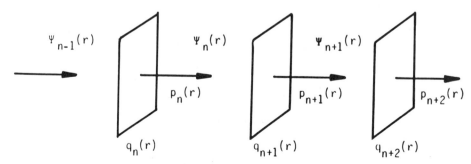

Fig. 3. Schematic representation of the multislice approach to dynamical scattering theory, in terms of object transmission functions $\Psi_i(r)$ and propagation functions $P_i(r)$.

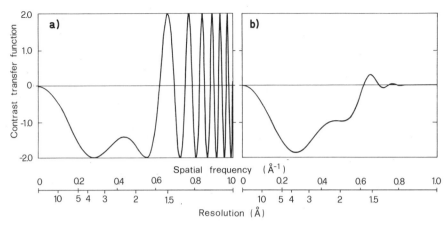

Fig. 4. Transfer functions for the objective lens (C_s=2.0mm) of a 500kV
HREM, axial illumination: (a) coherent illumination; (b) effect of
envelope functions, with focal spread of 100Å and incident beam
semi-angle of 0.4 mrad.

interval without phase reversals. This focus has the numerical value of
$(1.5C_s\lambda)^{1/2}$. Fig. 4a corresponds to the ideal case of fully coherent
illumination (i.e. assuming zero energy spread, zero beam divergence and no
high voltage or objective lens current fluctuations). The oscillatory
nature of such curves (usually termed the phase contrast transfer function
- PCTF - of the objective lens) reflect the angle-dependent phase changes
introduced by the spherical aberration of the objective lens. Electrons
scattered to different bands (corresponding to different resolutions)
obviously have different phases which can result in image artefacts.
Moreover, the strong focus-dependence of the PCTF should be noted since the
movement of the cross-over points with focus will also result in phase
reversals of particular image detail during the recording of a through-
focal series of images. In practice, the incident electron beam is not a
monochromatic plane wave. Fig. 4b corresponds to a more realistic
practical situation and illustrates the effects on the PCTF caused by
partially coherent illumination.

Envelope functions

It is possible to separate the effects of focal spread (temporal
coherence) and finite beam divergence (spatial coherence), and each can be
represented as envelope functions multiplying the ideal PCTF. The effect
of finite spatial coherence can be represented as:

$$B(k) = \exp[-\pi^2 s^2 k^2 (k^2-D)^2]$$ (11)

where s is a generalized coordinate representing the illumination
divergence. Note that this function depends on the defocus D; this
dependence is sometimes useful in determining the particular factor which
is limiting microscope performance. The effect of focal spread can be
represented as:

$$C(k)=\exp[-\pi^2 d_o^2 k^2/2]$$ (12)

where d_o represents the effective half-width of the focal spread
distribution, and includes contributions from the finite energy spread of
the beam, as well as high voltage and lens current instabilities.

35

Definitions and measurement of resolution

The position of the first zero crossover of the PCTF which occurs at the broad passband focus illustrated in Fig. 4, is variously described as the "interpretable", "point" or "structural" resolution limit of the particular microscope. It has the numerical value given by $\delta \sim 0.66$ $(C_s \lambda^3)^{1/4}$, and it effectively represents the limit to which intuitive image interpretation in terms of projected atomic structure is (sometimes) possible. The considerable reductions in λ which result from the use of higher accelerating voltages have a much greater influence on the value of δ than the slight increase in C_s which results from saturation of pole-piece material. The net result is that δ improves from ~2.8Å at 100kV, to ~1.7Å at 400kV and to ~1.3Å at 1000kV. Practical considerations, such as size and expense, as well as the deleterious effects of electron irradiation, generally serve, however, to restrict the maximum voltages available with most HREMs, with 300 to 400kV becoming a popular compromise. The "instrumental" or "information" limit of resolution is defined in terms of the attenuation produced by the envelope functions with exp(-2) (ie. 15%) normally taken as the cut-off for which a posteriori image processing might usefully be applied to deconvolute the effect of the PCTF and to retrieve specimen information. Finally, the "lattice fringe" resolution is the finest spacing visible in the one-dimensional fringes which can result from interference between two, or more, diffracted beams. Although these fringes reflect the instrumental stability of the HREM, they do not convey any useful information about local atomic arrangements.

There are two common methods used for determining the resolution limits of a microscope, as well as its other variable parameters. Images taken of amorphous objects are placed in an optical bench and the resulting ring patterns, called optical diffractograms, can be analyzed to determine, for example, Δf, C_s, the focal spread, the angle of illumination, and any residual image astigmatism. Alternatively, through-focal series from known test objects can be compared with image simulations as a means of establishing most of these parameters (Wilson, Spargo and Smith, 1982).

IMAGE SIMULATIONS

Weak-Phase-Object images

While WPO images may only represent a qualitative approximation to the actual HREM micrograph, it is, nevertheless, often very useful in practice to compute WPO images for the sample of interest as a function of image resolution. In such computations, the structure factors of the various diffracted beams are artificially given unit weight out to the desired resolution and zero thereafter. These WPO images are comparatively quick and simple to calculate, since the effects of beam divergence, focal spread and the lens transfer function are not included, but they provide a rough guide to the resolutions required to image particular features of the object. For example, the WPO images of the 15R polytype of SiC, shown in Fig. 5 (Smith and O'Keefe, 1983), indicate that a resolution of about 2.4Å would be required in order to distinguish the polytypic stacking sequence and that about 0.9Å would probably be needed in order to distinguish the separate Si and C atomic columns on the basis of intensity levels. In practice, slightly higher resolutions would be required because the WPO image simulations give too much weight to the higher order diffracted beams compared to the real transfer properties of the objective lens where the damping effects reduce these beams.

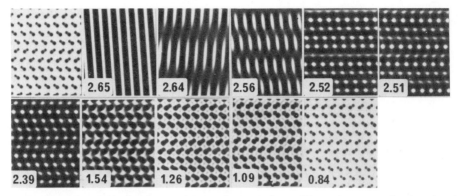

Fig. 5. Series of Weak-Phase-Object images from 15R polytype of SiC showing
the effect of increasing resolution.

Multislice simulations

The most universal method used for the simulation of high-resolution
images is based on the multislice algorithm (Goodman and Moodie, 1974),
derived from the theory of Cowley and Moodie (1957a) (see also Chapter by
Gibson). Computer packages have been available for some time (for example,
O'Keefe and Buseck, 1979), and some practical experience with these
programs is commonly regarded as an excellent means whereby the newcomer
can gain valuable insight into the extreme sensitivity of high-resolution
images to the specimen and imaging parameters. Nowadays, the programs are
fast, highly interactive and "user-friendly", making it a quick and simple
matter to adjust any one of a whole host of variables and almost
immediately view the consequences.

The use of image simulations must be considered as essential in any
quantitative high-resolution study. Figures 6 and 7 show, respectively,
examples of through-thickness and through-focal series which were generated
during an attempt (eventually successful) to determine the structure of
twin boundaries in tin dioxide (Smith, Bursill and Wood, 1983). In the
thickness series, note the differences in "extinction" effects of the
various diffracting beams: in [001], half-spacing detail is produced,
whereas in [100] and [111], the image details become predominantly one-
dimensional. In the focal series, note the reversals in contrast of the
major contrast features, particularly in [100] where the complete period of
reversal seems approximately equal to 900Å.

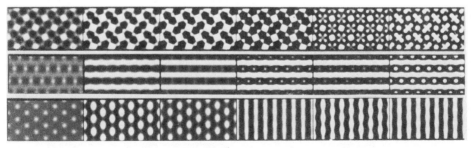

Fig. 6. Image simulations for SnO_2 in various orientations showing effect
of increasing crystal thickness (500kV; C_s=3.5mm; Δf=-865Å).
Top row, [001], 31.9Å, 63.7Å, 95.6Å, 127.5Å, 159.3Å and 191.2Å;
Middle, [111], 29.7Å, 59.4Å, 89.0Å, 118.7Å, 148.4Å, and 178.1Å;
Bottom, [100], 28.4Å, 56.8Å, 85.3Å, 113.7Å, 142.1Å and 170.5Å.

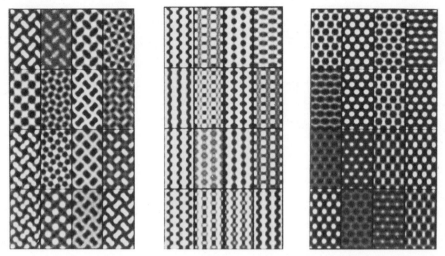

Fig. 7. Through-focal series of simulations for SnO_2 (500kV; C_s=3.5mm),
starting at -1400Å (top left), with +100Å steps (top to bottom),
and finishing at +100Å (bottom right). left: [001], 31.9Å thick;
middle: [111], 29.7Å thick; right: [100], 28.4Å thick.

Interference effects and image artefacts

In small unit cell materials, when relatively few diffracted beams
contribute to the image, identical images will recur periodically with
changes of defocus. These images, termed Fourier or "self"-images of the
lattice, with a period given by $2d^2/\lambda$ (Cowley and Moodie, 1957b), make it
impossible to determine the objective lens defocus unambiguously without
reference to a crystal defect or to the characteristic appearance of the
Fresnel fringe along the crystal edge. Another common and misleading
artefact, which often occurs in the vicinity of the first thickness
extinction contour, is the splitting of white spot contrast into two
separate white spots which, in the case of silicon and other tetrahedrally-
bonded semiconductors, can be mistaken for "atom-pair" images. These image
artefacts result from second-order interference effects between diffracted
beams which are visible when the transmitted beam is of low intensity.
Examples of "dumbbell" images for [110] gold and [100] SnO_2 are shown in
Fig. 8-in neither case do the white spots correspond to projections of atom
columns!

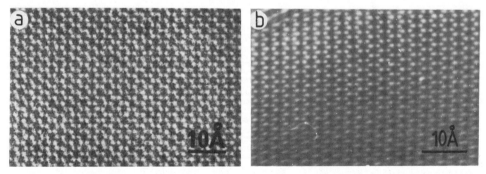

Fig. 8. Interference lattice images from (a) SnO_2 and (b) Au showing
apparent dumbbell structures.

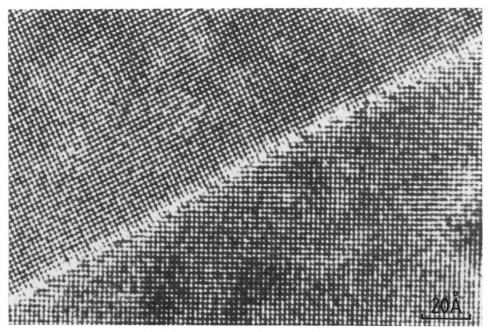

Fig. 9. Interface between two <001> crystals of nickel oxide.

APPLICATIONS OF ATOMIC-RESOLUTION IMAGING

It should now be apparent that, provided the region of interest is
sufficiently thin, it is comparatively straightforward with contemporary
HREMs to record electron micrographs containing atomic-resolution detail.
In order to interpret at least some of the image features, however, the
micrograph should be recorded close to the optimum objective lens defocus.
It may first be necessary to locate the more distinct focus corresponding
to minimum edge contrast, ie. zero Fresnel fringe, before offsetting by the
desired amount to attain the optimum conditions of Fig. 4.

The interface between two <001>-oriented crystals of nickel oxide,
shown in Fig. 9, was recorded under these circumstances (Merkle and Smith,
1987). From similar high-resolution images, it was possible to deduce, for
example, that crystallinity in the NiO was always retained right up to the
interface irrespective of the angular misorientation between the two
grains. The development of distinct facetting along the interface,
apparently as a means of minimizing the interface energy, was also
observed. Note, however, that the accurate location of atomic columns along
the interface, which is information required to substantiate theoretical
lattice statics calculations, can not be determined without careful
quantitative matching between image simulations and through-focal
micrograph series.

An interface between silicon/silicon carbide is shown in Fig. 10. In
this case, the SiC had been grown by a metallo-organic-chemical-vapor-
deposition process, after an initial carburization of the Si substrate, and
there was great interest in the quality of the SiC, in particular the
structure of any growth defects, as well as the nature of the interface and
any topotactical relationship between the two lattices (Nutt et al, 1987).
Qualitative information deduced directly from the high-resolution
micrographs provided adequate answers to these important materials'
questions, and there was no necessity for a comprehensive analysis.

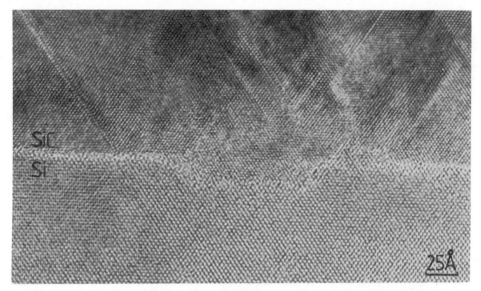

Fig. 10. Interface between crystal of Si and SiC. Note the apparent
continuity of some (111) planes across the interface.

Typical profile images from an extended Au foil are shown in Fig. 11.
Topographical information about the various Au surfaces, for example,
details of steps, terraces and surface twins, could be directly deduced
(Smith and Marks, 1984). However, quantifying the extent of any surface
contraction/expansion and the nature of any impurities, was a lengthy
process involving many computations (Marks, 1984). As shown in Fig. 12, it
was found that the "apparent" expansion at the surface, as it would be
measured from the high-resolution electron micrograph, was not the same as
the "true" expansion. It was also found that the real positions of the
atomic columns at the surface could be determined to within 5% (ie. ±0.2Å)
of the lattice parameter.

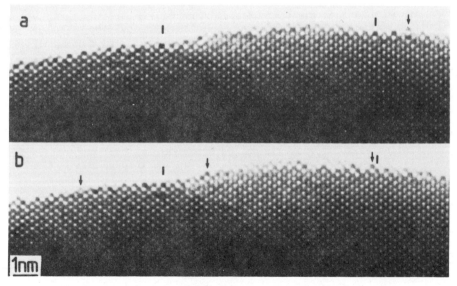

Fig. 11. Successive profile images from rough Au (110) surface, showing
changes in atomic column positions (from Smith and Marks, 1984).

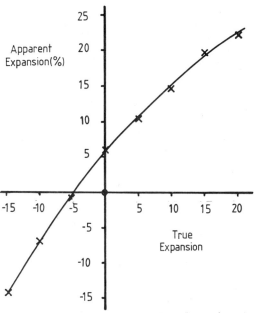

Fig. 12. Relationship between apparent and real relaxations at a gold (111)
surface (after Marks, 1984).

PERSPECTIVE AND OUTLOOK

Unlike the broad beam diffraction methods which provide information
averaged over large specimen regions, the technique of high-resolution
electron microscopy is unsurpassed as a means for characterizing local
disorder. Invaluable qualitative information about surfaces and interfaces
are comparatively easy to obtain. In general, however, there is no linear
relationship between image intensity and projected potential. Therefore,
in order to obtain quantitative information, for example about the accurate
location of atomic arrangements or the extent of relaxation at surfaces and
interfaces, usually requires more effort. Comparisons of image simulations
and through-focal image series are needed to establish the specimen and
microscope parameters before the detailed nature of any crystalline defects
or local atomic rearrangements can be determined unambiguously.

It should always be remembered that the high-resolution image
generally represents a two-dimensional projection of the specimen along the
beam direction (but see the later chapters on REM by Cowley). It is not
always valid to assume that the projected atomic structure of the sample is
uniform and, indeed, disorder along the beam direction is often apparent in
many electron micrographs. Moreover, the high-resolution technique is not
always the best choice of the available microscopy modes to solve some
interface or surface problems. For example, scattering from surface
(ad)layers is most apparent using "forbidden" reflections (Howie, this
volume) and Fresnel methods can provide precise data about bulk lattice
displacements at grain boundaries (Stobbs, this volume). The limited
vacuum available in most electron microscopes (10^{-6}–10^{-7} Torr) obviously
restrict the ranges of surfaces which can be imaged and the possibility of
reaction with residual gases induced by the incident electron beam should
be appreciated (Smith, this volume). Nevertheless, the power of electron
microscopy methods for characterizing surfaces should become apparent from
other contributions to this volume.

REFERENCES

Cowley, J.M., 1981, "Diffraction Physics", North-Holland, Amsterdam.

Cowley, J.M. and Moodie, A.F., 1957a, Acta Cryst., 10:609.

Cowley, J.M. and Moodie, A.F., 1957b, Proc. Phys. Soc., B70:486.

Goodman, P. and Moodie, A.F., 1974, Acta Cryst., A30:280.

Hawkes, P.W., 1978, in: "Advances in Optical and Electron Microscopy", R. Barer and V.E. Cosslett, eds., Academic Press, London, Vol. 7:101.

Marks, L.D., 1984, Surface Sci., 139:281.

Merkle, K.L. and Smith, D.J., 1987, Ultramicroscopy, 22:57.

Nutt, S.R., Smith, D.J., Kim, H.J. and Davis, R.F., 1987, Appl. Phys. Lett. 50:203.

O'Keefe, M.A. and Buseck, P.R., 1979, Trans. Am. Cryst. Assoc., 15:27.

Saxton, W.O., 1978, "Computer Techniques for Image Processing in Electron Microscopy", Academic, New York.

Smith, D.J., Bursill, L.A. and Wood, G.J., 1983, J. Sol. State Chem., 50:51.

Smith, D.J. and Marks, L.D., 1984, Ultramicroscopy, 16:101.

Smith, D.J. and O'Keefe, M.A., 1983, Acta Cryst., A39:139.

Van Dyck, D., 1978, in: "Diffraction and Imaging Techniques in Material Science", S. Amelinckx, R. Gevers and J. Van Landuyt, eds., North-Holland, Amsterdam.

Wilson, A.R., Spargo, A.E. and Smith, D.J., 1982, Optik, 61:63.

PROFILE IMAGING OF SMALL PARTICLES, EXTENDED SURFACES AND DYNAMIC SURFACE
PHENOMENA

David J. Smith

Center for Solid State Science and Department of Physics
Arizona State University, Tempe, Arizona 85287, USA

INTRODUCTION

     The transmission electron microscope (TEM) can be used in several
different geometries to provide atomic-level information about surfaces.
These operating modes include: conventional diffraction contrast imaging,
in both bright-field and dark-field; reflection electron microscopy (REM),
with the incident electron beam striking the surface of the sample at a
glancing angle; and profile imaging, when the TEM is operated with a large
objective aperture as though for high resolution and the surface profile is
imaged. For a microscope capable of atomic resolution, the profile image
can reveal the projected arrangement of atomic columns along the surface.
Topographical features such as steps and terraces are readily visible. The
profile imaging technique is also unsurpassed as a means for monitoring
changes in surface morphology occurring in real time. Motion of atomic
columns, surface rearrangements and a variety of surface reactions can be
observed. In this chapter, the electron microscopical approaches to
imaging surfaces are described and the various types of surface information
available are compared. The results of profile imaging studies of small
particles, surfaces and dynamic surface phenomena are reviewed and image
simulations and developments in instrumentation are briefly discussed.

SURFACE IMAGING METHODS

     The TEM geometries available for characterizing surfaces are
represented schematically in Fig. 1 and some common acronyms are listed in
Table I. In the case of the scanning microscopy modes, namely SREM and
STEM, the relative configurations are identical with the only differences
being the reversal of the electron beam direction and, obviously, the
location of the final image.

     The bright-field TEM mode, Fig. 1(a), where the transmitted
(unscattered) beam contributes to the image, is the configuration most
familiar to all microscopists. Depending on the particular material, and
the size of the objective aperture, certain surface features may be
visible. With the aperture size chosen to exclude the diffracted beams,
then the surface topography may be revealed by either focus-dependent phase
contrast or by diffraction contrast. In the former mode, the (crystalline)
sample is tilted away from any orientations where strong low-index

TABLE I
Common acronyms in electron microscopy

BF	Bright-Field
DF	Dark-Field
HREM	High-Resolution Electron Microscopy
REM	Reflection Electron Microscopy
SREM	Scanning Reflection Electron Microscopy
STEM	Scanning Transmission Electron Microscopy
TEM	Transmission Electron Microscopy
WBDF	Weak-Beam Dark-Field

reflections are excited. Provided that there is no underlying contrast from the specimen support, a situation which can be arranged by employing an appropriate holey film, then surface details are visible in either slight underfocus or overfocus conditions. Alternatively, the specimen is tilted to excite a particular diffracted beam, or a systematic row of reflections. Under these circumstances, the image contrast can be highly sensitive to thickness differences, particularly when the sample thickness is close to the condition for extinction of the bright-field beam where it has a minimum in intensity due to dynamical scattering. Although the objective aperture size may limit the lateral image resolution to 5Å or more, surface steps can be seen with appreciable contrast even when they are only of single-atom height (Lehmpfuhl and Warble, 1986). A great enhancement of contrast can sometimes be obtained by tilting the crystal normal well away from the incident beam direction while still retaining one systematic row of reflections in a strongly diffracting condition. Klaua and Bethge (1983) demonstrated this possibility in an Au foil which was estimated to have an average thickness of only 3.5 monolayers. Finally, another BF method which can be used to image surface monolayers, with lateral atomic resolution as well, relies on the weak scattering into normally forbidden reflections which occurs when (see Chapter by Howie) there are partially unfilled projected unit cells in the beam direction (eg. in <100> and <111>). The objective aperture must be large enough to pass the "forbidden" beams which are surface sensitive, but exclude the diffracted beams of the bulk crystal which are not. This method has been successfully demonstrated for [100] Ag foils by Takayanagi et al (1983).

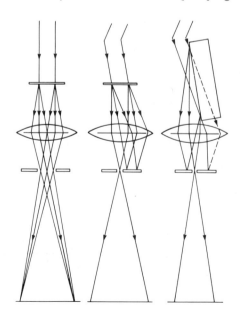

Fig. 1 Schematic showing the different geometries used for imaging surfaces in the electron microscope. (a) bright-field TEM; (b) dark-field TEM; (c) REM.

When the objective aperture is enlarged sufficiently to transmit the bulk-diffracted beams, then the surface detail is not normally visible above the high-contrast lattice image of the bulk crystal. However, under circumstances where the specimen is thin enough for structure imaging (see Smith, this volume), then the profile of the crystal is visible with high-resolution. The technique, now called surface profile imaging, was initiated with observations of CdTe by Sinclair and colleagues (1981) and then extended to observations of Au and Ag metal particles with atomic resolution by Marks and Smith (1983a). As will be shown later, profile imaging is ideally suited to following dynamic processes in real time with atomic resolution.

The dark-field TEM mode, Fig. 1(b), can provide surface images of very high contrast, particularly when advantage is taken of forbidden reflections to highlight variations in specimen thickness. Diffracted beam amplitudes are a serious limitation, however, and the exposure times of 60s or more seem to represent a serious deterrent to more frequent utilization of the technique. The method was initiated by Cherns (1974) who studied surface step topography in thin gold foils using forbidden reflections, but it has since been extended by several other worker (for example, Lehmpfuhl and Uchida, 1979) who have used the more conventional weak-beam DF mode. Note that the direction of the incident beam is usually tilted so that the diffracted beam used for imaging is located on the optic axis of the objective lens. Off-axis aberrations do not then degrade the image unnecessarily, though the image resolution is still limited by the aperture size.

In the REM mode, Fig. 1(c), the incident beam is tilted through angles of 1°-2° before striking the sample at glancing incidence angles. Discussions of imaging geometries, the theory of REM imaging, and examples of the technique which make it ideal for bulk samples, are not given here since they are described elsewhere in this volume by Cowley.

PROFILE IMAGE SIMULATIONS

With the latest generation of HREMs it is comparatively simple to obtain images, at least from thin specimen regions, which can be related in an intuitive manner to a two-dimensional projection of the atomic structure of the object (Smith, this volume). Thus, the features of profile images which have been recorded near the optimum defocus should be interpretable in terms of the atomic arrangements along the edge of the specimen. Nevertheless, because of the possibility of artefactual detail arising from (Fresnel) edge diffraction effects, and since some assessment of the reliability of profile images is required, image simulations are generally advisable. Practical difficulties can arise in the process of computing profile images because of the abrupt termination of material at the crystal edge, and the so-called "periodic continuation", involving an artificially large unit cell, is required to eliminate computing artefacts (Wilson and Spargo, 198x). Several workers have, however, reported successful simulation studies, so these problems are apparently not insurmountable.

The most comprehensive computer investigations have been documented by Marks (1984) who was concerned both to establish the validity of the original atomic-resolution profile images (Marks and Smith, 1983a), and to determine the accuracy with which the atomic columns along a reconstructed Au particle surface could be located. Good agreement with the experimental micrographs of Au surface profiles was obtained, even to the extent of determining, by careful comparison with images recorded at the reverse contrast ("white dot") focus for the bulk lattice, whether the Au surfaces were free of any carbon contamination layer. Further computations showed

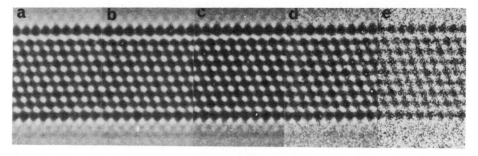

Fig. 2. The effect of shot noise on image clarity, with electron doses
(from 1. to r.) of ∞, 10^6, 10^5, 10^4 and 10^3 e$^-$/$\mathring{A}^2$ (from Marks,
1984).

that there were slight deviations between the apparent surface relaxation
as measured from images and the true relaxations (Marks, 1984). However,
armed with a suitable calibration curve (see Fig. 12, Smith, this volume),
it was possible to locate the surface atomic columns to within +5% of the
lattice parameter (i.e. ±0.2Å). Such high accuracy might, at first sight,
seem contradictory for an HREM having a structural resolution of about
1.8Å, but it should be realized that what are being determined are <u>relative</u>
positions. The signal-to-noise of the micrograph, which depends primarily
on the shot noise of the incident electron beam and the photographic
emulsion, has far more effect on this determination - see Fig. 2 from the
work of Marks (1984) - and for this reason it is important that the
experimental micrographs be recorded at an adequate magnification. It was
later shown by Saxton and Smith (1985), with the assistance of a through-
focal series restoration, that atomic columns might sometimes be located to
better than one-twentieth of the microscope resolution limit despite
apparent variations of atomic separations by 20% or more during a focal
series.

Recently, there has been a great deal of interest in the modes of
reconstruction of clean semiconductor surfaces. Image simulations predict,
for example, that different models for the Si {111} 7x7 surface recon –

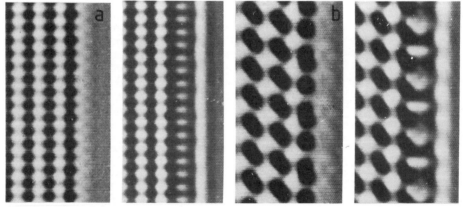

Fig. 3. Simulated GaAs profile images at 400kV of the vacancy-induced 2x2
reconstruction of Ga(111) surface: (a) [112] projection, thickness of 27Å,
defoci of -500Å(1), -1000Å(r); (b) [110] projection, 32Å, -500Å(1),
-1000Å(r).

struction could be distinguished using the profile imaging method (Krivanek and Wood, 1984), though it is not yet clear whether the Si crystal faces, which must be narrow along the beam direction, could support an extended reconstruction (Gibson, this volume). The 2x1 (110) surface reconstruction is common to many compound semiconductors (Duke, 1983). Fig. 3 shows simulations for the vacancy-induced 2x2 reconstruction of the GaAs (111) surface which again suggest that by using experimental micrographs it should be possible to discriminate between the likely structures provided that they are not too similar (Lu and Smith, 1985).

SMALL PARTICLES

A characterization of the microstructure of small metal particles is of fundamental importance, given the pivotal role that these particles play in such fields as heterogeneous catalysis and thin-film growth. The atomic-level detail which can be extracted by careful application of the HREM technique should therefore be extremely useful (Marks and Smith, 1983b; Marks, 1985; Iijima, 1986; Gai et al., 1986). As an example, Fig. 4 shows a high-resolution electron micrograph of a decahedral-shaped multiply-twinned particle of gold, recorded in this case with the reverse "white dot" contrast so that the positions of individual atomic columns appear white. The five twin boundaries present in this particle are clearly visible. Moreover, of more immediate relevance to our interest in surfaces, it is possible to map out most of the external surface profile of the individual tetrahedral segments. Facetting of the particle has occurred, with the development of extra (111) surfaces, leading to a more rounded equilibrium shape (Marks, 1985). The contrast from the underlying (carbon) support film can, however, obscure some details of the particle shape and edge definition (Gai et al., 1986). Details of surface micro-facetting can be imaged with greater visibility when the surface profile protrudes beyond the edge of the support.

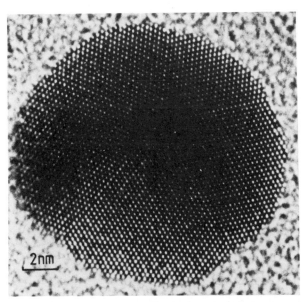

Fig. 4.   High-resolution image of a decahedral multiply-twinned particle of gold recorded at the reverse contrast position (atomic columns appear white). Note the microfacetting of the surface on each tetrahedral segment (from Marks and Smith, 1983b).

EXTENDED SURFACES

A major drawback to imaging the entire profile of a small particle is
obviously that overlapping contrast from the support must occur in some
places. For self-supporting materials such as thin metal films and bulk
semiconductor crystals, or even for large oxide particles supported on
holey carbon substrates, it is usually possible to locate extensive regions
where the images of the surface profiles are free from any support
interference. The "only" remaining problem is to ensure that the surfaces
are clean?!

Surface cleanliness has not proven to be a serious limitation in our
observations of oxide surfaces (Smith, Bursill and Jefferson, 1986). Our
normal technique of sample preparation for microscopy has usually involved
crushing crystals of the oxide under a solvent such as ethanol or
isopropanol, paying careful attention to using clean utensils and pure
solvents. In some cases, thin layers of surface contamination (<10°Å) might
still remain but these would invariably be etched away during the initial
observation periods. Figure 5 shows a crystal of uranium dioxide, imaged
in a [001]-type projection, which has been prepared in this manner: profile
images of the (100) and (110) surfaces at A and B are shown enlarged in
Figs. 6(a), (b), and it is clear that the surfaces are remarkably free of
any form of carbonaceous deposit. Note the high visibility of the metal
atom columns at the surface and the clarity of the surface steps and
ledges   (Bursill et al., 1985).

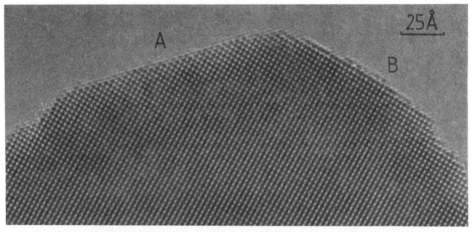

Fig. 5.   High-resolution electron micrograph of <001> uranium dioxide with
           (110) surface profile at A and (100) profile at B.

Fig. 6.   Enlargements from Figure 5 showing profile images of (a) (110)
           and (b) (100) surfaces of <100> uranium dioxide.

48

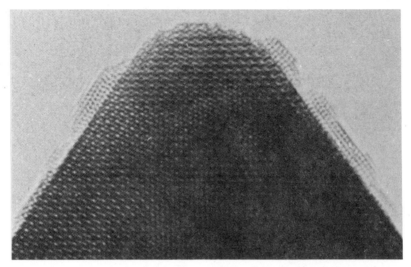

Fig. 7 Profile image from a crystal of ZnCrFeO$_4$ spinel catalyst showing
growth of small ZnO surface rafts (from Hutchison & Briscoe, 1985)

Many oxides represent good catalysts and there are strong reasons for
believing that surface morphology plays a large role in determining
relative catalytic activity. Hutchison and Briscoe (1985) have studied a
number of oxide spinels before and after their use as catalysts. The
profile image in Fig. 7 shows the presence of small rafts, believed to be
ZnO, which have formed on the (111) surface of a ZnCrFeO$_4$ catalyst
following a catalytic reaction.

Unlike the surfaces of oxides, it is much more difficult to obtain
clean surfaces of semiconductors. We have had limited success with
compound semiconductors, such as CdTe and GaSb, by crushing them with a
mortar and pestle under isopropanol and then transferring the holey support
film into the airlock of the microscope while it is still wet with the
solvent (Lu and Smith, 1987). In some cases, such as the crystal of <110>
CdTe shown in Fig. 8, the crystal profile is found to be free of
contaminants and surface reconstruction will then take place driven by the
excess energy associated with the dangling bonds at the surface. In
general, this does not occur and it is necessary to provide both in situ
specimen treat- ment facilities and an ultra-high-vacuum environment
(preferably $<10^{-9}$ Torr). Under such conditions, Gibson and colleagues
(1985) were able to obtain profile images from samples of Si, including a
novel and unexpected 1x1 structure on the reconstructed (113) surface.
Another possibility, as demonstrated by the studies of Ge by Nihoul et al
(1986) is to characterize the surface profile buried beneath an amorphous
surface layer, but genuine reconstruction is then unlikely to be observed.

Fig. 8 Profile image from a <110> CdTe crystal showing the development of
twinning on a clean [111] surface.

Profile imaging of metal surfaces presents a greater challenge to the microscopist since the projected separations between the close-packed columns of metal atoms are closer to the theoretical resolution limit of the HREM instrument. Moreover, it can be a serious problem to maintain surface cleanliness since metals such as palladium and silver are found to oxidize during electron irradiation under normal vacuum conditions (Lodge and Cowley, 1982; Ye and Smith, 1986). Nevertheless, in the case of gold surfaces, which remain clean once the overlying carbon film has been etched away, an interesting variety of surface structures have been observed. Examples include a partial reconstruction of the (110) surface into a 2x1 "missing-row" structure (Marks and Smith, 1983), development of hill-andvalley structures on (111) surfaces (Marks, Heine and Smith, 1984) and irregularities on (100) surfaces (Smith and Marks, 1985). Figure 9 shows the surface profile image of a large gold particle which was grown in situ by Hasegawa et al. (1986). Note the unusual corrugations on the (00$\bar{1}$) surface and an apparent slight relaxation on the (1$\bar{1}$1) surface.

DYNAMIC SURFACE PHENOMENA

A wide variety of dynamic surface phenomena, including diffusion, rearrangements and electron–beam–induced surface reactions, have been studied at the atomic level using the profile imaging technique (Smith, 1985). The first observations were by Sinclair et al (1981) who published a sequential series of micrographs which showed rearrangements along the surface of a CdTe sample. Dynamic rearrangements on extended gold surfaces were later reported by Smith and Marks (1985b), who found that different characteristic behaviour occurred, depending on the particular surface, even under identical imaging conditions. For example, extensive changes on (100) surfaces only took place in the vicinity of surface steps, whereas (111) surfaces were found to develop a hill–and–valley morphology (Marks et al., 1984), and (110) surfaces exhibited considerable movement of atomic columns between successive exposures. Subsequent observations of dynamic events in small gold particles took advantage of the development of high-quality online TV viewing and recording facilities. These made it possible, for example, to document the growth process, row by row, of small gold particles (Wallenberg et al., 1985), to observe near-surface phenomena

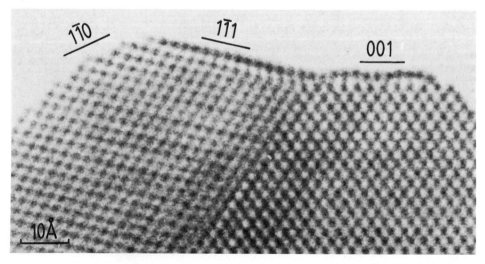

Fig. 9.   Profile image of a large Au particle showing (1$\bar{1}$0), (1$\bar{1}$1) and (001) surfaces. Note the corrugated structure of the (001) surface due to reconstruction (from Hasegawa et al, 1986).

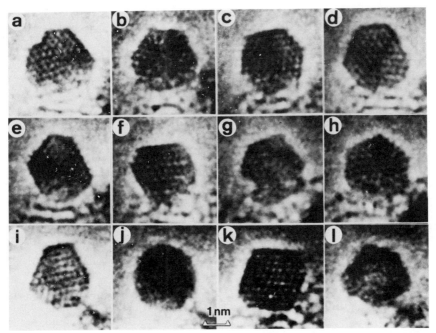

Fig. 10.   Series of electron micrographs showing various shapes of a small
          gold cluster as reproduced from single frames of a video-tape
          recording (from Iijima, 1986).

(Bovin et al., 1985) and to follow the fascinating structural
rearrangements in small metal particles which occur under intense electron
irradiation (Iijima and Ichihashi, 1986; Smith, Petford-Long et al., 1986).
As shown in Fig. 10, from Iijima (1986), small clusters can rapidly change
from single crystals to various twinned shapes including icosahedral and
decahedral. This motion has been shown to depend on such factors as the
particle size, the substrate material and the beam current, and it has been
suggested that similar structural changes could occur in small catalyst
particles when they are heated (Smith et al., 1986). This possibility
remains to be proven, and is also subject to recent debate by Williams
(1987), who believes that the dynamical particle behaviour can be explained
in terms of brief Auger excitation processes.

    Surface modifications induced by electron beam irradiation include
several different chemical processes so that the microscopist must always
remain aware of the possibility that the surface phase he has just detected
could have been caused by the electron beam. Mention has already been made
of the oxidation of silver and palladium metal foils which only seems to
occur once any overlying carbonaceous material has been etched away (Lodge
and Cowley, 1982; Ye and Smith, 1986). In the case of indium III-V
compound semiconductors it was observed (Petford-Long and Smith, 1986),
that irradiation led to the formation of irregular crystals of indium
oxide, In_2O_3, along the surface, a result which was attributed to the
initial loss of the anion species by some sort of radiolytic process
followed by oxidation of the residual metal. Irradiation of CdS led,
however, in some cases, to the accumulation of Cd metal, rather than the
oxide, along the surface (Smith and Ehrlich, 1986), presumably because the
latter is unstable under electron irradiation. It was also significant
that, even at 100keV, Cd was observed to form at the surface, since this
beam energy is below the threshold for direct knock-on displacement of
either Cd or S atoms.

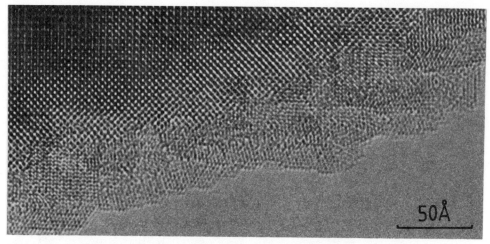

Fig. 11. Electron micrograph of TiO_2 showing the development of tetragonal titanium monoxide due to electron irradiation at 400keV.

Under intense irradiation ($>25A/cm^2$), considerably greater than normally used for high-resolution imaging, oxides of maximally-valent transition metals, such as Nb_2O_5, V_2O_5 and TiO_2, develop thin strips of fine fringes along the crystal edges. Using bulk lattice spacings as a reference calibration, these fringes have been identified as corresponding to the respective binary oxides (Smith et al., 1987), which are presumably formed as a result of electron-stimulated desorption (ESD) of oxygen from the near- surface region (Knotek and Feibelmann, 1979). Figure 11 shows an example of the development of tetragonal-shaped lattice fringes on the surface of a titanium dioxide crystal which match the spacings of titanium monoxide. The reduced oxides generally have a defective rocksalt structure and it is significant that they are also metallic in nature, meaning that there would then be a ready supply of conduction-band electrons which should stifle any further desorption process. It is interesting that previous Auger studies of ESD processes had reported depletion of oxygen from the surface, but not its complete removal (Lin and Lichtman, 1979).

DISCUSSION AND OUTLOOK

The application of profile imaging to the characterization of surface morphology and surface reactions has been a very recent development relative to most other techniques for surface analysis (with scanning tunneling microscopy being the notable exception). Nevertheless, the information about surfaces which is immediately available, such as the distribution of steps, the presence of reconstructions and the nature of surface phases, suggests that profile imaging should become a technique which will find increasing application in surface investigations. Moreover, it is unique in its ability to follow dynamic events in real time. So far, the electron- beam-stimulated processes which have been studied have taken place under poorly-controlled environmental conditions. The presence of unknown impurities could, for example, seriously influence the surface reactions which have been observed. Several groups are working towards truly ultrahigh-vacuum instruments (eg. Ponce et al., 1986; Takayanagi et al., 1986; Swann et al., 1987), which is no easy task given the very confined space within the region of the objective lens pole-pieces, and in situ cleaning and treatment facilities are also being made available. Results from these effectively prototype instruments are of special significance since they should settle the burning issue of whether electron microscopic methods will have a lasting impact on many surface science studies.

FURTHER READING/ACKNOWLEDGEMENT

In this brief review of profile imaging, the literature has been cited extensively so that the interested reader could easily identify original sources of material. Broader perspectives on the applications of EM methods to surface structure can be found elsewhere in this volume and, for example, in Takayanagi (1984), Cowley (1986), Yagi (1987) and Smith (1986). The last of these references also contains comprehensive bibliography of EM surface studies and lists the materials which had been studied at the time of writing, by the various TEM methods.

I am pleased to acknowledge numerous colleagues for their collaboration on some of the studies reported here, and I would also like to thank Prof. L.D. Marks, Dr. J.L. Hutchison, Dr. S. Iijima and Prof. K. Yagi for kindly providing the simulations and micrographs used in Figs. 2, 7, 9 and 10 respectively.

REFERENCES

Bovin, J.-O., Wallenberg, L.R. and Smith, D.J., 1985, Nature, 317:47.

Bursill, L.A., Peng Ju Lin, Smith, D.J. and Grey, I.E., 1985, in "Electron Microscopy and Analysis 1985", P. Goodhew, ed. (Adam Hilger, London and Bristol) pp. 467-470.

Cherns, D., 1974, Phil. Mag., 30:549.

Cowley, J.M., 1986, Progress Surf. Sci. 21:209.

Duke, C.B., 1983, J. Vac. Sci. Technol., B1:732.

Gai, P.L., Goringe, M.J. and Barry, J.C., 1986, J. Microscopy, 142:9.

Gibson, J.M., McDonald, M.L. and Unterwald, F.C., 1985, Phys. Rev. Letts., 55:1765.

Hasegawa, T., Kobayashi, K., Ikarashi, N., Takayanagi, K. and Yagi, K., 1986, Japan. J. Appl. Phys. 25:L366.

Hutchison, J.L. and Briscoe, N., 1985, Ultramicroscopy 18:435.

Iijima, S., 1985, J. Electron Microsc. 34:249.

Iijima, S. and Ichihashi, T., 1986, Phys. Rev. Letts., 56:616.

Klaua, M. and Bethge, H., 1983, Ultramicroscopy, 11:275.

Knotek, M.L. and Feibelmann, P.J., 1979, Surface Sci., 90:78.

Krivanek, O.L. and Wood, G.J., 1985, in: "Proc. 43rd Ann. Meet. Electron Microscope Society of America", G.W. Bailey, ed. (San Francisco Press, San Francisco) pp. 262-263.

Lehmpfuhl, G. and Uchida, Y., 1979, Ultramicroscopy, 4:275.

Lehmpfuhl, G. and Warble, C.E., 1986, Ultramicroscopy, 19:135.

Lin, T. and Lichtman, D., 1979, J. Maters. Sci., 14:455.

Lodge, E.A. and Cowley, J.M., 1982, Ultramicroscopy 13:213.

Lu, P. and Smith, D.J., 1986, in: "Proc. 44th. Ann. Meet. Electron Microscope Society of America", G.W. Bailey, ed. (San Francisco Press, San Francisco) pp. 388-389.

Lu, P. and Smith, D.J., 1987, Phys. Rev. Letts. 59:2177.

Marks, L.D., 1984, Surface Sci., 139:281.

Marks, L.D., 1985, Ultramicroscopy 18:445.

Marks, L.D., Heine, V. and Smith, D.J., 1984; Phys. Rev. Letts. 52:656.

Marks, L.D. and Smith, D.J., 1983a, Nature, 303:316.

Marks, L.D. and Smith, D.J., 1983b, J. Microscopy, 130:249.

Nihoul, G. 1986, in: "Proceedings 44th Ann. Meet. Electron Microscope Society of America", G.W. Bailey, ed. (San Francisco Press, San Francisco) pp. 384-387.

Petford-Long, A.K. and Smith, D.J., 1986, Phil. Mag., A54:837.

Ponce, F.A., Suzuki, S. Kobayashi, K., Ishibashi, Y., Ishida, Y. and Eto, T., 1986, in: "Proc. 44th. Ann. Meet. Electron Microscope Society of America", G.W. Bailey, ed. (San Francisco Press, San Francisco) pp. 606-609.

Saxton, W.O. and Smith, D.J., 1985, Ultramicroscopy 18:39.

Sinclair, R., Yamashita, T. and Ponce, F.A., 1981, Nature, 290:386.

Smith, D.J., 1985, J. Vac. Sci. Technol., B3:1563.

Smith, D.J., 1986, in: "Chemistry and Physics of Solid Surfaces VI", R. Vanselow and R. Howe, eds., (Springer, Heidelberg) Chap. 15.

Smith, D.J. and Ehrlich, D.J., 1986, J. Mater. Res., 1:560.

Smith, D.J. and Marks, L.D., 1985a, Ultramicroscopy 16:101.

Smith, D.J. and Marks, L.D., 1985b, Mater. Res. Soc. Symp. Proc., 41:129.

Smith, D.J., Bursill, L.A. and Jefferson, D.A., 1986, Surface Sci., 175:684.

Smith, D.J., McCartney, M.R. and Bursill, L.A., 1987, Ultramicroscopy, 23:299.

Smith, D.J., Petford-Long, A.K., Wallenberg, L.R. and Bovin, J.-O., 1986, Science, 233:872.

Swann, P.R., Jones, J.S., Krivanek, O.L., Smith, D.J., Venables, J.A. and Cowley, J.M., 1987, in: "Proc. 45th. Ann. Meet. Electron Microscope Society of America", G.W. Bailey, ed. (San Francisco Press, San Francisco) pp. 136-137.

Takayanagi, K., 1984, J. Microscopy 136:287.

Takayanagi, K., Kobayashi, K., Kodaira, Y., Yokoyama, Y. and Yagi, K., 1983, in: "Proc. 7th. Int. Conf. High Voltage Electron Microscopy", R.M. Fisher, R. Gronsky and K. Westmacott, eds. (Lawrence Berkeley Laboratory, Berkeley) pp.47-50.

Takayanagi, K., Tanishiro, Y. Kobayashi, H., and Harada, Y., 1986, in: "Proc. XIth. Int. Cong. Electron Microscopy, Kyoto, 1986", T. Imura, S. Maruse and T. Suzuki, eds. (Japanese Society of Electron Microscopy, Tokyo) 2:1337.

Wallenberg, L.R., Bovin, J.-O. and Schmid, G., 1983, Surface Sci., 156:256.

Williams, P.R., 1987, Appl. Phys. Letts., 50:1760.

Wilson, A.R. and Spargo, A.E., 198 , Phil. Mag. A46:435.

Yagi, K., 1987, J. Appl. Cryst., 20:147.

Ye, H.Q. and Smith, D.J., 1986, in: "Proc. XIth. Int. Cong. Electron Microscopy, Kyoto, 1986", T. Imura, S. Maruse and T. Suzuki, eds. (Japanese Society of Electron Microscopy, Tokyo) 2:959.

TRANSMISSION ELECTRON MICROSCOPY AND DIFFRACTION FROM

SEMICONDUCTOR INTERFACES AND SURFACES

J. Murray Gibson

AT&T Bell Laboratories
600 Mountain Ave
Murray Hill, NJ 07974, USA

INTRODUCTION: SEMICONDUCTOR INTERFACES

The semiconductor industry has provided strong impetus to the microscopic study of interfaces, since the electrical properties of these control the behaviour of semiconductor devices. Furthermore, because of the exceptionally high crystal quality, purity and uniformity of semiconductor crystals and thin films, meaningful structure-property relationships are possible from microscopic analysis of very small volumes. There are two modes in which interfaces or surfaces can be examined by transmission electron microscopy. These are referred to as plan-view ($\underline{n} x \underline{B} = 0$) and cross-section (or profile for surfaces), ($\underline{n}.\underline{B} = 0$), where $\underline{n}$ represents a vector normal to the interface and $\underline{B}$ is the electron beam direction. Plan-view specimens are often prepared by chemical etching whereas cross-sections usually involve ion-milling[1]. High-resolution transmission electron microscopy (TEM) is particularly useful in interface studies, utilizing the cross-section method for specimen preparation.

The Si/SiO_2 interface, for example, has been studied by several workers[2,3] and electron mobility has been correlated with apparent interface step density[4]. Figure 1 shows an image from a somewhat unusual Si/SiO_2 interface, formed by high energy O_2^+ implantation and annealing to form a buried oxide layer[5]. Nevertheless the image reveals the exceptional flatness typical of Si/SO_2 interfaces on (100) Si. The detailed interpretation of such images, taking into account projection effects and signal-to-noise limitations, will be discussed later.

The Si image in fig.1 is in the <100> projection in which single atomic columns can be resolved[6,7]. This was possible with a JEOL 4000EX HREM operating at 400 kV with point-to-point resolution $\approx$ 1.65 Å (see inset optical diffractogram to fig.1). Figure 2 is a stereographic projection of Si (which is appropriate for cubic semiconductors with lattice parameter 5.5 - 6.5 Å) illustrating the directions in which high-resolution microscopy can be performed with an instrument of resolution either ~3 Å or ~1.8 Å. One can see that zone axes of type <110> are accessible at 3 Å resolution and these have been most extensively used (e.g., fig.4c). However, in this projection at such resolution only pairs of atoms can be resolved and no two such zones are closer than 90°. With resolution of the

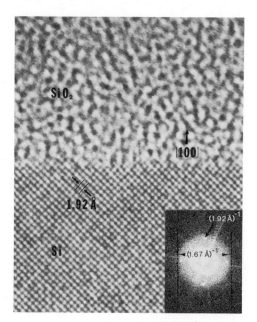

Fig. 1. An axial bright field image of a ~100 Å thick cross-section of a (100) Si/SiO$_2$ interface viewed in the [010] direction. This is the lower interface in a buried Si/SiO$_2$/Si structure created by high-energy O$_2$ ion implantation and annealing. This image reveals all individual atomic columns in projection and was taken with an instrumental resolution of ~1.7 Å near the Scherzer focus.

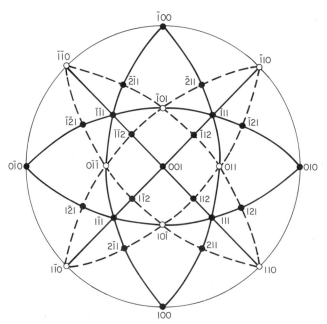

Fig. 2. A stereographic projection of the diamond Si structure which shows the planar spacings which can be resolved at 3 Å resolution (dotted lines) and 1.8 Å resolution (solid lines).

(220) planar spacing at 1.92 Å images in <100>, <111> and <112> zone axes also become possible, the first two representing true projections of the atomic structure[7]. In the study of interfaces in cross-section it is necessary that the interface be viewed edge-on, so zone axes are restricted to those for which $\underline{n} \cdot \underline{B} = 0$. For (100) interfaces, images in either <110> or <100> projections are possible and the same area may be viewed in both projections with a tilt of $45°$. For (111) interfaces, images with <110> and <112> can be obtained with a tilt of only $19°$. The capability to view the same area of interface in different projections is highly desirable, e.g., for studying steps[8] or rigid shift of lattices (see later). Of course, if an interface has uniform and periodic structure it may not be necessary to image the same area in different projections and two samples can be used.

In experiments with electron microscopy and semiconductor interfaces, high resolution images of cross-section samples should only be attempted after it has been confirmed by other techniques that the interface is sufficiently uniform to justify such study. This can be most easily performed with plan-view specimens or perhaps complementary techniques such as Rutherford ion Backscattering (RBS) and ion-channeling. The use of cross-section samples to study defect densities in thin films and at interfaces is a particularly poor example of this concept. For example, fig.3a shows a cross-section bright-field image from a 4 μm thick layer of GaAs on Si[9] from which one might be tempted to make the statement that the surface of the layer is "defect-free". However, a plan-view image, fig.3b, taken with the same diffraction conditions reveals a defect density at the surface in excess of 10^8 cm^{-2}. Of course, the reason for the apparent perfection in fig.3a is the statistics of sampling a very small interface/surface area. For example, in fig.3a the sampled interface area is [1 μm width x 1000 Å specimen thickness] = 10^{-1} μm^2 in contrast with 50 μm^2 in fig.3b. Thus at a defect density < 10^9 cm^{-2} it is likely that no defects will appear in single cross-section image. To re-emphasize this

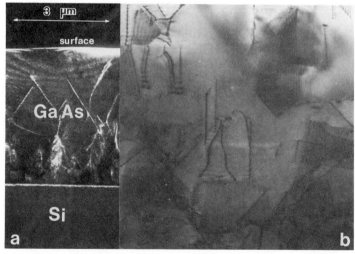

Fig. 3. Two images of a GaAs layer epitaxially grown on Si(100). In the cross-section image (a) threading dislocations do not appear to reach the surface whereas in a plan view image (b) the density of these is seen to be ~10^8 cm^{-2}. The difference occurs because of the poor sampling statistics of cross-section samples. Threading dislocations would be of concern in potential applications for epitaxially grown GaAs on Si.

trivial but poorly accepted lesson, one might reflect that the population density of New York City is ~10^5 km^{-2} and in a short random walk of 1 m with outstretched arms it would be very misleading to conclude that if you touched no-one, the population density of New York City is zero, yet the experiment is statistically similar to cross-section imaging.

It is profitable to divide up the discussion of TEM interface imaging by the nature of the boundary as follows: 1) Crystallographic - which generally implies crystal/crystal boundaries but could include crystal/amorphous boundaries; 2) Compositional - which may include crystal/crystal boundaries with little or no structural difference between the two materials and encompasses amorphous/amorphous boundaries. In the first case, high-resolution imaging can be pushed to its limits with useful results. In the latter case, signal-to-noise and projection limitations usually reduce the effective resolution considerably.

Crystallographic Interfaces

Thin films of metals on semiconductors present an interesting example of crystal/crystal boundaries which have important uses as Schottky barriers and contacts. In the study of microscopic properties of such interfaces, it is particularly interesting to examine epitaxial interfaces since the uniformity of these single-crystal/single-crystal interfaces allows direct microstructure/property correlations. In illustrating techniques for crystal/crystal interfaces, epitaxial thin films are used as examples since these have been the focus of the author's own work. Nevertheless, the ideas apply to general crystal/crystal interfaces which do not show long-range uniformity although often more work is required in these cases to identify the property determining aspects of microstructure.

The term "crystallographic interface" refers to a system in which the crystal structure or orientation of the two different components of the interface are different. These represent the simplest interfaces to study. For example, dark-field diffraction contrast images can be formed from either side of the interface, both in plan-view and cross-section. An example is shown in fig.4 where a 200 reflection from a Si/CoSi$_2$/Si sandwich heterostructure is used to image the silicide layer a) in cross-section and b) in plan-view.

Although CoSi$_2$ has the same crystal structure as Si, in this heterostructure grown by MBE[10], the silicide layer has a 180° rotated orientation with respect to the Si (this twinned orientation is referred to as type B, relative to type A with the same orientation as the substrate). This permits easy imaging of the crystallographic boundary either with separate or coincident reflections. (The 200 reflection which is not kinematically allowed in Si can also be used to image the silicide when it has the A or B orientation). Dark-field images such as fig.4 (a) and (b) are very useful in determining the uniformity of the silicide layer and its interfaces, the presence of misfit dislocations at the interfaces and other defects in the layers, e.g., pinholes through the silicide[11] which are important for applications. A necessary condition for most useful high-resolution imaging of a crystallographic interface is that both crystals have a zone axis parallel to the beam and perpendicular to the interface. This is a very restrictive condition, obviously satisfied by the silicide in the <110> direction as seen in fig.4(c), and is alleviated by higher resolution as can be seen in the stereographic projection (fig.2).

Given high resolution images such as fig.4(c), what is the procedure for interpretation? It is first necessary to perform multislice image simulations for the perfect crystal structures of the two materials involved in the interface. This allows one to confirm qualitatively the structure of the interface, identify appropriate parameters such as crystal thickness, defocus and possible misorientation to be incorporated in full interface image simulations (models for which can be guessed from this procedure) and make rigid shift measurements, consider interface coherence, etc.

The Multi-Slice Algorithm for Image Simulation. The weak-phase object approximation for high-resolution images has been described by Smith in this volume. This approximation is valid only if the phase shift in the specimen is much less than $\pi/4$ and propagation of the scattered electron wave normal to the incident beam direction can be neglected within the specimen. Both of these effects become significant in typical real specimens for electron microscopy, which are of minimum thickness 50 Å. The propagation effect represents the breakdown of the so-called "column approximation". The large phase-shift is characteristic of multiple scattering of electrons and in contrast to the propagation error, is much greater for crystalline oriented specimens than for amorphous ones. A simple prescription for calculating images and diffraction patterns from thick specimens without making either of these assumptions was given by

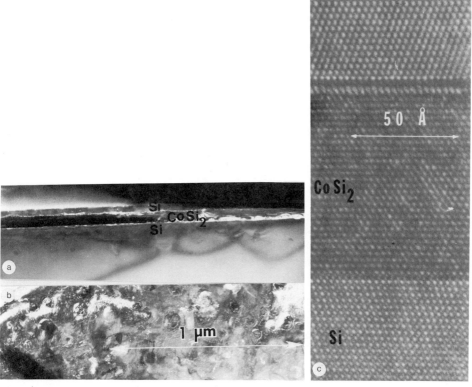

Fig. 4. Images in cross-section a) and plan-view b) of a crystallographic boundary in a Si/CoSi$_2$/Si structure, taken using a 200 Bragg reflection which is forbidden in the Si structure only. c) is a <110> high-resolution cross-section image of a similar interface.

Cowley and Moodie[12]. In their multi-slice algorithm, the specimen is divided up into weak-phase objects by projecting the potential onto slices normal to the incident beam direction. Propagation through vacuum between slices is then assumed. The wavefunction $\psi_{n+1}(\underline{r})$ emergent from the (n + 1)th slice is given from that for the nth slice $[\psi_n(\underline{r})]$ convolved with the propagator $P(\underline{r})$ and then multiplied by the phase-shift from the (n+1)th slice, $\exp[i\sigma\phi_{n+1}(\underline{r})\Delta z]$

$$\psi_{n+1}(\underline{r}) = [\psi_n(\underline{r})\odot P(\underline{r})]\exp[i\sigma\phi_{n+1}(\underline{r})\Delta z]$$

where the propagator $P(\underline{r}) = \exp[ik(x^2 + y^2)/2\Delta z]$, k is the electron wavevector and Δz is the slice thickness. This algorithm can accurately calculate images arising only from elastic scattering, provided the slice thickness and number of beams included in the calculation are correctly chosen. (Details on how to calculate with this algorithm are given by Goodman and Moodie[13] and the program is readily available in FORTRAN from Arizona State University). Cowley and Moodie showed[12] that the method is a complete solution of the Schrödinger equation in its infinitesimal limit. The matrix-method[14] can also be used to simulate multiple scattering and can be modified to exclude the column approximation[15]. The latter algorithm has the attraction for the column approximation that its result is thickness-independent since it involves Bloch-wave solutions to Schrödinger's equation. However, in thin crystals where very large numbers of beams are necessary, the multi-slice algorithm is most efficient, i.e., for high resolution images.

For example, fig.5 shows the results of multislice simulations for a <110> Si specimen at 200 kV. The first results of the multi-slice computation to examine are the beam intensities and phases versus thickness (fig.5a,b). From these it can be seen that the 000 beam has zero intensity at 140 Å. This dark "pendellosung fringe" can be seen usually in experimental images and from this the specimen thickness can be deduced. Note that this thickness will generally be different in the other material abutting the interface. The phases of the beams as a function of thickness are very useful in predicting the nature of the image.

To understand this let us consider the imaging obtained with the three beams $(\underline{K}_0 - \underline{g})$, $\underline{K}_0$, $(\underline{K}_0 + \underline{g})$ in an HREM bright-field image of a centro-symmetric axially aligned crystal. For generality, which we will require later, we will not assume that the microscope is properly aligned so that the beam phase-shifts from the Scherzer aberration function $\gamma(\underline{K})$ are not necessarily $\gamma(\underline{K} + \underline{g}) = \gamma(\underline{K} - \underline{g})$ or $\gamma(\underline{K}_0) = 0$.

For three beams:

$a_0 \exp[i(\phi_0 + \underline{K}_0 \cdot \underline{r})]$

$a_g \exp\{i(\phi_g + [\underline{K}_0 - \underline{g}] \cdot \underline{r})\}$

$a_g \exp\{i(\phi_g + [\underline{K}_0 + \underline{g}] \cdot \underline{r})\}$

$I(\underline{r}) = a_0^2 + 4a_0 a_g \cos[\phi_g - \phi_0 - \gamma(\underline{K}_0) + \varepsilon_1]\cos[\underline{g}\cdot\underline{r} + \varepsilon_2] +$

$\qquad 2a_g^2\cos[2\underline{g}\cdot\underline{r} + 2\varepsilon_2]$ $\qquad\qquad\qquad\qquad\qquad\qquad$ (1)

where

$\varepsilon_1 = [\gamma(\underline{K}_0 + \underline{g}) + \gamma(\underline{K}_0 - \underline{g})]/2$

$\varepsilon_2 = [\gamma(\underline{K}_0 + \underline{g}) - \gamma(\underline{K}_0 - \underline{g})]/2$

for axial alignment $\underline{K}_0 = 0$ therefore $\varepsilon_1 = Y(\underline{g})$ and $\varepsilon_2 = 0$ so that

$$I(\underline{r}) = a_0^2 + 4a_0a_g\cos[\phi_g - \phi_0 + Y(\underline{g})]\cos[\underline{g}.\underline{r}] + 2a_g^2\cos[2\underline{g}.\underline{r}] \qquad (2)$$

The second term in equation 2 represents the lattice image $\cos(g.r)$ which is transferred with amplitude $\cos[\phi_g - \phi_0 + Y(g)]$. For very thin objects (weak phase objects) $\phi_g - \phi_0 = -\pi/2$ so that the transfer function is $\sin Y$. However under dynamical conditions, as we can see from the curves in fig 5b, the phase of diffracted beams change. If $Y(g) = -\pi/2$ (typical of Scherzer) then the lattice frequency image will be extinguished when $\phi_g - \phi_0 = n\pi$. (Note that the third term gives an image with twice the lattice frequency and this is readily seen at such thicknesses). This extinction is expected for 111 beams at t = 130 Å. Below this thickness the image of silicon will represent a true projection of the projected potential (provided the microscope's transfer function is appropriate).

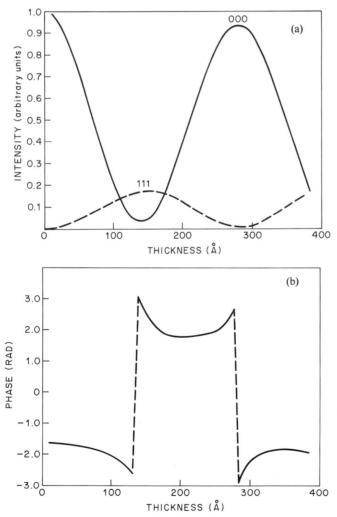

Fig. 5. Multislice calculations of <110> Si at 200 kV involving 75 beams at 100 kV: a) Beam intensities versus thickness; b) Beam phases versus thickness.

Above this thickness the image will reverse contrast and the atoms will be bright and not dark. The thickness at which this image reversal occurs can be seen on the image and often (but not always in the presence of many beams) corresponds to the first minimum in main beam intensity. From image simulations of the perfect crystals one can thus determine the parameters and guess the structure of the interface for subsequent calculations. It is also a good idea to calculate multi-slice beam intensities as the crystal is slightly misaligned from the exact zone axis to investigate these effects. This possibility is particularly important at interfaces because there is quite commonly a very small misorientation from exact alignment of zone axes in both crystals.

Rigid Shift Measurement. From image simulations of the perfect crystals on either side of a crystal/crystal interface several indirect determinations of interface properties can be made. A simple example might be the identification of the interface habit plane. One very useful example of indirect structure determination is the measurement of the rigid shift between the lattices on either side. This measurement is based on the location of atom positions relative to the lattice image of the "perfect" crystal regions on either side of the interface. This technique has been successfully applied to several semiconductor interfaces including $CoSi_2/Si$[16], $NiSi_2/Si$[17,18] and CaF_2/Si[19]. It is easy to obtain an accuracy of a very small fraction of the lattice spacing in such measurements and the theoretical basis of this has been discussed in detail by several authors[20-22]. The accuracy is based on "interferometry" in which the phase of a quasi-monochromatic wave can be determined quite accurately. There are several sources of error which shall be briefly discussed here.

For example, it is not possible to correctly align the incident beam direction $\underline{K}_O$ exactly coincident with the optical axis of the objective lens $[\gamma(\underline{K}_O) = 0]$. In this case it can be seen from equation 1 that a slight phase shift

$$\Delta = \varepsilon_2 = \{\gamma(\underline{K}_O + \underline{g}) - \gamma(\underline{K}_O - \underline{g})\}/2$$

is introduced in the relative location of the lattice image and the atomic positions. The magnitude of this shift is ~0.3 Å for a 3 Å lattice spacing and a 1 mrad beam misalignment. With great care it may be possible to align the microscope to better than 1 mrad accuracy[20] but this gives an order of estimate for accuracy. Other effects which can lead to similar inaccuracy are relative crystal misorientation[21] and the disturbance of the perfect crystal image by the interface. One obvious method to minimize random errors is the simple experimental technique of making several different measurements from different samples. Of course, this does not remove all systematic errors such as small relative crystal tilt and some interface image effects.

Another source of error has recently been exposed in this context in thin crystal films containing strain: the elastic relaxation of strain in thin samples[23-25]. A thin film which is coherently strained by magnitude ε_O in both the x and y directions will exhibit a tetragonal distortion in the free surface direction z of magnitude

$$\varepsilon_{zz} = \varepsilon_O(1 + \nu)/(1 - \nu)$$

where ν is the Poisson's ratio. If one makes a thin film only one unit cell in thickness, then the stress in one direction, say y, is relieved and this causes the tetragonal distortion to collapse to

$$\varepsilon_{zz} = \varepsilon_O(1 + \nu)$$

62

(the unit cell is now actually orthorhombic). In a real thin sample this collapse occurs as a function of distance from the interface, causing a continuously varying rigid shift. The simplest way to deal with this is by expanding the stress in a Fourier series[24], making it particularly suited to the study of superlattices[25]. The strain amplitude from a single sinusoidal component of the bulk strain wave of magnitude $\varepsilon_0 \sin(\alpha x)$ ($\alpha = 2\pi/\Lambda$, Λ is the strain wavelength) in a thin sample of thickness t is[24]

$$\varepsilon_{xx} = -2\varepsilon \left(\frac{1 + \sigma}{1 - \sigma}\right) \cdot \tag{3}$$

$$\left[\frac{(\frac{\alpha t}{2})\cosh(\frac{\alpha t}{2})\cosh(\alpha z) - (\alpha z)\sinh(\alpha z)\sinh(\frac{\alpha t}{2}) - (1-2\sigma)\sinh(\frac{\alpha t}{2})\cosh(\alpha z)}{\sinh(\alpha t) + \alpha t} - \frac{1}{2}\right]$$

In order to estimate the effect of relaxation on rigid shift measurements we first describe the strain in a superlattice of period Λ_0 where the difference in strain between the two layers is 2δ, as a Fourier series:

$$\varepsilon_{xx} = -\frac{2\delta}{\pi} \sum_{m=0}^{\infty} \frac{1}{2m + 1} \sin[(2m + 1)\alpha x] \tag{4}$$

This represents a strain $\varepsilon_{xx} = -\delta$ for $-\Lambda_0/2 < x \leq 0$ and $\varepsilon_{xx} = \delta$ for $0 < x \leq \Lambda_0/2$. The case of a single interface can be obtained from the limit $\Lambda \to \infty$. Each sinusoidal component of strain in the superlattice relaxes as given by eq.3 above. The effect of the relaxation on rigid shift measurement can be approximated by the value of $\varepsilon_{xx}(z = 0)$. The rigid shift for each component ($R_x = \int \varepsilon_{xx} dx$) is then summed using the Fourier series of eq.4 and the result is subtracted from the fully relaxed result (on the assumption that $\Lambda_0 \gg t$) to give the error in rigid shift measurement

$$\Delta R_x = \left[\frac{-4(1 + \nu)}{1 - \nu}\right] \cdot \tag{5}$$

$$\sum_{m=0}^{\infty} \left[\frac{1}{(2m + 1)^2} \left(\frac{\mu_m \cosh\mu_m - (1 - 2\nu)\sinh\mu_m}{\sinh 2\mu_m + 2\mu_m} - \frac{1}{2}\right) + \frac{1}{2}(1 - \nu)\right]\cos[(2m + 1)\alpha_0 x]$$

where $\mu_m = (2m + 1)\alpha_0 \cdot t/2$. The result of this summation, computed numerically for infinite Λ_0 and therefore appropriate to a single interface, is shown in fig.6. The error in rigid shift measurement, relative to that expected from a fully relaxed system [in which the lattice parameters in the interface normal direction still differ by $2(1 + \nu)\varepsilon_0$] has a maximum of $\sim 0.15\varepsilon_0 t$ at a distance from the interface of $\sim 0.3t$. The error can be minimized by measuring displacements farther from the interface (e.g., $d \gg t$) but other considerations such as projector lens distortions and thickness variations restrict this also. For the case of a 1% mismatch in a foil of thickness 100 Å, the error in rigid shift could be as large as 0.15 Å due to this effect. This is probably a major source of error in measurements on strained epitaxial layers where the accuracies proposed[20] of $\ll 0.1$ Å, based on only instrumental errors, are not practically attainable.

Nonetheless, this analysis suggests that an accuracy of 0.3 Å should be readily obtainable in rigid shift measurements at interfaces with strain less than 2% and this has been successfully used to identify interface structure in several systems. Detailed image simulations can be very useful in confirming interface structure although the accuracy in atomic positions will never rival that obtained with rigid shift measurements. The reader is referred to previous references for examples of these.

63

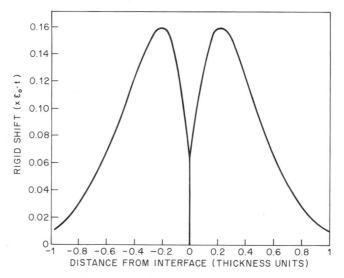

Fig. 6. The magnitude of the rigid shift introduced due to stress relaxation as a function of position near a single interface between two materials with lattice parameters different by $2\varepsilon_0$.

In making image simulations from interfaces, it is of course necessary to make assumptions of artificial periodicity so that one is effectively always calculating the image from a superlattice. Artefacts from this procedure could be easily studied from superlattices themselves.

It should also be noted that at much smaller values of strain, stress relaxation can have a very significant effect on diffraction contrast. This can be seen in fig.7 which shows two dark-field images of a $Ge_{0.05}Si_{0.95}$ superlattice in which the strain modulation amplitude is only 0.1%. The image taken with 220 parallel to the surface plane shows only structure factor contrast (dilatation effects can be ignored at this level of strain) which is, as expected, very weak. On the other hand the image taken with a reflection (400) parallel to the growth direction shows strong contrast due to stress relaxation. The reason why this shows up at such low values of strain is the bending of lattice planes resulting from the surface relaxation[24] and the sensitivity of diffraction contrast to lattice plane rotation.

Compositional Interfaces

Compositional interfaces present a different type of problem in interpretation. For example, fig.8 shows a single interface in a superlattice of $Ge_{0.5}Si_{0.5}/Si$ superlattice grown by molecular beam epitaxy on (100) Si and thinned by ion-beam thinning in <110> cross-section. In such images the contrast arises only from the difference in scattering factor between Ge and Si. It is possible through dynamical scattering effects and defocus optimization[26] to enhance the contrast of these interfaces but the fundamental resolution becomes limited by noise.

Before discussing noise, there is another aspect not necessarily unique to compositional interfaces which limits interpretation in cases like fig.8. Because the free energy of this interface is low, it is not likely to be atomically flat. In contrast the high free energy of interfaces such as $CoSi_2/Si$ is likely the reason why these interfaces are atomically flat over large regions (e.g. fig.4c). Returning to the case in

64

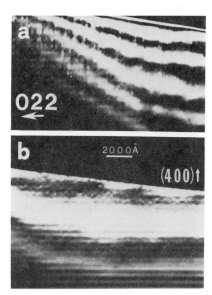

Fig. 7. Diffraction contrast images of a very low strain modulation wave in
a $Ge_{0.05}Si_{0.95}$/Si superlattice taken with reflections parallel a),
and perpendicular b), to the growth direction. The latter shows
only weak structure-factor contrast whereas the former reveals the
strong effects of strain relaxation on diffraction contrast even at
the 0.1% level.

Fig. 8. A high-resolution image in cross-section of a (100) $Ge_{0.5}Si_{0.5}$/Si
superlattice of period 100 Å grown by molecular-beam epitaxy.

hand, this likely absence of interface flatness leads to problems related to projection in HREM images, which typically plague amorphous materials[27] and their interfaces[4]. This makes it difficult to distinguish between interface roughness and interdiffusion. Careful quantitative measurements, e.g. by integration over rectangular slits of varying size during photodensitometry[28], may allow these cases to be distinguished and the Fourier components of roughness determined.

Returning to the problem of signal-to-noise, it is found that in practice the major contribution to noise arises from specimen preparation[29]. We have measured the magnitude of S/N from a series of samples made from the same Si<110> wafer by a variety of techniques[29]. The noise level was measured by Fourier transforming images and integrating the noise spectrum in the band 0.2 to 0.4 $\AA^{-1}$. This frequency band is appropriate to measurements of the variation in intensity near the lattice frequency, i.e. the probability distribution of bright dots in high-resolution images. The results of this study were that an ion-thinned specimen (3 keV, 20° incidence Ar ions) had noise of standard deviation 8% compared with a lattice image of intensity less than 40%. Cleaved specimens had S/N ~4% and in-situ cleaned specimens ~3%, which was the limit set by shot noise due to grain in the photographic plate. (This noise level can be reduced by longer exposures and higher magnification). These figures are quite sobering as, for example, the presence of 20% Ge in Si causes only 4% intensity change so that the position of an interface cannot be determined with atomic resolution in typical ion-tinned specimens because of S/N problems. The effect of limited S/N ratio on point defect imaging is discussed elsewhere[29]. Clearly, things can often be improved by attention to specimen preparation and in some cases in-situ specimen preparation which is necessary for clean surface studies may also aid bulk studies of buried interfaces and defects.

Surface imaging and diffraction in the transmission electron microscope

A single unit cell of Si(111) scatters 1.2% of the beam intensity of a 100 kV electron beam. Therefore it is possible to obtain diffraction patterns and images from surface layers in a transmission electron microscope, particularly when signal-to-noise is high as occurs on removal of surface-contaminating films. Figure 9 is a graphic example of the diffraction pattern from a Si(111) surface which exhibits a 7x7 reconstruction of the surface layers. The remainder of this chapter involves discussion of TEM studies from surfaces and buried surfaces. Examples will be drawn from the author's own work.

Cherns[30] demonstrated for the first time the diffraction from surface layers of thin Au(111) films using the kinematically forbidden 1/3 422 reflections. These reflections can be viewed as the Fourier transform of a single (111) layer of the Au crystal. If stacked in a hexagonal close-packed sequence ABABAB.... this reflection becomes the strong 100 reflection. However, in f.c.c. Au, stacking sequence ABCABCABC..., the reflection has zero structure factor for an integral number of unit cells. As a result it oscillates in intensity as a single atomic layer is added to the crystal foil and dark field images can reveal atomic steps[30]. The 1/3 422 reflection is very strong from the clean 7x7 reconstructed surface in fig.9. It is, however, not generally observed in typical TEM samples except for noble metals. The reason for this is almost certainly due to the lack of clean and controllable surface conditions in electron microscopes. The advent of UHV technology in the 1950's revolutionized surface science for it permitted surfaces to be cleaned and remain clean during experiments. Only recently and in a very few laboratories have such conditions become available in TEMs.

Fig. 9. The diffraction pattern from a 7x7 reconstructed Si(111) surface
taken at 100 kV in-situ in a UHV transmission electron microscope.
The 1/3 422 diffraction spot is the (1,0) beam from a single-atomic
layer of Si(111) and the 7-th order beams lie between (h,k) surface
beams.

The Si(111) 7x7 surface reconstruction

There are many surface characterization techniques which have the
capability of diffraction or microscopy and it is important at this stage
to identify the relative advantages and disadvantages of TEM from surfaces
relative to others, many of which are covered in this book. The Si(111)
7x7 surface provides an excellent case study for this purpose, since it was
discovered in the early days of UHV by Low Energy Electron Diffraction
(LEED)[31] and has been studied by many techniques. The currently accepted
model of the structure of the (111) 7x7: the Dimer-Adatom-Stacking Fault
(DAS) model was proposed to fit the TEM data by Takayanagi et al.[32] and so
its solution involved all the techniques including TEM. In fact the
multitude of studies of this surface probably had their greatest future
contribution in the development of new surface science techniques.

LEED has been used most extensively in the study of the 7x7. LEED is
of course very sensitive to the last layer or two of a surface and to
periodic distortions of atomic positions both in the plane and surface
normal directions, typical of surface reconstructions. Furthermore it is a
simple technique to implement and is nondestructive. However at low
energies electrons interact very strongly with single atoms so that weak
scattering approximations are completely invalid (although they are
commonly used for qualitative interpretation of LEED). More sophisticated
modeling is now available, but LEED alone was unable to solve the structure
of the 7x7 although its complexity allowed the spawning of many models[33].
However, LEED contributed the information that the 7x7 structure is a
triangular mesh, from the fact that only the (n/7,0) spots are very
intense[33]. Photoemission[34] and infrared spectroscopy[35] gave evidence of
dimerized atoms in the structure from the pairing of dangling bonds. The
scanning tunneling microscope gave conclusive evidence for the adatoms[36]
which had previously been proposed from LEED. However, ion channeling data
which gives information on atomic displacements beneath the surface,
indicated that the 7x7 is a "deep" reconstruction, propagating several

layers below the surface. From this data it was suggested that a stacking fault of some kind must exist in the structure[37]. Hydrogen atom adsorption[38] and Scanning Tunneling Microscopy (STM)[36] also showed the existence of a unique hollow-site at the corner of the unit cell. Thus we have all the components of the final model: dimers, adatoms and stacking faults. However, the complexity of the reconstruction was such that its solution had to await the development of ultra-high vacuum TEM. The first published TEM diffraction patterns were by Petroff and Wilson[39] but a thorough analysis of the Tokyo data by Takayanagi et al.[32] led to the currently accepted structure. High-energy transmission electron diffraction from superperiodic surfaces offers the great attraction that it is weak scattering (multiple interactions can occur with bulk reflections however) and it provides a projection of the structure, i.e., it has penetration to examine sub-surface layers. These two features are the most practically useful of TEM in surface science and we will see that their application is great in the study of adsorbates and thin film growth on surfaces.

TEM Surface Imaging

The example of the Si(111) 7x7 provides weight to the transmission electron diffraction technique in surface science and we shall see more. However, a major advantage to high-energy electron diffraction is that high-resolution optics can be readily achieved and so microscopy and diffraction are available in the same instrument.

There are two modes in which transmission electron images of surfaces have been achieved: plan-view and profile[40]. In the profile mode the surface normal $\underline{n}$ is perpendicular to the beam direction $\underline{B}$ whereas in plan-view mode $\underline{n}$ is parallel to $\underline{B}$.

The profile mode can give very descriptive images with information at high-resolution both parallel to and perpendicular to the surface. For example, fig.10 shows an image from a (113) facet on a Si<110> sample which was cleaned in-situ. The sawtooth structure of this surface can be easily modeled with multi-slice simulations as shown in fig.10(b)[41]. The profile view suffers from some major limitations, however. The most serious is that the "surface" being studied is in reality a strip of surface just ~100 Å wide. Since most surface cleaning involves high temperature processing, one would not expect this strip to be necessarily representative of flat surfaces. This problem is particularly exacerbated when adsorbates or deposited atoms are involved as these could readily diffuse to or from other surfaces if they desired. However, turning this problem on its head, the technique is highly suited for the study of naturally curved surfaces, such as the important class of small catalytic particles, as first suggested by Marks and Smith[40]. The second difficulty lies in the problem of interpretation beyond the resolution limit. For example, apparent surface relaxations can occur due to the presence of contamination or incorrect imaging conditions[42].

In contrast, the plan-view mode of imaging gives very little information concerning the dimension parallel to $\underline{n}$, although all sample detail is included in projection. However, flat specimens are examined which can be representative of flat surfaces and could be simultaneously characterized by other techniques. It is certainly possible to achieve and make use of atomic level resolution in plan-view samples although most later examples will not utilize this. Nevertheless fig.11 shows an example of the high-resolution imaging of the Si(111) 7x7 in plan-view which reveals its large scale interaction with SiC precipitates.

The electron microscope is a powerful tool in revealing the irregularities in an apparently perfect surface, and fig.11 is an excellent example of this. The STM has revolutionized the surface scientist's real space impression of surfaces but not sufficiently, in the author's opinion. The STM scans are generally < 1000 Å in dimension and have a low "success" rate so they do not give a complete picture of surface structure. For

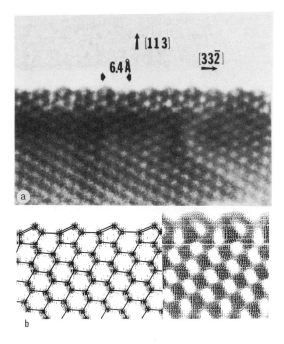

Fig.10. a) A profile view of the reconstructed Si(113) surface formed by high-temperature annealing of a Si<110> sample in-situ in a UHV TEM. b) A simulated image appropriate to the imaging conditions of fig.10 a): defocus -950 Å, thickness 60 Å, spherical aberration 1.2 mm and electron wavelength 0.025 Å.

Fig.11. A plan-view image of the Si(111) 7x7 surface showing individual unit cells of the reconstruction. Particles of epitaxial β-SiC are seen to be coexistent with the reconstruction.

example, the area viewed in fig.11 could give rise to good 7x7 STM images because the spacing between SiC particles is >> 1000 Å. Nevertheless, the presence of SiC "boulders" at a density of 10^8 cm^{-2} may impact adversely such behaviours as surface step motion, diffusion and epitaxial growth. Surface cleaning is in itself an important area of surface science and SiC has been well known as a potential problem on heated Si surfaces with an inevitable initial trace of carbon coverage. However, misleading results on the behaviour of carbon can be obtained from Auger spectroscopy, the commonly employed chemical characterization technique, due to the limited escape depth of Auger electrons (~15 Å). As the surface is heated, the diameter (r) of SiC particles increases as a result of ripening, while their number decreases. Even though the total amount of carbon may remain constant, the amount within the Auger escape depth appears to fall like 1/r, giving a misleading impression of improved surface cleanliness.

In order to fully clean Si surfaces we have found it necessary to anneal at ~1200°C for a few minutes. Experiments with low temperature annealing (~800°C), e.g, with the Shiraki chemical technique[43] have not shown considerable reduction in the SiC concentration (although size of particles and terrace width increases, possibly responsible for the apparent improvement in Auger and LEED). This issue bears more detailed investigation. In the meantime, the high temperature cleaning operation offers another unique advantage for TEM - it can be used to regenerate thin specimen areas. After annealing at 800°C and desorption of SiO$_2$ by the process SiO$_2$ + Si → 2SiO, the thinnest specimen areas are consumed on the wedge-shaped plan-view specimens. (Note that this effect may be partly responsible for the greater surface mobility on desorbing a sacrificial Shiraki oxide). On subsequent higher temperature annealing, evaporation and surface diffusion cause even more rounding of the once thin edges. However, above 1100°C, surface facetting occurs which causes new thin areas to appear in previously thick parts of the specimen, e.g., fig.12. These thin areas are typically atomically flat and several μm^2 in area, with heavily stepped edges. They invariably show good 7x7 (111) patterns on cooling to room temperature and very little sign of SiC precipitates.

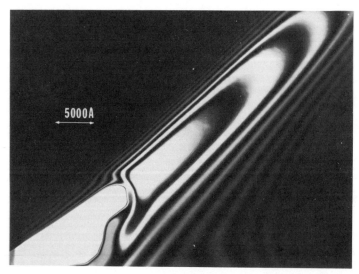

Fig.12. A low magnification image showing the regeneration of thin flat specimen areas in a wedge-shaped specimen by evaporation and surface diffusion at > 1000°C.

In-situ TEM of Epitaxial Growth

One of the major applications of TEM in surface science will probably be the study of epitaxial growth and deposited films. Epitaxy has been studied for many years[44] but in the past decade or more it has taken on great practical significance in the semiconductor industry. Molecular Beam Epitaxy (MBE) has been developed for exceptional control on the atomic level of crystal growth, utilizing clean surfaces. However, even when control can be exerted on the fluxes of deposited atoms, surface diffusion can cause redistribution, chemical reactions can be involved and defects can be introduced. The study of such phenomena is greatly facilitated by in-situ TEM monitoring of growth. We provide some examples of our own studies in this field.

Co on Si. Cobalt disilicide is a cubic metal which grows epitaxially on (111) Si to yield high-quality single crystals with a twin (type-B) orientation relative to Si. The most effective way of growing $CoSi_2$ is by room temperature deposition of Co on clean Si followed by in-situ annealing to 500-600°C. Previous studies of $CoSi_2$ growth have either concentrated on the reaction of "thin" (1000 Å) films by annealing at 800-900°C or have been in-situ studies without microscopic and penetrating power. Our studies have shown that intermediate phases do form on annealing Co on Si under clean conditions[45]. Figure 13 shows a series of selected-area diffraction patterns from a 17 Å thick Co films deposited on Si(111) and annealed in-situ in a TEM. The instrument in which these experiments were performed is described in a separate publication[46]. The ambient vacuum in the instrument is 1×10^{-9} Torr and metals are evaporated on the specimen in the viewing position by thermal evaporation from a coated Ta filament. Thicknesses are measured a posteriori by RBS.

The 100 kV transmission electron diffraction patterns show that even at room temperature an epitaxial phase is formed. Figure 13(a) reveals an extra diffraction spot parallel to each 220 Si corresponding to a planar spacing of 2.04 ± 0.02 Å. The symmetry of this pattern is apparently hexagonal (extra spots occur from double diffraction). It does not simply correspond with any published phase of Co_xSi_{1-x}; however, with just a small distortion of the <100> projection of the orthorhombic Co_2Si structure, its diffraction pattern could exhibit hexagonal symmetry and we propose this explanation[47].

The structure of the room temperature Co on Si film has been the subject of several in-situ studies which provide apparently conflicting information. For example, photoemission indicates that at these high coverages, unreacted Co remains on the surface, whereas there is evidence for immediate reaction to form $CoSi_2$ at coverages less than ~10 Å[48]. Other evidence for unreacted Co comes from Auger spectroscopy on thicker layers[49]. However, up to the maximum thickness (25 Å) which we have investigated, we see only evidence for the reacted "Co_2Si" phase. Perhaps this can be reconciled from the observation that in the <010> direction, the projected atomic structure of Co_2Si has close to hexagonal packing of Co atoms, with a difference of 13% from the spacing in h.c.p. Co. The Si atoms are stacked in alternate sites between the Co layers. As a result it seems plausible that the structure could accommodate large deficiency in Si concentration at large distances from the interface. This situation is very likely to occur due to the limited amount of room temperature interaction. The in-situ diffraction experiments are the best in confirming crystallographic data, yet the photoemission and Auger spectroscopy gives more accurate chemical and electronic information. We also note that some remnant of the "Co_2Si" phase occurs even at temperatures above that at which $CoSi_2$ grows, indicating its stability.

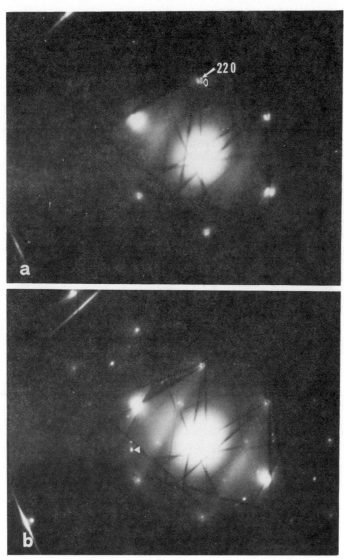

Fig.13. Diffraction patterns from in-situ deposited Co on Si(111). a) Showing at room temperature the existence of an epitaxial phase believed to be a hexagonally-distorted form of Co_2Si; b) Epitaxial cubic CoSi, another intermediate phase which forms at 350°C in the same sample.

Some ex-situ microscopic experiments have been attempted on the room temperature structure of Co on Si[50], but with such reactive films and the high temperatures which may be involved in specimen preparation, it is hardly surprising that the results are at odds with ours.

At 350°C we observe (fig.13b) the growth of epitaxial CoSi, which can be simply indexed as the <010> orientation of this cubic structure. Because of its small lattice parameter, the silicide grows with <100> parallel to <110> Si and thus there are three orientation variants.   At

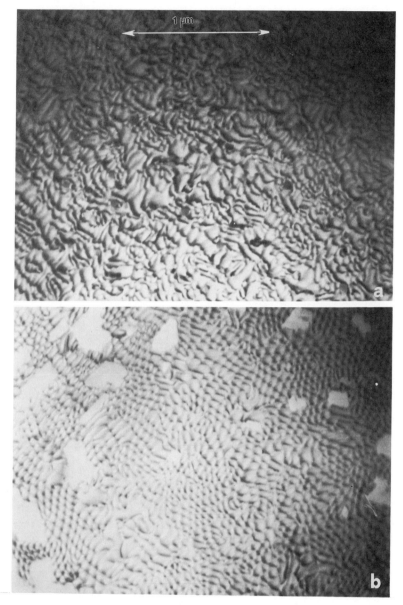

Fig.14. Two images of CoSi$_2$ on Si(111), deposited in-situ (17 Å Co) and annealed to 475°C (a) and to 550°C for 5 minutes (b).

450°C we observe transformation to CoSi$_2$. Dynamic imaging study of the transformation from CoSi to CoSi$_2$ reveals that the growth process is lateral, i.e. in the plane of the surface. This is in marked contrast to the behaviour of so-called "thin" films, in which diffusion limits the growth of phases to the vertical direction, with successive Si rich phases nucleating at the interface. The higher temperatures observed for silicide growth in 1000 Å thick films is thus attributable to bulk diffusion coefficients, which are usually much lower than surface diffusion. We observe the same phenomena in the transition from "Co$_2$Si" to CoSi. Lateral growth is likely to lead to misleading impression of stoichiometry from surface-averaging probes[49].

Another interesting observation in this study has been the origin of pinholes in thin CoSi$_2$ films. These holes had been observed by plan-view TEM and found to be important in electrical properties of heterostructures involving CoSi$_2$[11]. Figure 14 shows two bright-field images of CoSi$_2$ formed in-situ: a) at 475°C and b) at 550°C after 5 minutes. As can be seen, holes have opened up in the same area after the higher temperature anneal. These holes are a symptom of the "unwetting" of the Si substrate by the CoSi$_2$ film. It clearly is not in stable equilibrium as a uniform layer because of a high interface energy. By careful study such as this, pinhole densities can be minimized in metastable films.

Ni on Si. In recent in-situ studies of Ni deposition on clean Si surfaces and annealing to form NiSi$_2$, we have found additional evidence for the importance of metastable epitaxial phases[51]. The NiSi$_2$/Si system is a fascinating one, for it has proven possible by MBE to grow either of two orientations (type-A or type-B) on Si(111). The type-A orientation has crystallographic axes aligned in both the cubic silicide and the substrate, whereas type-B has a 180° rotation about the (111) growth direction. Single-crystals of either orientation may be grown dependent only on the initially deposited Ni thickness prior to a 450°C anneal[52]. This unique phenomenon has the important consequence that the physical properties of these two interfaces can be studied and the Schottky Barrier heights have been found to be different[53].

In-situ study has shown that the thickness dependence of the orientation of the nickel disilicide is a consequence of intermediate metastable phases. The metastable hexagonal phase, θ-Ni$_2$Si, grows for Ni coverages between 15 and 20 Å. Although this phase is only found in bulk Ni/Si alloys above 850°C, its excellent fit to Si(111) reduces the free energy to make it stable for low coverages of Ni. Imaging studies in-situ show that the existence of this phase correlates with the subsequent growth of type-A NiSi$_2$, whereas type-B NiSi appears to grow immediately for very low Ni coverages. The Ni/Si system has been studied by many surface science techniques, but the combination of diffraction, imaging and penetration power of transmission electron microscopy has proved unique in elucidating the microstructure of reactions on surfaces involving more than one monolayer.

In conclusion, the transmission electron microscope has proven very useful in studies of semiconductor interfaces and surfaces because of its unique combination of imaging, diffraction and penetrating power. The future of such studies is facilitated by improved vacuum conditions and greater attention to quantification, such as improving signal-to-noise ratio.

The assistance and collaboration of J.L. Batstone, R. Hull, R.T. Tung, J.C. Bean, M.L. McDonald and F.C. Unterwald in the work described here is gratefully acknowledged.

REFERENCES

1. T.T. Sheng and R.B. Marcus, J. Electrochem. Soc. 127:737 (1980).
2. O.L. Krivanek, T.T. Sheng and D.C. Tsui, Appl. Phys. Lett. 32:437 (1978).
3. C. D'Anterroches, J. Micros. Spect. Elec. 9:147 (1984).
4. Z. Liliental, O.L. Krivanek, S.M. Goodnick and C.W. Wilmsen, Mat. Res. Soc. Proc. Ser. 37:193 (1985).
5. A.E. White, K.T. Short, J.L. Batstone, D.C. Jacobson, J.M. Poate and K.W. West, Appl. Phys. Lett. 50:19 (1987).
6. A. Ourmazd, K. Ahlborn, K. Ibeh and T. Honda, Appl. Phys. Lett. 47:685 (1985).
7. A. Bourret and J. Pennison, JEOL News 25E:2 (1987).
8. R. Hull, S.J. Rosner, S.M. Koch and J.S. Harris Jr., Appl. Phys. Lett. 49:1714 (1986).
9. S.J. Pearton, S.M. Vernon, C.R. Abernathy, K.T. Short, R. Caruso, M. Stavola, J.M. Gibson, V.E. Haven, A.E. White and D.C. Jacobson, J. Appl. Phys. 62:862 (1987).
10. R.T. Tung, J.C. Bean, J.M. Gibson, J.M. Poate and D.C. Jacobson, Appl. Phys. Lett. 40:684 (1982).
11. R.T. Tung, A.F.J. Levi and J.M. Gibson, Appl. Phys. Lett. 48:635 (1986).
12. J.M. Cowley and A.F. Moodie, Acta Cryst. 10:609 (1957).
13. P. Goodman and A.F. Moodie, Acta Cryst. A30:280 (1973).
14. P.B. Hirsch, A. Howie, R.B. Nicholson, D.W. Pashley and M.J. Whelan, "Electron Microscopy of Thin Crystals", R.E. Krieger Publ. Co., Malabar, Florida, USA (1977).
15. A. Howie and Z.S. Basinski, Phil. Mag. 17:1039 (1968).
16. J.M. Gibson, J.C. Bean, J.M. Poate and R.T. Tung, Appl. Phys. Lett. 41:818 (1982.
17. D. Cherns, J.C.H. Spence, G.R. Anstis and J.L. Hutchison, Phil. Mag. A26:849 (1982).
18. J.M. Gibson, R.T. Tung and J.M. Poate, Mat. Res. Soc. Proc. 14:395 (1983).
19. F.A. Ponce, G.B. Anderson, M.A. O'Keefe and L.J. Schowalter, J. Vac. Sci. Tech. B4:1121 (1986).
20. D.J. Smith, W.O. Saxton, M.A. O'Keefe, G.J. Wood and W.M. Stobbs, Ultramic. 11:263 (1983).
21. J.M. Gibson, Ultramic 14:1 (1984).
22. W.O. Saxton and D.J. Smith, Ultramic. 18:39 (1985).
23. R.F. Cook and A. Howie, Phil. Mag. 165:641 (1969).
24. M.M.J. Treacy, J.M. Gibson and A. Howie, Phil. Mag. A51:389 (1985).
25. J.M. Gibson, R. Hull, J.C. Bean and M.M.J. Treacy, Appl. Phys. Lett. 46:649 (1985)
26. R. Hull, J.M. Gibson and J.C. Bean, Appl. Phys. Lett. 46:179 (1985).
27. J.M. Gibson and A. Howie, Chemica Scripta 14:109 (1978-9).
28. R. Hull, A.T. Fiory, J.C. Bean, J.M. Gibson, L. Scott, J.L. Benton and S. Nakahara, Proc. 13th Int. Conf. on Defects in Semiconductors, p 505 (1985).
29. J.M. Gibson and M.L. McDonald, Mat. Res. Soc. Proc. 82:109 (1987).
30. D. Cherns, Phil. Mag. 30:549 (1974).
31. R.E. Schlier and H.E. Farnsworth, J. Chem. Phys. 30:917 (1959).
32. K. Takayanagi, Y. Tanishiro, M. Takahashi and S. Takahashi, J. Vac. Sci. Tech. A3:1502 (1985).
33. E.G. McRae, Surf. Sci. 124:106 (1983).
34. H.D. Hagstrum and G.E. Becker, Phys. Rev. (1973).
35. Y.J. Chabal, Phys. Rev. Lett. 50:1850 (1983).
36. G. Binnig, H. Rohrer, Ch. Gerber and E. Weibel, Phys. Rev. Lett. 50:120 (1983).

37. R.J. Culbertson, L.C.Feldman and P.J. Silverman, Phys. Rev. Lett. 45:2043 (1980).

38. J.A. Appelbaum, H.D. Hagstrum, D.R. Hamman and T. Sakurai, Surf. Sci. 58:479 (1976).

39. P.M. Petroff and R.J. Wilson, Phys. Rev. Lett. 51:199 (1983).

40. L.D. Marks and D.J. Smith, Nature 303: 315 (1983).

41. J.M. Gibson, M.L. McDonald and F.C. Unterwald, Phys. Rev. Lett. 55:1765 (1985)

42. J.M. Gibson, Phys. Rev. Lett. 53:1859 (1984).

43. A. Ishizaka, K. Nakagawa and Y. Shiraki, Proc. 2nd Int. Symp. on "MBE and Related Clean Surface Techniques", p 183 (1982).

44. D.W. Pashley, in: "Epitaxial Growth", J.W. Matthews, ed., Academic Press, New York (1975).

45. J.M. Gibson, J.L. Batstone and R.T. Tung, Appl. Phys. Lett. 51:45 (1987).

46. M.L. McDonald, J.M. Gibson and F.C. Unterwald, J. Sci. Inst. to appear.

47. J.M. Gibson, J.L. Batstone and R.T. Tung, Appl. Phys. Lett. 51:45 (1987).

48. C. Pirri, J.C. Peruchetti, G. Gewinner and J. Derrien, Phys. Rev. B29:3391 (1984).

49. F. Arnaud D'Avitaya, S. Delage, E. Rosencher and J. Derrien, J. Vac. Sci. Tech. 2:770 (1985).

50. C. D'Anterroches, Surf. Sci. 168:751 (1986).

51. J.M. Gibson, J.L. Batstone, R.T. Tung and F.C. Unterwald, Phys. Rev. Lett. 60:1158 (1988).

52. R.T. Tung, J.M. Gibson and J.M. Poate, Phys. Rev. Lett. 50:429 (1983).

53. R.T. Tung, Phys. Rev. Lett. 52:461 (1984).

THE TRANSMISSION ELECTRON MICROSCOPY OF INTERFACES AND MULTILAYERS

W. Michael Stobbs

Materials Science and Metallurgy Department
Pembroke Street
Cambridge, CB2 3QZ, U.K.

INTRODUCTION

The diversity of the types of structural boundary which have been studied by TEM is as wide ranging as are the methods available to the microscopist for their characterisation. The choice of the best technique to use, as well as an understanding of the limitations of each, can often thus be as important as the recognition of the specific characteristic of a given interface which is of most relevance in relation to the material properties investigated. Sometimes it is not even clear that the local heterogeneity of structure or chemistry of interest can actually properly be described as being delineated on the atomic scale by a discrete plane. Cases in point can be found in the study of amorphous structure, spinodal decomposition, short range order and the characterisation of icosahedral materials. We will examine in this area not only the relative usefulness of techniques such as high resolution microscopy and the weak beam method, but also come to realise that the method used can allow useful definitions for discreteness which parallel the modelling of the origins for the structural inhomogeneity.

Man-made multilayers, in both semiconductor and metallic fields, can show periodic changes of both composition and lattice parameter with the wave form in each case exhibiting anything from square to simple sine wave character. Thus the problems alluded to above, in the characterisation of the class of interface involved, are equally relevant here. Furthermore a simple rigid body displacement at a grain boundary is necessarily associated with a local change in scattering potential in a way which can be very difficult to distinguish uniquely from that caused by a local change in composition. Clearly the quantitative structural and compositional characterisation of even a single interface is as difficult as it is rewarding.

The approach taken in this paper will follow the theme introduced above and represents a personal view of the area rather than a broad review. I will deal firstly with the choice of specimen geometry, the specimen preparation methods required and the more basic type of structure which can be examined. Secondly I will look at examples of problems for which definition of the character, or indeed the presence, of the boundary is central. The specific atomic scale characteristics of discrete boundaries which can be evaluated will then be discussed as will be the difficulties in obtaining a unique model for the locally variable displacement fields associated with intrinsic boundary defects. With this background I will finally be able to discuss current work on the characterisation of the different classes of man-made heterostructures and multilayers so increasingly important in the semiconductor industry.

SPECIMEN PREPARATION AND CHOICE OF GEOMETRY WITH EXAMPLES OF SIMPLE CHARACTERISATION PROBLEMS AND TECHNIQUES

Thinning. When examining homogeneous materials or alloys containing discrete second phase particles it is rare that the experience of the last thirty years does not provide an efficient electrochemical or chemical thinning technique. Unfortunately the nature of most specimens containing boundaries separating phases of different composition forces the use of ion beam thinning methods. By comparison with the specimen changes which can occur during thinning by the former methods the state of a specimen prepared using ion beam techiques can be grossly unsatisfactory unless particular care is taken. For example a layer at the surface is generally amorphised and dilated, this causing localised bending in the thinner regions. Disparities in the relative thinning rates of different chemical species can however be limited both by cooling and by a variety of masking techniques and care with the remanent atmosphere in the ion beam thinner can pay dividends. Perhaps the most awkward interface to thin is that between sapphire and niobium, the relative thinning rates being very dependent upon the partial oxygen pressure near the specimen. However, adequate specimens for high resolution electron microscopy can be obtained for even this interface (e.g. ref.1) and the characterisation of metal/ceramic interfaces is increasingly important in materials areas as disparate as the evaluation of potential packaging systems in the semiconductor industry and the optimisation of metal matrix fibre composites (examples of the observational techniques which can be used in this latter field are discussed in ref.2).

An "edge-on" specimen geometry is nearly always required for high resolution work and is generally useful in gaining a topographic understanding of the spatial relationships between different phases as they are formed relative to, for example, a reaction front. The preparation problem associated with this geometry is that the interface line is very often fragile: haematite can often be blown off the underlying magnetite on outward growing oxides on certain iron alloys. This both precludes the use of glues such as araldite in preparing the initial blocks or rods from which specimen discs containing an interface are to be cut and necessitates the protection of the free surface to be examined by the initial evaporation of a metal and/or by electroplating. The choice of plate can be important too, it being sensible to use a metal with a similar ion-beam thinning rate to that of the specimen to be examined.

The general specimen preparation techniques required have been discussed by a number of authors (e.g. ref.3), but a specific feature of all those methods which can be used with a high success rate is that the interface of interest is constrained at an early stage of the specimen fabrication process in a ring which increases the strength of the foil as a whole. This allows a simplified approach to disk handling during grinding and dimpling as well as when the final foil is placed in the microscope.

Geometry Choice. If an interface is to be viewed in projection, as is a necessity if an atomic resolution image is to have any pretence to a representation of a relevant projection, the edge-on geometry is the only

one that can be used.   In general, when examining the historical
development of reaction fronts, such specimens are also useful but often
not the only ones needed.   If only rather thin (<10 nm) second phase
regions develop, the edge-on geometry precludes the easy use of selected
area diffraction methods and limits the accuracy of EDX and EELS analysis
because of beam spreading.   Arguably a convergent beam diffraction approach
can then allow crystallographic characterisation, and specialised small
probe (<2 nm) EELS methods in thin areas will usually enable composition
measurements to be made.   In general however "plan-view" specimens, best
prepared by off-centre thinning on either side of a disc containing the
interface transversely, repay the investment of time in their fabrication.
In such specimens successive regions at different distances from the
interface can be viewed separately side by side and furthermore chemical or
electrochemical methods can often be used to thin the material on at least
one side of the boundary region to be examined.   The sort of situation
where this approach can be used beneficially is exemplified by the
characterisation of the interaction front between a metal and silicon,
where it allows the ready characterisation of the localised nature of the
silicides which are often formed (e.g. ref.4).

Technique Choice.   The choice of the method used in the evaluation of
boundary structures often seems to be dominated by, and even a
justification for, the money spent on a modern HREM or small probe STEM.
As we will see below such methods usually require very careful appraisal
and if the real problem to be solved is kept in mind, the relatively
qualitative application of very standard methods can reap substantial
benefits once an appropriate specimen foil has been made.   For example when
examining oxidation microstructures in-situ (e.g. ref.5,6), surface changes
due to deformation (e.g. ref.7), or chemical attack generally, it is rare
that the thoughtful use of two beam and weak beam imaging methods coupled
with reasonably quantitative EDX analysis techniques do not provide the
data required.

THE EXISTENCE OF THE BOUNDARY

In discussing methods for the examination of interfaces and multilayer
"superlattices" of such interfaces it might well seem perverse to examine
situations in which the discrete existence of a boundary is questionable.
It should however be remembered that some of the first man-made metallic
superlattices were prepared in an attempt to model spinodal decomposition[8]
where changes in both composition and lattice parameter are continuous and,
by definition, no discrete boundaries exist.   Equally it is in the
quantitative evaluation of the compositional and planar spacing wave forms
in such metallic systems that the most stringently demanding application of
modern methods is required[9].

The amorphous structure is characterised by the existence of three
dimensional heterogeneity without boundary formation.   While argument
continued for many years on the differentiation of microcrystalline from
truly amorphous models for a variety of materials, it is noteworthy that
high resolution axial bright-field methods even at current resolutions
(~0.15 nm) are rarely usefully applicable (because of the projection
problem) while dark field techniques can provide a differential
characterisation with ease.   If dark field stereo-images of "speckle"
cannot be obtained, then a microcrystalline model is inappropriate[10].
Furthermore this differentiation can be made using the electron microscope
rather elegantly down to a limit in the size of the locally disparate
regions of a microcrystalline specimen at which any attempts at further
differentiation would cease to be physically meaningful (~two unit cells).

The above situation is precisely paralleled in the analysis of short range order as being either of microdomain or of cluster type in a material and working definitions of the physically relevant differences between the two descriptions were given some time ago[11]. Nonetheless the debate on this subject continues (e.g. ref.12,13).

A more interesting area where the evaluation of local variation in contrast in dark field images is totally central to the characterisation of a structure lies in the study of icosahedral materials[14]. It is now generally recognised that a three dimensional Penrose tiling approach can, even if not uniquely, describe a framework for this type of structure. However it is depressing that, judging by the current literature (e.g. ref.15,16), electron microscopists are not going to learn from the experience gained in the study of amorphous and short range ordered materials and, in general, still concentrate on the use of high resolution techniques.

High resolution images (see figs 1a and 1b) of this type of material are very beautiful and the characterisation of the "long-short" golden section related planar interrelationships, as well as the irregularities in these associated with "phasons" (such as that arrowed in fig.1a) is useful in delineating the framework. But how is this framework atomically decorated? Those used to a regular lattice framework seek to answer this question by the modelling of high resolution images and yet a glance at the effect of a change of deviation parameter on a dark field image (c.f. figs 2a and b) is sufficient to guarantee that such an approach is not going to be successful. Fascinating images such as those shown here characteristically exhibit "speckle" on a non-aperture size limited scale (~2-3 nm) and yet show strong variation with small changes of tilt which demonstrate that we are not dealing with "microcrystalline" like regions of locally well coordinated composition. As discussed elsewhere[17] we are dealing here with variations in local scattering potential indicative of fluctuations in the local composition probably caused by changes in the coordination inherently required in the Penrose framework. This explanation for the dark field

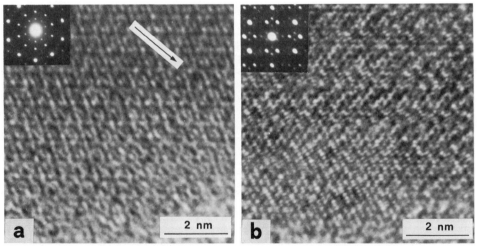

Fig. 1. High resolution images of an Al-Mn icosahedral alloy with inset diffraction patterns for each normal. a) the 5 fold axis, b) the 2 fold axis. Note the "phason"[15] disruption, as arrowed, which is not associated with the compositional inhomogeneity demonstrated by fig.2 (Courtesy of K.M. Knowles).

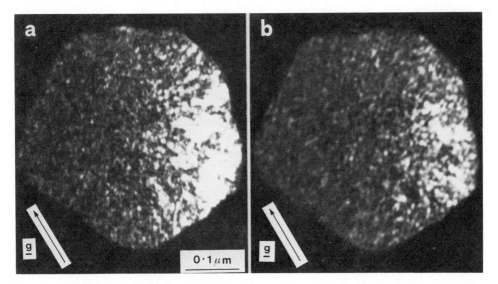

Fig. 2a,b. A pair of dark field images, taken at two different deviation
parameters, using a characteristic reflection of the icosahedral
stucture. The images are in this case of an Al-V phase. The
important point is that the "speckle" is characteristically similar
to that exhibited by an amorphous material (no stereo images
possible at high deviation parameter) but not aperture limited in
size (Courtesy of K.M. Knowles).

contrast is supported by the only weakly dynamic nature of the scattering
observed (thickness fringes exhibit poor contrast). Furthermore recent
observations of similar contrast in materials such as Al-Cu-LI alloys[18],
which can contain icosahedral regions despite being slowly cooled, suggest
that we are dealing with a characteristic feature of the structural density
of these materials rather than with an accident of the kinetics of their
formation.

SINGLE INTERFACE CHARACTERISATION

There are three main sets of measurements which are required for the
characterisation of an interface. These are:
1) the quantification of the rigid body displacement for a homophase
   boundary;
2) the measurement of the concentration of any segregant at the
   boundary;
3) the characterisation of the local relaxation structures when
   intrinsic boundary defects are present.

Unfortunately, as suggested earlier, the solutions of these individual
problems are not in general independent of one another and it is often
difficult to distinguish between the effects of segregation and those of a
rigid body displacement. Since the separation of these parameters is
fundamental to the unique solution of any boundary characterisation
problem, it is fortunate that individual techniques can usually be found
which are sufficiently independent of the two parameters not being measured
to yield a value for the other. The simplest way forward will be to

examine the approach best taken for each of the three measurements listed above.

The Rigid Body Displacement. Advances in high resolution electron microscopy have for a long time allowed the recognition of spacings well below 0.2 nm so this problem might at first sight seem to present little difficulty. However for a weak phase object, let alone a thicker specimen, a local change in spacing across a plane (even if this is an increase) is not reproduced faithfully without substantially improved resolution. The problem is related to the Fourier representation of a square well. The direct interpretation of small displacements using the comparison of modelled simulations with experimental image series is thus exceptionally difficult because the critical image changes with defocus depend on the transfer of amplitudes with rapidly varying phases at (and beyond) the normally accepted limits of the transfer function.

Fortunately, at least for a limited range of boundaries which exhibit a common reflection across them when appropriately orientated, there is no need to analyse the complex details of the image simulations. In any given image the interference fringe positions within about 2 nm of the boundary will still be non-uniformly spaced. However relatively far from the boundary, as long as the imaging conditions are identical on either side of the interface, then whatever the relationship of the fringes to the atomic plane positions this will be identical, so that the trans-boundary image shift will accurately represent the Coincident Site Lattice (C.S.L.) displacement. Such shifts can be measured to far higher accuracies than the microscope resolution because we simply need to know the relative positions of two large regions with a large number of "measurement markers". This approach has been used to evaluate $\Sigma 3$ boundary displacements for copper and gold to an accuracy about a hundred times better than the microscope resolution[19,20]. Interestingly in neither this "ruler" method nor in related moiré techniques[21,22] need the boundary plane itself be in projection. While there are many boundaries which cannot be treated using these methods[22], the requirement of a common, or at least parallel, reflection across the boundary is very often met in multilayers.

The Boundary Composition. As has been noted, a rigid body displacement is necessarily associated with a change in scattering potential which can easily be misinterpreted as being due to a local composition change. If, however, the rigid body displacement has already been determined by a method such as that described above, then this can be included in high resolution image simulations and the boundary composition adjusted to provide simulation matches with image series. Fortunately the high resolution approach is not needed: the Fresnel contrast at the edge of a foil has long been of use in the correction of astigmatism and the assessment of beam coherence but the contrast can also be related to the magnitude of the change in scattering potential. Thus the intensity of the fringed images of a boundary at large defoci can be related to the magnitude of the composition fluctuation provided only a single known element is segregated[23]. The method is surprisingly accurate and given care can yield results limited by the accuracy of foil thickness measurements to about ± 5%. Of course the nature of the segregating element has to be determined using EELS but the Fresnel approach is ideally complementary to the latter technique since the sensitivity of the contrast to the composition change increases as the thickness of the boundary layer decreases. The method is still being improved and it appears that the form of the local composition gradients can be determined. Furthermore it has been suggested[24] that a comparison of the Fresnel effects in dark field images with those obtained in the axial beam should allow the determination of both the rigid body displacement and the composition change; if this proves to be the case, the range of boundary classes which can be quantitatively characterised will be increased substantially.

The Localised Relaxation Structures. From the above it will be clear that the situation is still more open to misinterpretation when the problem is to characterise the local form of the intrinsic defects associated with a given boundary. The problem is nonetheless being tackled by many microscopists and a large number of very beautiful high resolution micrographs have been obtained. Of recent work it is noteworthy that significantly useful information can be obtained about the relative energy of different geometrically possible defects for a given boundary geometry without recourse to too detailed a comparison of a series of images and simulations[25]. It has however been suggested[26,27] that the detailed nature of a theoretical boundary defect structure can be related to the local image form, after digital image processing. While the uniqueness of such data-fits is perhaps questionable (e.g. ref.28) they are clearly achieved. That this is so is less surprising when it is remembered that the localised image distortions arising when the original image is formed, and emphasised further on Fourier filtering, are those that tend to even up local variations in the spacings. However such spacing changes are of course those that usually lower the energy of an unrelaxed boundary defect structure (e.g. ref.29).

In the end I would expect that this challenging area for the electron microscopist will require the quantified development of techniques requiring the comparison of beam tilted images and simulations. Even so the uniqueness of a given model is always a problem and for this reason data obtained by the weak beam technique can be very useful. Attempts have been made to use the method quantitatively in a "mapping approach"[30] by means of which the unknown displacement of a boundary defect is assessed by comparison of its images under a range of conditions with those of matrix defects with known displacements fields. Further, more recent, analysis[31] of the technique is somewhat discouraging in that simplistic interpretations are rarely viable because of dynamical effects. While these need not be dominant at high deviation parameters, they can be sufficiently different for a boundary and a matrix defect at the relatively low deviation parameters actually required. Thus simulations are still needed.

MULTILAYER STRUCTURE CHARACTERISATION

What then are the extra complications associated with the quantification of the structural and compositional modulations associated with heterostructural multilayers?

Multilayer systems tend to be of two general types: those which are, or are nearly, lattice matched (e.g. GaAs-$Al_xGa_{1-x}As$) and those which exhibit both compositional and lattice spacing variations (e.g. GaAs-$In_xGa_{1-x}As$ or Cu-Ni_xPd_y). The features of such multilayers which are generally required for correlation with models of the physical properties tend to be:
1) the amplitude and form of the compositional modulation;
2) the layer thickness and its vicinality;
3) the form of any interface steps or dislocations.

Considering firstly systems as epitomised by GaAs-$Al_xGa_{1-x}As$, the most appropriate techniques available for obtaining the required data have been discussed, from an industrial viewpoint, elsewhere[32] and will only be summarised here given also that much of the approach can be inferred from the above description of the methods used for single interfaces.

The amplitude of the compositional modulation in this system can be measured by techniques which are of special relevance to III-V alloy systems in that the contrast in an 002 type reflection is, on a kinematic basis, sensitively related to the Ga:Al ratio (e.g. ref.33). The full quantification of the approach is extremely difficult because of the effects of inelastic scattering (which are coming to be understood to have considerable relevance in high resolution image characterisation generally[34]) but the systemic errors caused by multiple inelastic/elastic scattering can at least be characterised to allow the use of a "fudge-factor" approach for more simply analysed data[35]. Other techniques for composition measurements include the characterisation of the thickness fringe periodicities for a cleaved wedge of known thickness[36] and a refinement of the CBED technique[37] but none seems more accurate than the dark field approach for this specific industrially relevant system.

An interesting current application of the Fresnel technique as summarised above, lies in the measurement of the abruptness of the compositional changes in multilayers such as these. Qualitatively the lack of detail in the contrast generally seen, as discussed elsewhere[38], suggests either that the composition changes are not as abrupt in M.B.E. grown material as is generally supposed or that the relationship between the change in composition and scattering potential is non-localised.

The thickness of the layers and their vicinality can be readily assessed to near monolayer accuracy by a combination of dark field and high resolution methods but it is important in this context to obtain images at a series of specimen tilts using the 002 reflection approximating to the growth direction[35,39].

Once the vicinality has been determined the average step density can be inferred. However optical filtering of images, though this aids the visualisation of the positions at which steps occur, is of dubious value in determining their displacement fields. A variety of methods are currently being developed, including the centre dark stop field technique, in attempts to retain resolution and the contrast changes associated with the composition wave form[39,40] (until energy loss filtering becomes more generally available at high resolution[34]).

Considering now the characterisation of multilayer systems exhibiting periodic variations in both the composition and lattice plane spacing, the two semiconductor systems which are receiving increasing attention are $GaAs/In_xG_{1-x}As$ and Si/Si_xGe_{1-x}. The alarming tendency for the former system to exhibit the reverse contrast to that expected on simple structure factor grounds in an 002 reflection does not lend confidence to any simple approach, and it should be noted in general that the importance of treating the foil surface relaxation in a strained multilayer properly[41] cannot be understimated. Equally this is not the origin of the contrast anomaly described.

From a more optimistic viewpoint one metallic superlattice system has been fully quantified using TEM with interesting results. The system examined was $Cu-Ni_xPd_y$ at varying lattice parameter differences and modulation wavelengths and all the techniques described above had to be applied quantitatively. The pair of dark field images in fig.3 demonstrates the epitaxial nature of the sputter grown films[42] but the evaluation of the layer composition variations required the Fresnel approach described here. While axial images of Au/Ag layers, as shown in fig.4, were of sufficient resolution to allow quantification of the spacing variations at the cube normal, non-axial images had to be used for the

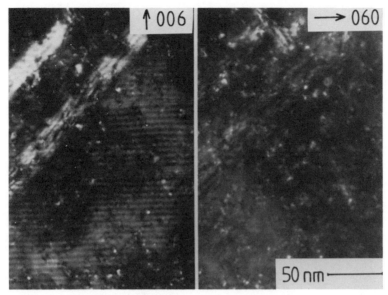

Fig. 3. A pair of dark images of a Cu-NiPd multilayer grown with 001
normal. The differences in appearance for 006 and 060 at a similar
deviation parameter demonstrate the epitaxially strained nature of
the system (Courtesy of C.S. Baxter).

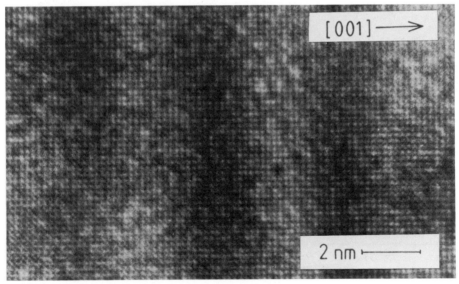

Fig. 4. A high resolution edge-on image of an Au/Ag multilayer grown with
001 normal (Courtesy of C.S. Baxter).

reduced spacings of the $Cu-Ni_xPd_y$ system (fig.5). It is noteworthy that
since simply spacing variations were required, rather than local atomic
position variations, the non-axial approach allowed the transfer of the
higher Fourier components of the diffracted amplitude, carrying the
relevant information[43]. Of course the fringe spacing variations (fig.6),
as determined from images such as that shown in fig.5, still have to be
related to the real changes in planar spacing by comparison with simulated
models incorporating the composition waveform (determined by the Fresnel
method). The work described allowed a structural interpretation of the
physical property anomalies associated with lower wavelength structures[9].

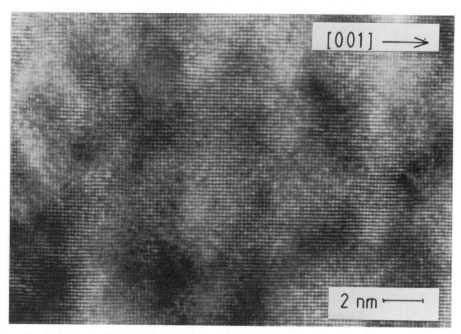

Fig. 5. A high resolution edge-on image of a Cu/NiPd multilayer of 001
normal. The image was obtained non-axially to enhance the accuracy
with which the spacing variations could be measured.

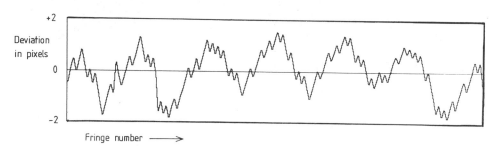

Fig. 6. Graph showing the deviation of successive fringe positions in fig.5
from those for an averaged uniform spacing, changes which (at ~9%)
are about 50% larger than the model spacings which best fit them in
simulations of the images of the type obtained.

CONCLUSION

I have concentrated here on giving a brief description of some of the newer methods which allow, or potentially allow, a full characterisation of a layered system. Many of these methods can be system specific but several, as exemplified by the Fresnel approach, would appear to have very general application. Because microscopes have such broad application it is important to develop methods which can be quantitative and yet not too time consuming. The full quantification of a heterostructure is only, for example, worthwhile, since it will nevertheless take months to complete it for a single specimen, when new physics is involved. In general, however, the current rapid expansion in the use of III-V layered systems technologically puts considerable pressure on the ability of the crystal growers to make systems of very precise forms. It is thus fortunate that most of the simpler data the grower needs to improve his methods, when specimens are not "providing the expected characteristics", can be obtained using rather simple TEM techniques in a matter of hours.

ACKNOWLEDGEMENTS

I am grateful to Prof. D. Hull for the provision of laboratory facilities and to both the SERC and a number of industries for support, including British Telecom, G.E.C., Philips, S.T.C. and Johnson and Matthey. I acknowledge discussion in the fields described with a number of co-workers in the Department including K.B. Alexander, C.S. Baxter, C.B. Boothroyd, E.G. Britton, P.E. Donovan, K.M. Knowles, S.B. Newcomb, F.M. Ross and E.J. Williams.

REFERENCES

1. W.M. Stobbs, in: "The Physics and Fabrication of Microstructures and Microdevices", M.J. Kelly and C. Weisbuch eds, Springer-Verlag Physics Series Vol.13, Berlin, p.136 (1986).
2. P.J. Withers, W.M. Stobbs and A.J. Bourdillon, J. Microsc. (1988) in press.
3. S.B. Newcomb, C.B. Boothroyd and W.M. Stobbs, J. Microsc. 140:195 (1985).
4. C.B. Boothroyd, W.M. Stobbs and K.N. Tu, Appl. Phys. Letts 50:577 (1987).
5. S.B. Newcomb, W.M. Stobbs and E. Metcalfe, Phil. Trans. Roy. Soc. (London) A319:191 (1986).
6. W.M. Stobbs, S.B. Newcomb and E. Metcalfe, Phil. Trans. Roy. Soc. (London) A319:219 (1986).
7. S.B. Newcomb and W.M. Stobbs, J. Mat. Sci. and Eng. 66:195 (1984).
8. M.E. Cook and J.E. Hilliard, J. Appl. Phys. 40:2191 (1969).
9. C.S. Baxter and W.M. Stobbs, Nature 322:814 (1986).
10. W.M. Stobbs, in: "Structure on Non-Crystalline Materials", P.H. Gaskell ed. p.253 (1976).
11. W.M. Stobbs and J.P. Chevalier, Acta Metall. 26:233 (1978).
12. G. Van Tendeloo, S. Amelinckx and D. De Fontaine, Acta Crystallogr. B41:281 (1985).
13. W.M. Stobbs and S.H. Stobbs, Phil. Mag. B53:537 (1986).
14. D. Shechtman, I. Blech, D. Gratias and J.W. Cahn, Phys. Rev. Lett. 53:1951 (1984).
15. References in: J. de Physique, 47 Coll. C3 (1986).
16. References in: Journal of Microscopy, 146 pt III June (1987).
17. K.M. Knowles and W.M. Stobbs, Nature 323:313 (1986).

18. K.M. Knowles, in: "Electron Beam Imaging of Non-Crystalline Materials", K.M. Knowles ed., I.O.P. Short Meeting Series No 11, Inst. of Phys. Bristol, p.99 (1988).
19. G.J. Wood, W.M. Stobbs and D.J. Smith, Phil. Mag. A50:375 (1984).
20. W.M. Stobbs, G.J. Wood and D.J. Smith, Ultramicrosc. 14:145 (1985).
21. J.W. Matthews and W.M. Stobbs, Phil. Mag. 36:373 (1977).
22. R.C. Ecob and W.M. Stobbs, J. Microsc. 129:275 (1985).
23. J.W. Ness, W.M. Stobbs and T.F. Page, Phil. Mag. A54:679 (1986).
24. C.S. Baxter: Private Communication (1987).
25. K.N. Merkle and D.J. Smith, Ultramicrosc. 22:57 (1987).
26. W. Krakow, J.T. Wetzel and D.A. Smith, Phil. Mag. A53:739 (1986).
27. W. Krakow and D.A. Smith, Ultramicrosc. 22:47 (1987).
28. W. Wunderlich and M. Rühle, Proc. Int. Conf. Electron Micr. Kyoto 1986, T. Imura, S. Maruse and J. Suzuki eds, p.1335 (1986).
29. A.P. Sutton and V. Vitek, Phil. Trans. Roy. Soc. (London) A309:1 (1983).
30. P.E. Donovan and W.M. Stobbs, J. Microsc. 130:361 (1983).
31. P.E. Donovan and W.M. Stobbs, Ultramicrosc. 23:119 (1987).
32. E.G. Britton, K.B. Alexander, W.M. Stobbs, M.J. Kelly and T.M. Kerr, GEC Journal of Research 5:31 (1987).
33. P.M. Petroff, J. Vac. Sci. Technol. 14:974 (1977).
34. W.M. Stobbs and W.O. Saxton, J. Microsc. (1988) in press.
35. E.G. Britton, Ph.D. Thesis, University of Cambridge (1987).
36. H. Kakibayashi and F. Nagata, Jap. J. Appl. Phys. 25:1644 (1986).
37. D.J. Eaglesham and C.J. Humphreys, Proc. Int. Conf. Electron Micr. (Jap. Soc. E.M. Tokyo) p.209 (1986).
38. W.M. Stobbs, in IIIrd M.S.S. Conference Montpellier, J. de Physique C5, 48:33 (1987).
39. C.B. Boothroyd, E.G. Britton, F.M. Ross, C.S. Baxter, K.B. Alexander and W.M. Stobbs, in: "Microscopy of Semiconducting Materials" 1987, A.G. Cullis and P.D. Augustus eds, Inst. of Phys., London p.195 (1987).
40. K.B. Alexander, C.B. Boothroyd, E.G. Britton, C.S. Baxter, F.M. Ross and W.M. Stobbs, in: "Microscopy of Semiconducting Materials" 1987, A.G. Cullis and P.D. Augustus eds, Inst. of Phys., London p.15 (1987).
41. M.M.J. Treacy, J.M. Gibson and A. Howie, Phil. Mag. A51:389 (1985).
42. R.E. Somekh and C.S. Baxter, J. Cryst. Growth 76:119 (1986).
43. D.J. Hall, P.G. Self and W.M. Stobbs, J. Microsc. 130:215 (1983).

SURFACE MICROANALYSIS AND MICROSCOPY BY X-RAY PHOTOELECTRON SPECTROSCOPY (XPS), CORE-LOSS SPECTROSCOPY (CLS) AND AUGER ELECTRON SPECTROSCOPY (AES)

Jacques Cazaux

Faculté des Sciences
F-51062 Reims Cedex
France

1. INTRODUCTION

X-ray photoelectron spectroscopy (XPS, or ESCA: Electron Spectroscopy for Chemical Analysis) and Auger Electron Spectroscopy (AES) are presently two of the three most popular techniques (with SIMS: Secondary Ion Mass Spectroscopy) used for the elemental identification of the species composing the first atomic layers of a surface. Core-loss Spectroscopy (CLS or ILS: Ionization Loss Spectroscopy) is far less popular than XPS and AES. Nevertheless it is considered here because this electron spectroscopy is based on a mechanism similar to that of XPS, it can be developed in parallel with AES and it allows also the surface elemental identification with an information depth in the nanometer range.

The atomic mechanisms involved in these spectroscopies have been known for a long time (from the work of E. Rutherford[1] for XPS, P. Auger[2] for AES and G. Ruthemann[3] for CLS) but their use for surface analysis only started at the end of the sixties (by K. Siegbahn[4] for XPS and L.A. Harris[5] for AES). Because their concepts have been exploited commercially in ultra high vacuum systems, the pioneering papers have been followed rapidly by a proliferation of papers devoted to surface studies not only from the fundamental point of view but also for solving practical problems having industrial applications.

In the next section I will try to show why the chemical composition of the few uppermost layers of a surface or an interface has considerable importance from the practical point of view. The principles of (and the chemical information that can be deduced from) the electron spectroscopies of interest are described in section 3. Their surface sensitivities and use for non-destructive and destructive depth profiling are the subject of the section 4. The minimum detectable concentration and the accuracy with which the concentration can be obtained are indicated in section 5. Section 6 is devoted to the lateral resolution problem associated with the corresponding microscopies while the additional structural information (nearest-neighbour atomic distances) they are able to give, is indicated in section 7.

2. EXAMPLES OF INDUSTRIAL APPLICATIONS

General considerations

The fact that the most modern industries, such as the semi-conductor industry, use highly sophisticated techniques to solve problems of increasing complexity with increasing device density is not surprising. It is an industry for which the gap between fundamental research and applications is quite indistinguishable.

On the other hand, the motor-car industry is representative of a rather old and well established industry for which the use of modern techniques does not seem to be necessary. In the next section I will try to show that this is not the case but before entering into details, a naive question: why is stainless steel not oxidizable?

Stainless steel cannot be oxidized (under standard conditions) because its surface is protected by a natural chromium oxide layer, 1-2 nm thick[6] (see also Stobbs' contribution, this volume). But when it is heated (500 °C; 75 minutes), the sulfur and phosphorus impurities (initial concentration <0.1%) migrate towards the surface, leading to a surface segregation concentration of ~25% for S and ~8% for P and breaking the continuity of the chromium layer.

Similar mechanisms occur for nickel in an acidic medium. The formation of a passive NiO layer a few monoatomic layers thick prevents any kind of further oxidation (passivation process) except when sulfur impurities (coming from the metal or the electrolyte) are concentrated at the surface leading to corrosion (dissolution process)[7,8].

Examples taken from the motor car industry

Figure 1 indicates some parts of a motor car for which knowledge of the surface and interface compositions at the monolayer scale is very useful.

Steel tyre cords. In modern steel-belted radial tyres, the steel tyre cords are strongly bonded to the natural rubber because they are plated with 200nm brass which gives strong bonds (at the atomic scale) with the sulfur atoms contained in natural rubber. The knowledge of the characteristics of this bond and the effect of the brass parameters (Cu-Zn alloy) on the bond strength and durability is important for optimizing the performance of the tyre[9,10].

The body. The painted body of the car offers another example of metal-polymer adhesion studies. Between the steel and the paint there is an interfacial phosphate conversion layer (iron and zinc phosphate) which, from the adhesion standpoint, involves two interfaces as well as a possibility of cohesive failure within the phosphate layer. From the corrosion standpoint the possible chemical reaction of the additive paint pigments with the phosphate layer has to be investigated at the corresponding interface by using surface sensitive techniques.

Lubricant oils. Modern lubricating oils contain pure zinc dibutyl-dithiophosphate (ZDP). The influence of this and other additives on the wear between surfaces in relative motion requires the identification of the elements present at the edge of the worn surfaces: a subject for tribological research[11,12].

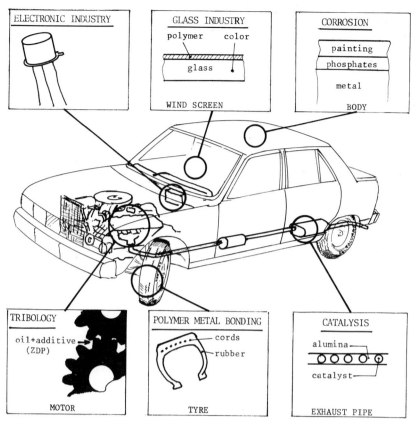

Fig. 1. Some selected elements of a modern motor car in which the surface elemental analysis is involved. The plated elements, the battery, as well as the plastics, are also relevant for surface analytical technique.

The windscreen. The windscreen can be tinted, covered by a thin plastic foil (for safety) or by transparent conductive layers (for deicing). A good understanding of the adhesion mechanism between these layers and the windscreen can only be obtained by accurate analysis of a limited number of atomic layers composing the corresponding interface. This example illustrates the increasing use of surface microanalytical techniques (mainly XPS) in the glass industry[13].

The exhaust pipe. For environmental reasons, quality catalyst converters in the exhaust pipes have to be used. But there is a need to improve their lifetime by studying the poisoning mechanism of catalysts. This surface effect is investigated by means of electron spectroscopies not only for this specific problem but for all the fundamental research used in development of catalysts in research laboratories and in the petroleum industry[14].

Other examples

In fact, surface chemical analyses are now made on a large variety of advanced materials (e.g. semiconductors, polymers, metals, oxides, glasses) after fabrication and at various times during their service life for process optimization, failure analysis and quality control (see for instance relevant books[15-17], the excellent review paper of Powell[18] and the proceedings of a recent congress[19]).

For example, in the semiconductor industry, modern microanalytical techniques are used i) to identify the compounds formed at the interfaces by diffusion and dopants and contaminants at the various stages of processing, ii) to analyse the failures in lead bonding, hermetic seals and metallization adherence (see fig.2) and iii) in the establishment of adequate cleaning procedures[18].

The strong financial need for finding the causes of failure can be easily understood if one knows that the reliability and yield in the manufacturing of MOS/LSI (metal oxide semiconductor devices with large scale integration) is less than 10% of the final assemblies; 40% of device failures being attributed to surface related defects and close to 20% to metallization, oxide and bond defects.

Conclusion for industrial applications

The few examples given above illustrate the importance of the surface and interface characterization from the technological point of view. This importance results from the fact that the composition of the outermost atomic layers of a surface always differs from that of the bulk and this compositional change strongly influences the following properties (see

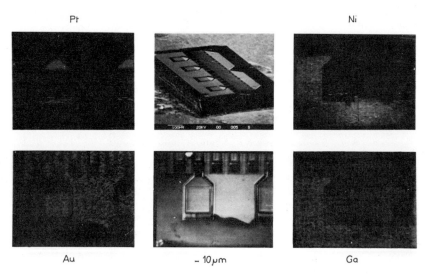

Fig. 2. Example of failure analysis of a microelectronic component by Auger electrons (courtesy of Thomson-CSF, Corbeville, France). The microelectronic devices are composed of multilayers for optimizing their electrical and mechanical properties, but a failure in adhesion between two adjacent layers may occur. The diagnosis is facilitated by obtaining Auger maps relative to the various elements (Au, Ga, Ni, Pt). The two central pictures are secondary electron images of the device at two different magnifications.

Table I. Number of monolayers and corresponding thicknesses affecting the technologically important properties indicated

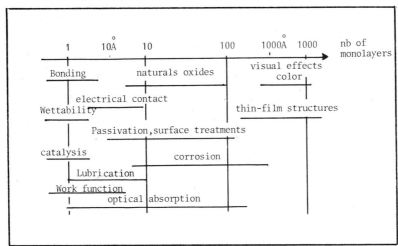

table 1 for an order of magnitude for the surface thicknesses involved):
i) electrical properties (e.g. integrated circuits),
ii) adhesion and bonding,
iii) catalytic activity,
iV) plating and decoration,
V) lubrication,
Vi) corrosion,
Vii) grain boundary segregation,
Viii) biocompatibility.

Finally, the technological importance of surface characterization can also be measured by the fact that the related surface chemical analysis are involved in the VAMAS project-area 2 (Versailles project on Advanced Materials and Standards[21]).

3. CHEMICAL INFORMATION OBTAINED FROM ELECTRON SPECTROSCOPIES

Basic mechanisms

The left hand side of fig.3 shows the electronic levels of 3 different solids: pure aluminium (Z=13), pure silicon (Z=14) and SiO$_2$. These diagrams illustrate the well-known fact that the binding energy of an atomic electron is strongly dependent on the atomic number of the atom in which this electron is involved (see the change from Al to Si). It illustrates also the fact that this binding energy slightly depends on the chemical surrounding of the atom of interest (see change from Si to SiO$_2$). This is the chemical shift effect [Δ(c.s) $\leq$ 10 eV]. The right hand side of fig.3 shows the mechanisms involved in electron spectroscopies. In XPS, monochromatic X-ray photons (for example AlKα: hν $\cong$ 1486 eV) are absorbed (X-ray absorption) causing photoionization of the atom and ejection of an inner-shell electron. The energy analysis of the photoelectron, E_K, permits the determination of the binding energy E_B of the inner-shell electron from the relationship:

$$E_K = h\nu - E_B \tag{1}$$

the energies being referred to the Fermi level).

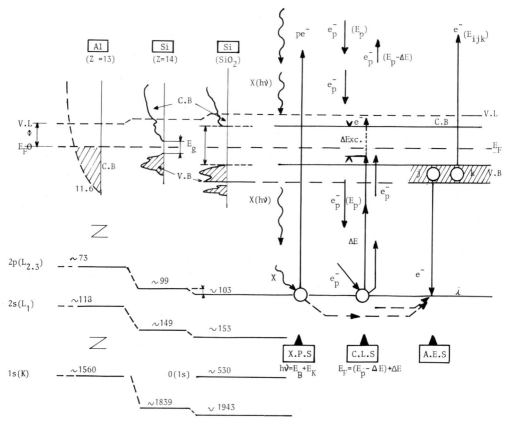

Fig. 3. Left hand side: Schematic representation of the electronic levels of Al, Si (pure) and SiO$_2$. Being the result of a compilation from various papers, the binding energy values (referred to the Fermi level) are only indicative. Right hand side: Schematic represent-ation of the excitation mechanisms involved in the spectroscopies of interest.

In CLS, monochromatic incident electrons (primary energy E_p) interact with inner shell electrons. An inner shell electron may gain the energy ΔE to reach an allowed final (empty) state above the Fermi level in metals. ΔE is deduced from the kinetic energy measurement of the outgoing electron ($E_p - \Delta E$). C.L.S. is nothing other than Electron Energy Loss Spectroscopy (EELS) but using rather slow energy incident electrons (keV range) on surfaces instead of high incident energy electrons (100 keV) through thin foils.

Such ionization processes (created by incidents photons or electrons) are followed by two de-excitation processes: X-ray photon and Auger electron emissions but the Auger emission probability, a_{ijk}, is greater than the fluorescence yield, ω_{ij}, when the initial binding energy is $E_B(e) < 10$ keV.

The result is that (X-ray induced) Auger electron lines can be detected when an XPS spectrum is acquired as well as CLS signals can be detected when an (electron induced) AES spectrum is acquired. The simple way to distinguish between Auger lines and true XPS or CLS lines is that Auger lines do not change positions when the energy of the incident particles (hν or E_p) is changed: this explains that Auger spectroscopy does not require in principle monochromatic incident beams.

An XPS instrument is mainly composed of an X-ray source (AlKα or MgKα), a sample holder and an electron analyser (concentric hemispherical analyser in most of the cases). An X-ray monochromator is often interposed between the X-ray source and the sample. The counting mode is always used for the signal acquisition.

An AES instrument is mainly composed of an electron gun, a sample holder and one of several types of electron analyser. The spectra of the electrons emitted from the sample can be obtained directly [N(E) mode by pulse counting or beam blanking] or with the use of a lock-in amplifier to obtain the first, N'(E), and the second derivatives, N"(E). These two modes are useful to suppress the background on which the Auger (and core loss) signals are superimposed but they distort the physical information contained in the N(E) spectra; for example they annihilate some unstructured but valuable information and they overestimate sharp structures of low intensity. The core loss peaks are expected to be edge shaped (see fig.3 and fig.4a) but they look like Auger peaks in the N'(E) and N"(E) modes.

All the XPS and AES instruments have in common the kinetic energy range being explored by the analyser (0 to ~2 keV), the need of a very good vacuum and some additional systems such as an ion gun for cleaning the sample.

Chemical information obtained

XPS is mainly based on a one electron level process, CLS on a two-electrons level process and AES on a three-electrons level process but in all the cases the kinetic energy of the characteristic electrons reflects (more or less directly) the binding energy of an inner shell electron before its excitation.

The elemental identification property of the spectroscopies results from the above remark. The measurement of the kinetic energies of the characteristic electrons allows the determination of the elements that have emitted the characteristic signal (by calculation or by using published charts for the microanalytical use: see Problem 1 and references therein or ref.16 for details). All the elements can be identified in this way except H and He in AES (see fig.3c); (in fact these two elements are also difficult to identify in CLS and XPS because of their weak ionization cross sections and of their low binding energies, leading to overlap with plasmons losses of the substrate in CLS and valence bound signals of the same substrate in XPS).

The additional chemical information that can be obtained is:

The chemical shift effect. The amplitude of this shift differs from one spectroscopy to another. The considerable advantage of XPS is that the kinetic energy change, $\Delta E_K(i)$, is the direct result of the change in the binding energy, $\Delta E_B(i)$:

$$\Delta E_K(XPS) = -\Delta E_B(i) \qquad (2)$$

(see ref.16a). This ignores the possible change of the work function and of the vacuum level by charging effects.

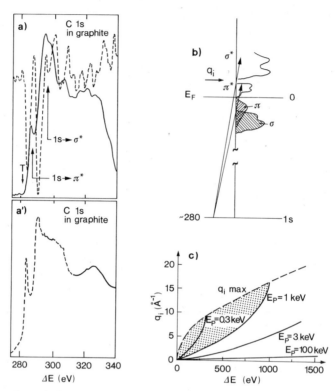

Fig. 4. Comparison between the near edge structures of graphite obtained by
CLS (a) and by EELS (a'), from ref.24. The spectrum (a) has been
obtained in the first derivative mode by reflection of E_p = 1 keV
incident electrons (dotted lines) and then integrated numerically
(full line). The spectrum (a') has been obtained by transmission
of E_p = 100 keV electrons and the background has been subtracted.
The two main peaks correspond to 1s → π* and 1s → σ* transitions
respectively, as indicated in (b). From the theoretical point of
view the main difference between CLS and EELS is related to the
fact that the minimum value of the inelastic momentum transfer,
q_i(min), increases when E_p is decreased (c), leading to a
relaxation of the dipole selection rules. The differential cross-
section for a core ionization process can be written in the form
(see ref.23 and A. Howie, this book):

$$d^2\sigma/dqdE \propto (1/q^3)|\langle f|e^{i\underline{q}\cdot\underline{r}}|i\rangle|^2, \quad \text{where} \quad e^{i\underline{q}\cdot\underline{r}} = 1 + i\underline{q}\cdot\underline{r} - (\underline{q}\cdot\underline{r})^2.$$

The first term is = 0 by orthogonality, the second term represents
the dipole transition ℓ' = $\ell \pm 1$ and the third term is the
quadrupole transition ℓ' = $\ell \pm 2$, ℓ' = ℓ.

In CLS, the measured shift reflects also the shift in the final state (for example the opening of the band gap when going from a metal to its oxide). Also, in insulators the excited electron is poorly screened by the valence electrons from the core hole, spelling the downfall of the one electron model. A core exciton final state may be formed lowering the excitation energy by Δ_{exc}[22]:

$$\Delta E(CLS) \simeq \Delta E_B(i) + \Delta E_g/2 - \Delta E_{exc} \qquad (3)$$

The advantage of CLS is that the measured shift is neither influenced by the work function nor the charging effect. It is similar to the shift observed in X-ray absorption spectroscopy.

In AES, the shift of the Auger line (involving the levels i, j and k) is given by[16a]:

$$\Delta E_{ijk}(AES) \simeq \Delta E_K(XPS) + 2\Delta R(i^+) \qquad (4)$$

where $\Delta E_K(XPS)$ is given by eq.2 and $\Delta R(i^+)$ is the change in the relaxation energies (predominantly extra atomic) in the final singly ionized state (i vacancy).

Equation 4 indicates that the influence of the work function change and the charging effects are the same in XPS and AES.

Density of states. The shape of the valence band density of states can, in principle, be obtained from the shape of the corresponding XPS line, but the related photoionization cross section values lead to rather weak intensities when the usual photon energies are used. This difficulty can be overcome by lowering the incident photon energy (YMξ or ZrMξ) down to the ultra-violet photoelectron spectroscopy regime.

The shape of the core loss peak corresponds to the so-called near edge structures in electron energy loss spectroscopy (transmission through thin foils) or in X-ray absorption spectroscopy because the initial and final state are quite the same. The most important differences are i) the fact that the dipole selection rules are relaxed when the primary beam energy is decreased (see caption of fig.4); ii) that CLS can be developed on bulk samples (instead of thin foils) but iii) the incident electron also suffers elastic events (before or after) the inelastic event of interest.

In AES, when transitions involving the valence band are concerned (CCV: core core valence; CVV: core valence valence transitions), it is possible to extract the corresponding valence density of states (DOS) from the Auger line shape but in the CVV case, this shape is not a simple self convolution of the DOS. It remains that this line shape contains chemical information that is the subject of considerable recent study[25].

Correlation studies

Although all the spectroscopies of interest here have the same elemental identification property, they differ from each other by some specific information (previous section) and the advantage of gathering all the possible information from the investigated sample needs no emphasis. Figure 5 shows an example of an XPS spectrum in which true XPS lines coexist with (X-ray induced) Auger lines. Their exact positions are influenced in the same manner by charging effects (when they occur in poorly conducting materials) leading to a false chemical shift. This difficulty is overcome by plotting the (modified) Auger parameter[27]:

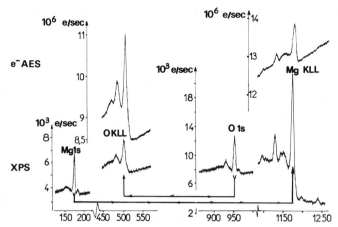

Fig. 5. (After ref.26). Bottom: XPS spectrum of a MgO sample showing the
O1s and Mg1s photolines associated to their parent (X-ray induced)
Auger lines, O(KLL) and Mg(KLL), which result from the de-
excitation process. Top: (electron induced) AES spectrum of the
same sample, for a comparison of the signal and the background
intensities. All these spectra have been obtained in the counting
mode [E.N(E) mode].

$$\alpha' = E_{ijk} + E_B(i) \tag{5}$$

This parameter is independent of the work function, the photon energy
and the surface potential. When various compounds of the same element are
investigated, the accuracy of α' (± 0.1 eV) is better than the dispersion of
the independent peaks position measurements and the change in the
relaxation energy (eq.4) can be accurately known. A good example of the
use of this parameter comes from the work of West and Castle on silicate
minerals[28]. In (electron induced) Auger analysis, it is possible to obtain
not only the Auger lines but also the elastic peak, the plasmon losses and
the core losses, in order to correlate the information they give.
Surprisingly, it is very scarcely done (to the author's knowledge). For
doing this, the need is only to lower the primary beam energy (relative to
its usual value in AES: 5-20 keV) down to an energy range accepted by the
electron analyser (see fig.6 for an example).

One advantage of this correlation (among others) is that the CLS peak
position is insensitive to the charging effects.

Another type of correlation can also be obtained by using coincidence
techniques when two events occur quite simultaneously. This is the case of
the photoelectron emission and the ejection of the Auger electron (related
to the subsequent de-excitation process), in XPS. Haak et al[30], succeeded
in this very fine experiment, elucidating definitively the assignment of
the Cu LMM Auger line (see fig.7).

A similar coincidence experiment can be imagined in AES by analysing
the core loss signal and the subsequent Auger process, but this experiment
has not yet been performed.

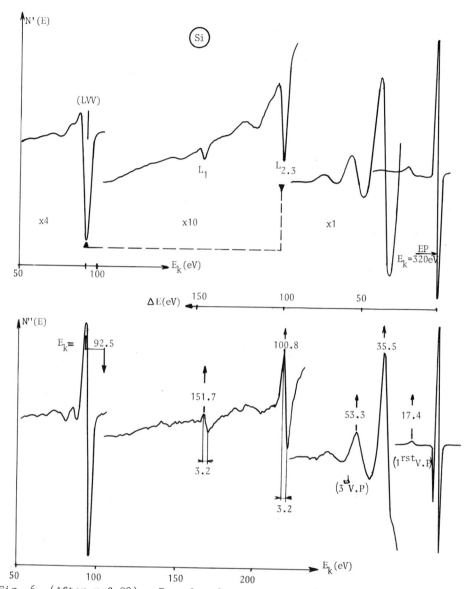

Fig. 6. (After ref.29). Example of spectra obtained on a silicon sample by using an electron bombardment ($E_p = 320$ eV).
Top: first derivative mode. Bottom: second derivative mode. Going from right to left is the elastic peak (EP) followed by three plasmon peaks (at 17.4, 35.5 and 53.3 eV). In the middle, the $L_{2,3}$ and the L_1 core loss peaks are clearly visible. The $L_{2,3}$ ionization process is followed (see dotted line) by the $L_{2,3}$ VV Auger emission (on the left). The useful scale for the loss structures is the ΔE scale starting from the elastic peak. The useful scale for the Auger line position is the usual kinetic energy scale. The depth of modulation and the channeltron voltage have been kept unchanged during the data acquisition, except for the elastic peak acquisition (to prevent channeltron saturation). Core loss peaks allow the accurate determination of the threshold energy positions ($L_{2,3}$ and L_1 lines) keeping in mind that, in the N(E) mode, core loss shapes are "edges", so that the exact position of the threshold energy, indicated by T in fig.4a, is ~3.2 eV before the maximum of the 2nd derivative.

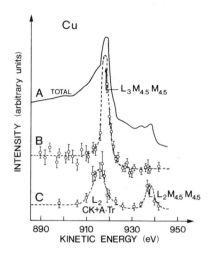

Fig. 7. (After ref.30). Auger photoelectron coincidence experiments in
copper. The curve marked in A is the normal Auger spectrum. B is
the Auger spectrum obtained in coincidence with the $2p_{3/2}$ photoelectrons and C is the Auger spectrum obtained in coincidence with
the $2p_{1/2}$ photoelectrons.

4. SURFACE SENSITIVITY

General

The characteristic (photo Auger) electrons carrying the chemical
information have kinetic energies, E_K, being in the 50 eV to ~2 keV range
to be identified by Auger and XPS equipments. Due to this rather low
kinetic energy, these electrons may suffer strong inelastic interactions
(loosing their chemical information) when they are issued from deep
atoms. In surface microanalysis, it is assumed that the electron created
at the depth z has the probability proportional to $e^{-z/\delta}$ to escape into the
vacuum without suffering one inelastic process. The information depth δ is
given by:

$$\delta = \lambda(A.L)\sin\theta \tag{6}$$

[$\lambda(A.L)$ = attenuation length; θ; take off angle of the analyser]. $\lambda(A.L)$
is typically between 3 and 50 Å for electrons having E_K: 50 eV < 2 keV (see
Dobson, this volume, for the compilation of Seah and Dench).

This is the reason for which XPS, CLS and AES are surface sensitive
techniques. But inherently this extreme surface sensitivity is paid for
by: i) the need for UHV attachments (10^{-6} torr x 1s = 1 langmuir = 1 monolayer coverage in 1 s if the sticking coefficient is one); ii) the great
sensitivity of the signal intensities to topographic effects (fig.8).

The correlation between $\lambda(A.L)$ and the interactions of electrons in
solid is indicated in the following subsection. The applications of
electron spectroscopies for non destructive and destructive depth profiling
are next pointed out (see subsections on Non-destructive depth profiling
and Destructive depth-profiling).

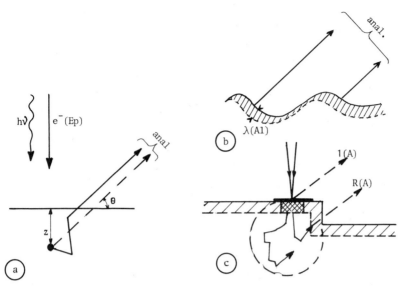

Fig. 8. a). A characteristic (Auger or photo) electron created at the depth z may have a zig zag path (full line) before escaping into the vacuum and entering the electron analyzer. The inelastic mean free path is associated with this real path (full line) while the attenuation length is associated to an ideal straight line (dotted line).

b). The topography of a surface influences the measured signal intensity because the mean thickness, from which the electrons come from, changes from place to place (hatched area).

c). By scanning the incident electron beam along a surface one may obtain an increase of the Auger signal when the incident beam is close to a step because of the reinforcement of the backscattered contribution R(A) due to the vertical part of the step.

Attenuation length and inelastic mean free path

The inelastic interactions of electrons (kinetic energy E_K) in a solid can be described by using the complex dielectric theory: $\tilde{\varepsilon}(\omega, q)$ (A. Howie: this volume and ref. 31). The differential inelastic cross section for the energy loss $\Delta E = \hbar\omega$ and the inelastic momentum transfer q_i in an infinite medium is:

$$\frac{d^2\sigma}{d(\Delta E)dq_i} = \frac{4\pi m e^2}{N\hbar^2 E_K} \text{Im}[\frac{-1}{\tilde{\varepsilon}(\omega, q_i)}] \frac{1}{q_i} \qquad (7)$$

where N is the density of atoms or molecules. The main energy loss mechanism is the plasmon excitation of valence electrons and $\text{Im}[(-1/\tilde{\varepsilon}(\omega, q_i)]$ can be, for simple solids, either evaluated theoretically or deduced from dielectric data (optical or characteristic loss experiments).

The integration of eq.7 over q and ΔE gives the total inelastic cross section σ_T and the corresponding value of the inelastic mean free path from:

$$\lambda[\text{IMFP}] = [N\sigma_T]^{-1} \tag{8}$$

Following this method, various expressions have been proposed[31], such as that of Szajman and Leckey[32]:

$$\lambda(\text{IMFP}) \text{ in } \AA \simeq 1.8 \ \overline{E} \ E_K^{3/4}/E_p^2 \tag{9}$$

where $\overline{E}$ is the centroid of the energy loss function and E_p the bulk plasmon energy (eq.9 can be simplified by assuming $\overline{E} \simeq E_p$ for Problem 2). Equation 9 shows the dependence of λ on the kinetic energy, E_K, of the electrons and on the material constants, through the plasmon energy.

For practical purposes (microanalysis), $\lambda(\text{IMFP})$ is often taken as $\lambda(\text{A.L})$ (see Problems 1 and 2, expression of Sjazman and Leckey). Nevertheless it must be kept in mind that $\lambda(\text{IMFP})$ may exceed $\lambda(\text{A.L})$ because of the elastic events suffered by the excited (photo or Auger) electron before entering the electron analyser (see fig.8c). Theoretical efforts are in progress to take into account the zig-zag paths of the electrons by solving transport equation or using Monte Carlo simulation[33,36].

Fortunately the result is that the exponential decay is kept in most of the cases but $\lambda(\text{A.L})$ is now a function of $\lambda(\text{IMFP})$ and of $\lambda(\text{T.E})$ (transport mean free path for elastic scattering). Without an analytical expression for $\lambda(\text{T.E})$ (which strongly influences the signal intensity in CLS), it remains possible to overcome these difficulties by determinating experimentally $\lambda(\text{A.L})$ (overlayer method in XPS and elastic peak intensity in AES[37]).

Non destructive depth profiling

When the surface thickness to be investigated, t, is $t \leq \lambda(\text{A.L})$, the electron spectroscopy of interest can be used for non-destructive depth profiling, the information depth being varied by changing θ (eq.6) or by changing E_K (eq.9). Depth information can also be obtained from the shape of the signal.

Emission angle dependent XPS measurements on layers of known thicknesses are frequently performed to obtain the experimental values of $\lambda(\text{A.L})$. In turn, they can also be used to obtain the thickness and composition of the overlayers observing that the signal intensity (of a given element at a given angle) is proportional to:

$$I(\theta) \propto \int_0^\infty C(z) \cdot \exp(-z/\lambda\sin\theta)dz \tag{10}$$

or to

$$I(\theta) \propto \sum_1^\infty C_i \cdot k^{i-1} \tag{11}$$

depending on the model used for describing the attenuation: continuous model or layer-by-layer model (see Problem 2: Gallon's model).

By using eq.10, the concentration profile, $C(z)$, is given by the inverse Laplace transform of the measured angular intensity, $I(\theta)$.

By using eq.11, the concentration of the ith layer, C_i, is the solution of a set of linear equations having coefficients $k = \exp(-a/\sin\theta)$ that can be changed by changing θ (a = thickness of one layer).

Due to inevitable measurement errors and the possible influence of topographic and crystallographic effects (there are assumed not to change when θ is changed) this procedure is of limited accuracy in the general case. Nevertheless it remains that the variation of the take off angle in XPS permits one to distinguish between a thin overlayer of an element A on a substrate B and an homogeneous surface of composition $A_X B_Y$[38,42] (see Problem 2).

An alternative solution is to change E_K (eq.9). In AES, this is possible for elements having two peaks at (rather) different energies. In XPS, it is possible by changing the primary radiation energy (twin anode or synchrotron radiation source[42]). In this field, the advantage of CLS is that the primary energy E_p can be easily changed continuously, changing the information depth of the core loss signal (through $E_p - \Delta E$). This idea, proposed independently by several authors[43-45] is illustrated in fig.9. Here again the advantage of correlating the information obtained by two (or more) techniques in the same equipment and on the same sample is obvious: in this case it will be the information depth given by the photolines compared to that given by the (X-ray induced) Auger lines in XPS or that given by the core loss signal compared to that given by the (electron induced) Auger lines in conventional AES.

For extracting in depth composition information, a similar philosophy consists in collecting not only the signal intensity in AES but also the background intensity and combining the two [46] or the exact shape of the Auger signal after a correct background substraction [47]. In XPS also, the exact shape of the photoelectron line (including intrinsic plasmons, shake up and off satellites) and the background intensity is expected to allow a fast non-destructive extraction of depth composition information[48]. In each case the basic idea is that electrons originally excited at some depth in the solid, lose energy on their way out to the solid surface. The peak shape in the measured energy spectrum is, as a result, distorted and the distortion depends on the path length travelled by electrons.

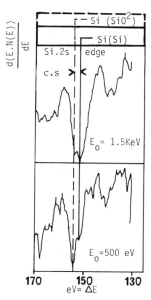

Fig. 9. (After ref.44). CLS spectrum of SiO_2 on Si; the 2s loss in pure silicon is indicated by the full line; the 2s loss of Si in SiO_2 is indicated by the dotted line. The (energetic) distance between the two corresponds to the chemical shift effect. For a primary beam energy of ≃ 0.5 keV (bottom) the signal related to the overlayer is more intense than that of the bulk, while the situation changes (top) when the primary beam energy is increased (1.5 keV).

Destructive depth profiling

When the composition depth profile concerns depths greater than $\lambda(A.L)$ (> 50 Å), the need is to use destructive methods to allow an initially buried layer to become a surface layer. For this, the most popular technique is to combine AES or XPS with the sputter etch removal of overlayers by ion beams (the cleaning of the sample by an ion beam is always needed before analysing it by electron spectroscopies). Ion etching is in general use and the readers are refered to the papers devoted to the potential and limits of this technique[40,49-51]. Mechanical angle lapping and ball cratering have also been used for metallurgical and semiconducting samples[52-54]. Nevertheless for semiconducting compounds the use of a softer treatment such as chemical etching seems to be better because they are well known[55-57]. The thermal nitridation study of SiO_2/Si has been done in this way[55]. Figure 10 illustrates the concentration profiles (of all the elements) at an InP/GaAs interface determined by Auger analysis of a chemical level[57].

5. QUANTIFICATION AND MINIMUM DETECTABLE CONCENTRATION

Another interesting parameter for characterizing the performance of a microanalytical technique is its minimum detectable concentration and the precision with which the concentration of a given element is known. For this the need is to establish the relationship between the signal intensity and the concentration of the corresponding element.

Signal intensity

For a binary alloy A_XB_Y, homogeneous in composition (inside the analyzed volume), investigated with the spectroscopies of interest (mainly XPS and AES), the signal intensity (in counts per second) can be written:

$$I_A = I_P \; \beta(A) \; T \qquad\qquad (12)$$

where I_P is the number (per second) of incident particles (e^- or X-ray photons); β is the quantum yield of the process of interest (i.e. the number of the characteristic particles created in the analyzed volume per incident particle) and T is the fraction of these characteristic particles

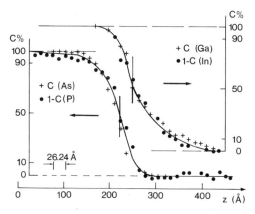

Fig.10. ([After ref.57). Concentration gradient that can be obtained by Auger analysis along a chemical bevel, exemplified with the study of a GaAs-InP interface. This is also the solution of Problem 2, question 3.

that is collected by the electron analyzer and detected (effect of solid angle, energy window of the analyzer and detective quantum efficiency of the detection unit).

In XPS:

$$\beta^X(A) \simeq N^A . \sigma^X(A_i, h\nu) . \lambda \qquad (13)$$

while in e^- (AES):

$$\beta^e(A) \simeq N^A . \sigma^e(A_i, E_p) \lambda . a_{ijk} . r \qquad (14)$$

where $N_A = N°[x/(x+y)]$; $N°$ is the atomic concentration of the sample $(5.10^{22} \text{ at/cm}^3)$, σ^X is the photoionization cross section (of atomic shell i of element A) by the radiation $h\nu$ while σ^e is the ionization cross-section by an incident electron (energy E_p), λ is the attenuation length of the characteristic electron of interest (resulting from the integration from 0 to ∞ of eq.10). In eq.14 a_{ijk} is the Auger probability of the process involving the i,j,k levels; "r" is the Auger backscattering coefficient describing the fact that the Auger signals result not only from the ionization process created by the incident electrons penetrating into the sample but also from the (incident) electrons that are backscattered by the substrate and ionize surface atoms before returning into the vacuum.

All the quantification procedures are based on the use of expressions similar to eq.12 and 14 (see ref.16b for details).

These expressions implicitly assume that the following effects can be neglected or are corrected or compensated:

Isotropy of the characteristic emission and non crystalline effects. For example in XPS the photoelectron emission is given by

$$d\sigma_{n\ell}/d\Omega = (\sigma_{n\ell}/4\pi)\{1 + (1/2)\beta(n,\ell)[(3/2)\sin^2\alpha - 1]\} \qquad (15)$$

where α is the angle between the incident photon and the outgoing electron and β is the asymmetry parameter. This emission is not isotropic except for "s" photoelectrons but this difficulty can be overcome by knowing β[58] or by setting α to be the magic angle of $\simeq 54°7$ (for the term in the bracket to become unity). Expressions (6), (9), (12)-(14) are only valid in amorphous materials; nevertheless they are often (and incorrectly) applied to microanalysis of crystalline specimens.

Flat surface (at the atomic scale). Topographic effects have been extensively studied in XPS[59,60] as well as in AES[61,62]. They can be taken into account if the surface topography is known. If not, Prutton et al[63] have found that the ratio S/(S+2B) (S: signal intensity; B: background intensity) is not too sensitive to topographical variations in SAM (Scanning Auger Microscopy).

Even when these effects are compensated (or taken into account) it remains to obtain the intrinsic line shape from the deconvoluted experimental spectrum in order to evaluate the exact peak area. This leads to take into account the influence of various instrumental and physical parameters such as the spectral broadening of the analyzer or the energy loss effects (for the background subtraction). See for example ref.64 for XPS.

Quantification

There are two main ways (and a combination of them) for quantifying the surface analysis experiments[16b]. Performing the ratio of the signal intensity relative to the two elements or to compare the signal intensity of the element A to that of a pure standard. In the first case (in Auger) one obtains from eq.14:

$$\frac{I(A)}{I(B)} = \frac{x}{y} \cdot \frac{\sigma^e(A_i)}{\sigma^e(B_{i'})} \cdot \frac{\lambda_{AB}(A_{ijk})}{\lambda_{AB}(B_{i'j'k'})} \cdot \frac{r^A_{AB}(E_p)}{r^A_{AB}(E_p)} \cdot \frac{T(E^A_{ijk})}{T(E^B_{i'j'k'})} \cdot \frac{a_{ijk}}{a_{i'j'k'}} \tag{16}$$

In the second case, one obtains:

$$\frac{I(A)}{I^0(A)} = x \cdot \frac{\lambda_{AB}(A_{ijk})}{\lambda_{A^0}(A_{ijk})} \cdot \frac{r^A_{AB}}{r^A_{A^0}} \tag{17}$$

By setting $a_{ijk} = 1$ and $r = 1$, similar expressions are obtained for XPS. The advantages (and disadvantages) of each method can be established if one knows that:

i). σ^x and σ^e are known with good accuracy. The most frequently used values are those published by Scofield σ^x for XPS and those deduced from the Gryzinski expression for AES (see Problem 1 and references therein), even if, in this last case, other expressions[65] or calculations[66] may give better results.

ii). Tabulated values of a_{ijk} also exist[66].

In fact, except for topographic effects on the one hand, and the response function of the analyser, on the other hand, the common key difficulty of the two approaches is related to the λ and r values (λ values alone in XPS) because these values depend on the sample composition which is unknown.

Most of the present efforts for improving the quantification are devoted to the evaluation of these matrix effects[19]. The r value can be evaluated experimentally[68] or deduced from Monte Carlo simulations[69]. The same remarks hold for $\lambda(A.L)$ despite the difficulties indicated in Section 4, (subsection on Attenuation length). For this, the advantage of eq.9 (over any kind of universal curves) is the fact that the matrix effect can be included in this expression by inserting in it the plasmon energy determined experimentally (plasmon loss peaks following the elastic peak in AES or the elastic photoline in XPS).

Ignoring these matrix terms may lead to errors reaching 30% (or more) in the composition of a binary alloy[16b]. One can measure the considerable effort that has to be made to reach a precision of ~1%.

Minimum detectable concentration

In pulse counting, it is generally admitted that a signal can be identified and detected if it is at least three times greater than the statistical fluctuations of the background surrounding it[70,71]. For the minimum detectable concentration x_m (atom/atom) of XPS and AES, this leads to[71,72]:

$$x_m \simeq 3 \, [S^o(A)/B]^{-1/2} \, [I_p \beta(A) T \tau]^{-1/2} \tag{18}$$

where $S^o(A)/B$ is the signal-to-background ratio for a pure element investigated by the spectroscopy of interest; τ is the time of measurement per channel for the signal. The advantage of this (approximate) expression is that it gives a good order of magnitude for the sensitivity of XPS ($x_m \simeq 10^{-3}$-10^{-4}) and AES ($x_m \simeq 10^{-2}$-10^{-3}) under standard conditions (leading to the usual sensitivity: a fraction of monolayer $\simeq x_m \delta$). See problem 3.

It allows also the study of the various factors influencing the sensitivity in order to optimize it (keeping in mind that x_m has to be less than unity in order to detect at least, a pure element). This is the optimum choice for the energy window of the analyzer, which corresponds nearly to the FWHM of the peak to be detected (in order to optimize T) and is conflicting with the use of a high energy resolution (for obtaining details on the peak shape). It also allows one to choose the best energy range for AES (10 - 20 keV for optimizing x_m alone).

When eq.18 is modified to take into account the lateral resolution "d" of a technique, one easily obtains:

$$x_m \simeq \frac{3[S^\circ(A)/B]^{-1/2}}{d}(\frac{\tau/t}{D_\beta T})^{1/2} \qquad (19)$$

where D is the dose (number of particles per unit area) received by the sample: $D \simeq I_p t/d^2$, and t is the minimum irradiation time.

Equation 19 shows the advantage of using a parallel recording system ($t/\tau \simeq 1$) instead of the sequential mode of acquisition ($t/\tau \gg 1$) (see Problem 3). The most important result that can be easily deduced from eq.19 is that the improvement of the lateral resolution (a decrease of d) has to be paid for by an increase of the dose received by the sample, x_m remaining a constant. This increase leads to 3 limits for improving d: i) the optical limit (related to the electron optical or X ray-optical aberrations; ii) the dose limitation given by the electron or the ray source; iii) the radiation damage limit for sensitive samples. Added to these limits, there is also the possible physical limitation due to electron scattering in the specimen such as the backscattering broadening in AES. See Section 6.

For sensitive samples eq.19 allows the evaluation of the best resolution that can be obtained (inserting the critical dose D_c they are able to suffer) when all the other parameters (T; τ/t; $S^\circ(A)/B$) are optimized. Taking into account the microscopic causes[72] as well as the macroscopic causes of radiation damage, (such as charging effects[73]), eq.19 allows one also to estimate the advantage of XPS over e⁻AES (from the minimum detectable concentration point of view). This advantage is of one or two orders of magnitude when the two techniques are allowed to have the same lateral resolution (instead of eight orders of magnitude when d(XPS) $\simeq$ 1 mm and d(Auger) $\simeq$ 1 μm)[73,74].

In this section, the possibilities of CLS have not been examined because this technique is at its beginning from the quantification point of view. The expected difficulties are clear: the need to obtain the signal in the N(E) form (for a correct evaluation of the signal intensity) leads to background substraction over several tens of eV, very similar to that developed in EELS[70]. Added to the complication related to the influence of the elastic events on this signal intensity, the quantification of CLS seems difficult but it remains that it will be developed in parallel with AES for the additional information it gives and its intrinsic advantages (Sections 3 and 4).

6. LATERAL RESOLUTION AND MICROSCOPIES

General

There are two different approaches for improving the lateral resolution of a technique and performing the corresponding microscopy[75]. It is the microprobe approach for which the first goal is to obtain a small spot on the sample by focusing the incident (electron or X-ray) beam on it. The various signals and spectra it emits may be then collected and the corresponding images obtained in the scanning mode (like in STEM). The second approach, the selected area approach, consists of irradiating the sample by a rather large incident particle beam and the first goal is to obtain a magnified image with the particles emitted by the sample (like in TEM). Various images may be obtained by changing the nature of the incident particles and, at least, a spectrum of the emitted particles can be obtained by selecting and analyzing a small feature in the image plane.

The three limitations (optical limit, critical dose limit, insufficient incident intensity limit) described in the previous subsection are independent from the approach being chosen and the minimum detectable concentration that can be expected can be evaluated by using eq.19, inserting in it the specific values relative to the instrument and to the sample. The obtained result for x_m has to be $x_m < 1$.

Examples of such estimates can be found in ref.72 and 75. More accurate evaluations, based on the same philosophy, have been given by Bauer and Telieps in this volume for AES.

Scanning Auger microscopy (SAM)

Apart from the attempt of Bauer and Telieps (this volume), all the other types of Auger microscopy are based on the microprobe approach and the images are obtained in the scanning mode. In this mode, an important question is the following: in SAM, does the resolving power depend upon the incident spot size or is it governed by the backscattered electron broadening (like in Electron Probe Microanalysis)? Many diverging opinions have been raised and various conflicting criteria have been proposed (including the change in opinion of the present author[75]). If one chooses the resolving power (i.e. the minimum distance between two points that can be distinctly separated by the optical system) of general use in Optics, one has to deal with the radial distribution of the Auger electron directly created by the incident beam penetrating into the sample, $j_1(r)$, plus the radial distribution of the Auger electrons due to the backscattering effects, $j_{b.s}(r)$. See problem 4.

The result is that the lateral resolution is mainly governed by the incident spot size, independent of the resolution criterion being used: either Rayleigh (two points can be distinguished if there is a 26.5% dip in intensity at the mid points between them) or better Sparrow (same as Rayleigh but with a 0% dip in intensity). The result is quite different if one deals with the edge spread function (Auger profile obtained across the chemical edge) and next applies the criterion δ (10%-90%) [the distance between the two points giving 10 and 90% of the maximum Auger signal]. In this last case, the lateral resolution is governed by the lateral extend of the backscattered contribution (see Problem 4, for a naive illustration). It seems that there is a problem of terminology (resolving power $\neq$ localisation $\neq$ detectability and quantification) on the one hand and an experimental difficulty to solve (how to evaluate the lateral resolution from the practical point of view) on the other hand.

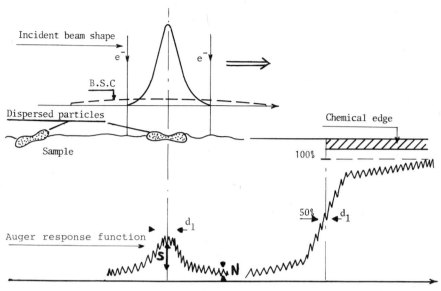

Fig.11. In SAM, the Auger response function is mainly the convolution of
the point spread function (PSF) by the surface distribution of the
element being detected. This response function may be distorted by
topographic effects and blurred by the statistical fluctuations of
the signal and background (noise). Dispersed particles having
dimensions less than the incident spot size, d, (left) can be
detected (detectability) when their signal is $S > 3N$; they may also
be localised (accuracy of localisation d_ℓ) with $d_\ell < d$. This
accuracy depends on the noise and on the shape of the incident beam
(through the PSF). Like in EELS, one trend in SAM will be the
knowledge of the PSF and of the incident beam [75,76]. Even when a
chemical edge profile is obtained (right) the localisation of the
interface d_ℓ may be inferior to the incident probe size.

As illustrated in fig.11, the Auger profiles obtained in the scanning
mode across the given sample is nothing other than the convolution of the
Auger signal [point spread function: $j_1(r) + j_{b.s}(r)$] by the shape of the
detail to be detected (assuming it has a uniform concentration of element
A) but the profile is partly blurred by the noise and is sensitive to
topographic effects. The localisation property is mainly governed by the
statistical fluctuation of the noise and the position of a chemical edge
(as well as the spatial localisation of a detail) can be determined with an
accuracy greater than the incident spot size (fig.11). The situation is
similar to that of EELS where the abruptness of an interface has been
obtained with better than 0.2 nm for an incident beam of $d \simeq 0.5$ nm (see
fig.11 in ref.76). Also the detectability of a detail is equal to or less
than the incident spot size when the incident beam intensity is sufficient
(it is related to S/N in fig.11). The quantification problem is difficult
to solve in SAM but we have seen (subsection 5.2) that it is also difficult
in the Auger analysis of laterally infinite samples and to clarify a
situation one should not mix two different problems. For the practical
point of view, the need is to define a procedure for evaluating the
resolving power (sample composition?). In this way the edge profile method
seems to present advantages (simplicity) but in fact it requires a sample
flat at the atomic scale to prevent the topographic effects (see fig.8c).
This difficulty explains the fact that the literature contains much more
theoretical results devoted to the lateral resolution problem (this paper

for instance) than experimental results. This difficulty being overcome, why not deal with the line spread function which corresponds to the first derivative of the experimental edge spread function? The obtained result (spot size dependent) will be more realistic than the result obtained by applying the most stringent criterion directly to the edge profile. The evaluation deduced from this last procedure is in conflict with the experimental result of Imeson and Milne (who have detected details having a size comparable to the incident probe size; i.e. ≃10 nm, see A. Howie this volume) as well as the previous results by Venables and co-workers (<30 nm)[77] (see also references therein).

An important consequence of the incident probe size dependence of the resolving power is in the choice of the primary beam energy, E_p, for optimizing simultaneously x_m and d. If the best choice for optimizing x_m alone is the 10-20 keV range, the use of higher energies seems to be very promising due to the increase of the incident intensity (I is proportional to V_p^n with $n \simeq 1/4$ up to 3/4 depending of the probe size[78]; V_p being the relativistic primary voltage) for a given probe size. Figure 12 shows one of the first spectra obtained at $V_p \simeq 100$ kV, indicating the slight increase of the signal to background ratio when V_p increases (and an expected slight decrease of the signal to noise ratio)[79]. The experiments performed at such high voltages are too scarce to state a definitive law, but it is an exciting path to explore because the expected minimum number of detectable atoms (surface atoms on a bulk substrate) is expected to be comparable to (or less than) that obtained in EELS by transmission through

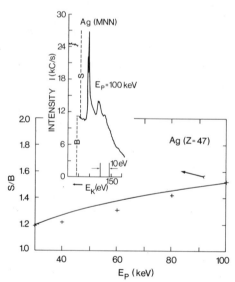

Fig.12. (From ref.79). Increase of the signal-to-background ratio as a function of the primary beam energy. Inset: Auger signal of silver obtained at E_p = 100 keV ($I_p \simeq 10$ nA). The FWHM of the Auger peak is ≤ 5 eV compared to ≥ 50-100 eV in EELS (and CLS). This advantage of SAM (over EELS) is counter-balanced by the isotropy of Auger emission (while in EELS, all the electrons carrying the signal can be collected in the forward direction). But for impurities on a surface, the minimum number of detectable atoms, Y_m in SAM, is expected to be comparable to that of EELS because $Y_m \simeq N.x_m$ (N number of atoms in the analysed volume) and N(AES) = $N^o d^2$(AES).δ, while N(EELS) = $N^o d^2$(EELS)·t (t, thickness of the sample and $N^o \simeq 5.10^{22}$ at/cm^3).

thin foils, because of the smaller minimum detectable concentration being required in the SAM (author's opinion).

One of the advantages of the scanning mode (over the conventional mode) is the signal processing facilities offered by digital acquisition. Recently this advantage has been pushed to a new step by El Gomati et al[80] by correlating, for each picture point, the signal relative to an element A and to and element B (instead of dealing with these two signals independently from each other to obtain two characteristic images). The interest (among others) of the scatter diagram technique is to derive true phase images ($A_x B_y$ images different from A_x,B_y, images) and to be applicable to other scanning imaging techniques such as EELS and Electron Probe Microanalysis. For additional details on SAM, the readers are referred to the references in papers 75,77,80. In the field of SAM, a line resolution of 8 nm has been recently obtained at 100 keV on a Si/GaAs chemical edge[81]. Pd particles of the same size have been analyzed leading to a detectable mass of 7×10^{-19} g and a minimum number of detectable atoms less than 4000[82]. For a new approach to the quantification of scanned images and quantitative imaging in SAM, the most recent papers are given in refs 83 and 84.

XPS Microscopy (and improvement of the lateral resolution of XPS)

The intrinsic advantage of XPS over AES is the chemical shift effect which is easier to interpret (subsection on Chemical information obtained); an additional advantage is its better S/B leading to a greater sensitivity x_m (for the same dose and spot size) or to less damaging effects (for the same sensitivity and spot size), see subsection on Minimum detectable concentration. These advantages are sufficient to predict that all the experiments presently performed by AES, would be performed by XPS if the same lateral resolution was achievable, for the same cost and facilities.

Unfortunately the present disadvantage of XPS over AES is its poor lateral resolution (in the mm range for almost all the conventional instruments). Most of the present efforts consist in optimizing the electron optical property of the transfer lens between the sample and the electron analyser in order to select a detail in the image plane and analyse it (selected area approach). Following the suggestion of Yates and West[85], the advantage of this solution is that the basic equipment of XPS remains quite unchanged and the inherent loss of intensity in the signal is compensated by using a parallel recording system. The same approach consists in using directly the double focusing properties of the spherical analyser to obtain XPS images[86]. The lateral resolution is presently at around 150-200 μm. Based also on the selected area approach, the attempt of Turner et al[87] is quite different because their instrument is a real microscope (and not a modified photoelectron spectrometer) in which superconducting coils are used to obtain a magnified image of the sample. The lateral resolution is governed by the cyclotron radius of photo-electrons spiralling in the diverging magnetic field this created. Under favourable conditions (low kinetic energy photoelectrons) a resolution of 10-30 μm can be reached. In principle this optical limit can be overcome with the Low Energy Emission Microscope of Bauer and Telieps (this volume), but the problem to solve is to illuminate the sample with a sufficient incident photon intensity.

Even if it seems very promising for the future, (synchrotron radiation + focussing elements[75,88]) the microprobe approach is presently less developed (than the selected area approach), except in the version marketed by SSL[89]. In this version, a bent crystal monochromator is used to focus incident X-rays on the sample and a resolution in the 100-200 μm range is attained. If a scanning photoemission microscope is presently designed to work at photon energy in the 500-800 eV range (synchrotron radiation) in order to reach a 75 nm lateral resolution by using zone plates as focussing elements[90], the simplest solution to obtain a small and scannable X-ray spot is to focus and to scan an incident electron beam on an anode in the form of a foil[88,91]. The present disadvantage of this solution is the need for the sample to be in the form of a foil (thickness in the μm range); this is the reason for which the present experiments concern scanning X-ray radiography (based on the same principle) but it is expected that this difficulty will be overcome in the near future.

CLS Microscopy

In principle CLS Microscopy can be developed in the same manner (same instrument) as Auger microscopy (SAM) but surprisingly only one attempt has been made (see ref.75), if plasmon loss microscopy is not considered[92] because the elemental identification property of plasmons is not general (this is the reason for which plasmon losses have not been considered in this article). The expected advantage of this microscopy is to image buried elements (by using the property described in the subsection on Non destructive depth profiling).

The difficulties are related to the need for using a rather low primary beam energy (in order for the loss to be analysed) and to the rather poor signal-to-background ratio in the N(E) mode. There is also the fact that elastic events have to be associated with the inelastic event of interest: in the scanning mode (microprobe approach), the lateral resolution will be deteriorated (it will be greater than the incident spot size) when the useful inelastic event is preceeded by many elastic ones while the selected area approach (LEEM: Bauer and Telieps, this volume) will suffer the same deterioration in the opposite situation (when the inelastic event is succeeded by many elastic ones).

7. ELECTRON ENERGY LOSS FINE STRUCTURES (EELFS) AND EXTENDED X-RAY FINE AUGER STRUCTURES (EXFAS)

The localisation property of a technique can be directly improved by improving the lateral resolution (see section 6) but also, in an indirect way, when it is able to give structural information on, for example, the nearest neighbour atomic distances. This is the case of Extended X-ray Absorption Fine Structures (EXAFS) following the threshold absorption edge in X-ray Absorption Spectroscopy[93]. Observing the similarities between X-ray Absorption Spectroscopy and Core Loss Spectroscopy (see Section 3 and fig.3), one can expect to obtain similar oscillations after the core loss edge in CLS. The great merit of de Crescenzi (and coworkers) is to have demonstrated that it is possible to obtain this structural information on a surface by using simple equipment instead of synchrotron radiation. Examples of such oscillations (EELFS) are shown on the top of fig.13 and compared to that of EXAFS (after 94). The accuracy in the interatomic distance is expected to be ± 0.05 Å and a recent theory suggest also a bond angle determination by using this technique[95] (see also ref.96 for some theoretical aspects). Recently again, de Crescenzi et al[97] have identified oscillations associated with the Auger line (and before this line instead of being after the core loss edge in the kinetic energy scale). These oscillations (EXFAS) have been attributed to some kind of core-conduction-valence Auger transition where the excited core electron, after being reflected by the nearest neighbour, participates to the Auger de-excitation process following its initial ionization process (see bottom of fig.13).

112

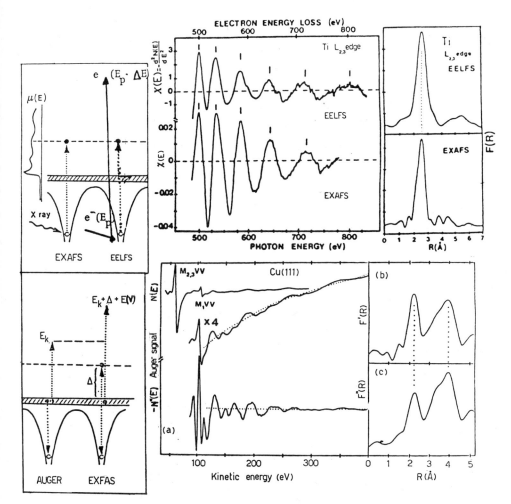

Fig.13. Top (After ref.94). Comparison between EXFAS and EELFS. Left: Basic mechanisms; middle: Experimental oscillations; right: Nearest neighbour distance (deduced from the experimental oscillations). Bottom (After ref.97): EXFAS. Left: Basic mechanisms of usual Auger process and EXFAS process; Middle: Experimental oscillations in the first and second derivative modes; Right: Deduced nearest neighbour distance.

CONCLUSIONS

In this paper, I have tried to describe the performance of XPS, CLS and AES with also an overview of their applications (from the industrial point of view). Each spectroscopy presents some specific advantage (over the others), this is, for example, the lateral resolution for AES, the chemical shift effect for XPS and the EELFS for CLS. If only one specific information is required, the use of only the corresponding (best) technique is sufficient. Nevertheless, in the general case, the need is to gather all the possible information that can be obtained on a given sample under the same experimental conditions. For this, I believe, the best is to develop all the possible techniques on the sample in order to correlate the information they give. This is one possible trend, in the near future, for research in this field. The other trends are also clear (even if the goals are difficult to reach): one is the improvement of the quantification procedure (to reach an accuracy of 1%?) in surface microanalysis, another is the improvement of the lateral resolution of XPS (to the submicron limit ?) and also the development of non destructive depth profiling techniques.

Due to their importance from fundamental research as well as practical applications, one can be optimistic for the future.

Final note. The following bibliography is far from being exhaustive. To be useful I have chosen to give the last references I know in this field. For reasons of space it has, unfortunately, been impossible to include references to all the significant work in the field.

REFERENCES

1. E. Rutherford, Phil. Mag. 28:305 (1914).
2. P. Auger, J. Phys. Radium 6:205 (1925).
3. G. Ruthemann, Naturwiss 29:648 (1941).
4. K. Siegbahn et al, Nova Acta Regia Scient. Upsal. 20:1 (1967).
5. L.A. Harris, J. Appl. Phys. 30:1419 (1968).
6. Phys. Electronic Industries Technical sheet n°7402 (1974).
7. J. Oudar and P. Marcus, Applic. of Surf. Sci. 3:48 (1979).
8. P. Marcus, J. Oudar, I. Olefjord, Mat. Sci. and Engineer. 42:191 (1980).
9. W.J. Van Ooij, T.H. Visser and M.E.F. Biemond, Surf. Interf. Analysis 6:197 (1984).
10. G. Marletta, S. Pignataro and G. Sancisi, Appl. of Surf. Sci. 17:390 (1984).
11. Y. Garaud and Tran Minh Duc, Analusis 9:231 (1981).
12. M.R. Philipps, M. Dewey, D.D. Hall, Vacuum 26:451 (1976).
13. J. Chenebaux, S.T. Gobain, Recherches: private communication.
14. T.L. Barr in ref.16 283 (1983).
15. L.A. Casper and C.J. Powell (Eds), Industrial Applications of Surface Analysis American Chemical Soc. Symposium. Series n°199 (1982).
16. D. Briggs and M.P. Seah, Eds, Practical Surface Analysis by Auger and X-ray Photoelectron Spectroscopy. J. Wiley, New York (1983);
    a) D. Briggs and J.C. Riviere 125;
    b) M.P. Seah 87.
17. H.W. Czanderna (Ed), Methods of Surface Analysis. Elsevier Scientific Publishing Co., Amsterdam, N.Y., (1975).
18. C.J. Powell, Appl. of Surf. Sci. 1:43 (1978).
19. Proceedings of ECASIA 85 (Veldhoven) special issue of Surf. Interf. Analysis 9 (1986).
20. D.K. Bakale and C.E. Bryson III, Surface Science Laboratory Technical Sheet 1982.
21. C.J. Powell and M.P. Seah, Surf. Interf. Analys. 9:79 (1986).
22. L.A. Grunes, Thesis Cornell University, Unpublished 1982;
    L.A. Grunes, R.D. Leapman, C.N. Wilker, R. Hoffman and A.B. Kunz, Phys. Rev. B25:7157 (1982).
23. S.E. Schnatterly, Solid State Physics. 34:273 (1985).
24. A.G. Nassiopoulos and J. Cazaux, Surf. Sci. 165:203 (1986).

25. H.H. Madden, Surf. Sci. 126:80 (1983);
    M.G. Ramsey and G.J. Russell, Phys. Rev. B32:3654 (1985);
    B. Carriere, J.P. Deville and P. Humbert, J. Micros. Spectros. Electr.
    10:29 (1985);
    D.E. Ramaker and F.L. Hutson, J. Vac. Scien. Tech. A4:1600 (1986);
    J.W. Rogers, J.E. Hutson and R.R. Rye, J. Vac. Sci. Tech. A4:1601
    (1986).
26. J. Cazaux, D. Gramari, D. Mouze, J. Perrin and X. Thomas, Inst. Phys.
    Conf. Series 61:425 (1982).
27. C.D. Wagner, H.A. Six, W.T. Jansen and J.A. Taylor, Appl. of Surf.
    Science 9:203 (1981);
    C.D. Wagner, L.H. Hale and R.H. Raymond, Anal. Chemistry 51:466
    (1979).
28. R.H. West and J.E. Castle, Surf. Interf. Anal. 4:68 (1982), see also
    S. Kohiki, S. Ozaki and T. Hamada, Appl. Surf. Sci. 28:103 (1987).
29. Spectrum obtained at NPL New Dehli (Dr J.K.N. Sharma Laboratory) by
    Drs B. Chakraborty and S.M. Shivaprasad during the 3-weeks stay of
    the author in this lab. (1987).
30. H.W. Haak, G.A. Sawartzki and T.D. Thomas, Phys. Rev. Letters 41:1825
    (1978).
31. C.J. Powell, Surface Interf. Analysis 7:256 and 263 (1985); J. Vac.
    Sci. Tech. A3:1338 (1985).
32. J. Szajman and R.C.G. Leckey, Electr. Spectr. Rel. Phenomena 23:83
    (1981).
33. A.L. Tofterup, Phys. Rev.B32:2808 (1985);
    A.L. Tofterup, S.Tougard and P. Sigmund, Surf. Int. Analysis 9:130
    (1986).
34. V.M. Dwyer and J.A.D. Matthew, Surf. Sci. 152/153:884 (1985).
35. M. Cailler, K. Barzine and J.P. Ganachaud, Surf. Sci. 154:548 (1985).
36. E.C. Goldberg and J. Ferron, Surf. Sci. 172:1523 (1986).
37. G. Gergely, Scanning 8:203 (1986);
    A. Jablonski, Surf. Sci. 151:166 (1985).
38. H. Iwasaki, R. Nishitani and S. Nakamura, Japanese Journal of Appl.
    Phys. 17:1519 (1978).
39. M. Pijolat and G. Hollinger, Surf. Sci. 105:114 (1981).
40. S. Hoffmann and S.M. Sanz in: "Topics in Current Physics" H. Oechsner,
    Ed., Springer Verlag, Berlin 37:141 (1984).
41. F. Ebel, Surf. Interf. Anal. 3:173 (1981).
42. M.H. Hecht, F.J. Grunthaner, P. Pianetta, L.I. Johansson and
    I. Lindau, J. Vac. Tech. A2:584 (1984).
43. K. Yoshimura and A. Koma, Proceed. Conf. Solid State Devices and
    Materials, Kobe 293 (1984).
44. A. Nassiopoulos and J. Cazaux, Surf. Sci. 149:313 (1985).
45. G. Gergely, M. Menyhard and A. Sulyok, Proc. of the 7th Czechoslovak
    Conference on Electronics and Vacuum Physics 457 (1985).
46. J.P. Langeron, L. Minel, J.L. Vignes, S. Bousquet, F. Pellerin,
    G. Lorang, P. Ailloud and J. Le Hericy, Surf. Sci. 138:610 (1984);
    Solid State Com. 49:405 (1984);
    Surf. Sci. 152/153:957 (1985).
47. D.C. Peacock, Surf. Sci. 152/153:895 (1985);
    M.M. El Gomati, J.A.D. Matthew and M. Prutton, Appl. Surf. Sci. 24:147
    (1985).
    D.C. Peacock and J.P. Duraud, Surf. Int. Anal. 8:1 (1986);
    M. Prutton et al., Proc. 4th Conf. on Quantitative Surface Analysis.
    Surf. Interf. Analysis 11:173 (1988).
48. S. Tougaard, Phys. Rev. B34:6779 (1986);
    Surf. Sci. 172:L503 (1986);

Surf. Interf. Anal. 8:257 (1986);
J. Vac. Sci. Techn. 1987 (to appear).

49. S. Hofmann, in: "Pract. Surf. Analysis", see ref.16b, page 141;
    S. Hofmann, Surf. Interf. Analysis 9:3 (1986).
50. H.J. Mathieu, in: "Thin film and depth profile analysis", H. Oechsner,
    Ed., Topics in current Physics 37:39 (1984).
51. H.W. Werner, Surf. Interf. Anal. 4:1 (1982).
52. M. Keenlyside, F.M. Scott and G.C. Wood, Surf. Int. Anal. 5:64 (1983).
53. J.M. Walls, I.K. Brown and D.D. Hall, Appl. Surf. Sci. 15:93 (1983).
54. A. Benninghoven, Thin Solid Films 39:3 (1976).
55. R.P. Vasquez, A. Madhuikar, F.J. Grunthaner and M.L. Naiman, J. Appl.
    Physics 60:226 (1986).
56. J.F. Bresse, J. Appl. Phys. 59:2026 (1986).
57. J. Cazaux, P. Etienne and M. Razeghi, J. Appl. Phys. 59:3598 (1986).
58. R.F. Reilmann, A. Msezane and S.T. Manson, J. Electr. Spectr. Related
    Phenomenae 8:389 (1976).
59. H. Ebel, M.F. Ebel and E. Hillbrand, J. Electr. Spectr. Related
    Phenom. 2:277 (1973).
60. C.S. Fadley, R. Baird, W. Siekhaus, T. Novakov and S.A.L. Bergstrom,
    J. Electr. Spectros. Related Phenomenae 4:93 (1974).
61. P.H. Holloway, J. Electr. Spectr. Rel. Phen. 7:215 (1975).
62. D. Wehbi and C. Roques Carme, Scann. Electr. Micros. 1:171 (1985).
63. M. Prutton, L.A. Larson and H. Poppa, J. App. Phys. 54:374 (1983).
64. D.D. Hawn and B.M. Dekoven, Surf. Interf. Anal. 10:63 (1987).
65. C.J. Powell, in: "Electron Impact Ionization", T.D. Mark and
    G.H. Dunn, Eds, Springer Verlag, N.Y. Chap. 6 (1985).
66. P. Rez, X-ray Spectrometry 13:55 (1984).
67. M.O. Krause, Phys. Rev. A22:1958 (1980).
68. A. Jablonski, Surf. Int. Anal. 1:122 (1979);
    J. Kirschner, Scann. Electr. Micros. I:215 (1976).
69. S. Ichimura, M. Aramata and R. Shimizu, J. Appl. Phys. 51:2853 (1980).
70. C. Colliex, in: "Advances in Optical and Electron Microscopy"
    R. Barer and V.E. Cosslett, Eds, Academic Press, London 9:65
    (1984).
71. J. Cazaux, Surf. Sci. 140:85 (1984).
72. J. Cazaux, Appl. of Surf. Science 20:457 (1985).
73. J. Cazaux, J. Micros. Spectros. Electr. 11:293 (1986).
74. L. Kirschner, in: "Springer Series in Optical Sciences"
    G. Schmahl and D. Rudolph, Eds, Springer Verlag 43:308 (1984).
75. J. Cazaux, J. of Microscopy 145:257 (1987).
76. M. Scheifein and M. Isaacson, J. Vac. Sci. Techn. B4:326 (1986).
77. J.A. Venables, D.R. Batchelor, M. Handbucken, C.J. Harland and
    G.W. Jones, Phil. Trans. R. Soc. London A318:243 (1986).
78. M. Troyon, 8th European Congress on Electron Microscopy), Budapest
    1984, Electron Microscopy I:11 (1984);
    European Workshop on the future of electron microscopy, Toulouse, 26
    Jan. 1986 and private communication.
79. J. Chazelas, A. Friederich and J. Cazaux, Surf. Interf. Anal. 11:36
    (1988).
80. M.M. El Gomati, D.C. Peacock, M. Prutton and C.G. Walker, J. of
    Micros. 147:149 (1987).
81. J. Cazaux, J. Chazelas, M.N. Charasse and J.P. Hirtz, Ultramicroscopy
    25:31 (1988).
82. J. Chazelas, J. Cazaux, G. Gillmann, J. Lynch and R. Szymanski,
    Proceedings ECASIA 88. To appear in Surface and Interface Analysis
    1988.
83. C.G.H. Walker, EMAG 87, Inst. Phys. Conf. Ser. n°90 p.189, 1987.
84. M. Prutton, M.M. El Gomati and C.G. Walker, EMAG 87, Inst. Phys. Conf.
    Ser. n°90 p.1 (1987).
85. K. Yates and R. West, Surf. Interf. Anal. 5:217 (1983).

86. N. Gurker, M. Ebel and H. Ebel, Surf. Interf. Anal. 5:13 (1983);
    N. Gurker, M.F. Ebel, H. Ebel, M. Mantler, H. Hedrich and P. Schön, Surface Interface Analysis 10:242 (1987).
87. D.W. Turner, J.R. Turner and H.Q. Porter, Phys. Trans. Royal Soc. A318:219 (1986).
88. J. Cazaux, Ultramicroscopy, 17:43 (1984).
89. R.L. Chaney, Surf. Interf. Anal. 10:36 (1987).
90. H. Ade, J. Kirz, H. Rarback, S. Hulbert, D. Kern, Y. Vladimirsky, Proc. Int. Symp. on X-ray Microscopy, Brookhaven August 1987, to appear in Springer Series in Optical Sciences, J. Kirz and H. Rarback, Eds, Springer Verlag (1988).
91. J. Cazaux, Scanning Electron Microscopy III 1193 (1984).
92. A.J. Bevolo, Scanning Electron Microscopy IV 1449 (1985);
    M.M. El Gomati and J.A.D. Matthew 147:137 (1987).
93. B.K. Teo and D.C. Joy, Eds, "EXFAS Spectroscopy" Plenum Press, N.Y., (1981).
94. M. de Crescenzi, Surf. Science 162:838 (1985).
95. F. Mila and C. Noguera, J. Phys. C, 20:3863 (1987).
96. C. Noguera, Scanning Electron Microscopy II 521 (1985).
97. M. de Crescenzi, E. Chainet and J. Derrien, Solid State Com. 57:487 (1986).

PROBLEMS

Problem 1. Elemental identification.

a. Find the elements A and B (see fig.1). Identify the atomic levels[1] associated with the A_1, A_2, A_3 and B_1 lines. Suggest an explanation for the lines C and D.

b. The spectrum has been obtained by using MgKα radiation ($h\nu$ = 1254 eV). With the help of the expression of Sjazman and Leckey[2]: $\lambda(\text{Å}) = 1.8\ E_K^{3/4}\ E_p^{-1}$, evaluate the mean free path λ associated with the lines A_1 and B_1 ($E_p \simeq 20$ eV).

c. The photoionization cross-sections are (Scofield[3]): $\sigma^X(B_1) \simeq 4.26$; $\sigma^X(A_1) = 5.3$ (unit = 22 200 barns/atom). Taking into account that $FWHM(A_1) \simeq 1.5\ FWHM(B_1)$ and neglecting the analyzer response function, evaluate the ratio x/y. Propose the chemical formula for the compound.

d. The sample is now investigated by e⁻AES at E_p = 5 keV. Enumerate the expected Auger lines, label them and give their approximate kinetic energies[4] (50 eV < E_K < 2 keV). From the Gryzinski's expression[5]:

$$\sigma_{n\ell}\ E^2_{n\ell} = 6.51 \cdot 10^{-14}\ Z_{n\ell}\ g(U)\ cm^2 eV^2,$$

evaluate the electron cross-sections associated with the B_1 and C lines. U is the reduced energy: $U = E_p/E_{n\ell}$; $Z_{n\ell}$ is the number of electrons in the $n\ell$ subshell, and

$$g(U) = \frac{1}{U}\ (\frac{U-1}{U+1})^{3/2}\{1 + (2/3)(1 - \frac{1}{2U})\lg[2.7 + (U-1)^{1/2}]\}.$$

e. The primary beam energy is now Ep = 1 keV. Find the energy position of the expected core-loss lines. How it is possible to distinguish the Auger lines from the C.L. lines?

1. J.A. Bearden and A.F. Burr, Rev. Mod. Phys. 39:125 (1967).
2. J. Szajman and R.C.G. Leckey, J. Electr. Spectros. Related Phenomena, 23:83 (1981); C.J. Powell, Scanning Electron Micros. IV 1649 (1984).

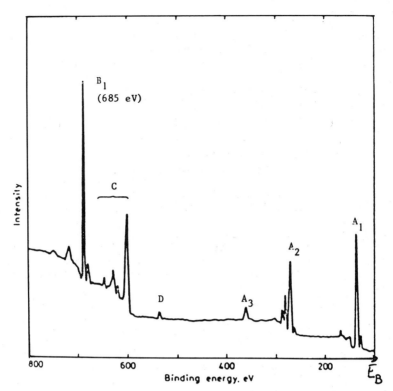

Fig. 1 (Problem 1). Showing the XPS spectrum of a binary compound A_xB_y.

3. J.H. Scofield, J. Electr. Spectros. Related Phenomena 8:129 (1976) and Kharitonov and Trzhoskivskaya, Atomic Data and Nuclear tables 23:444 (1979).
4. L.E. Davis, N.C. McDonald, P.W. Palmerg, G.E. Riach and R.E. Weber, Handbook of Auger Electron Spectroscopy (1978), Edited by Physical Electrons Industries.
5. M. Gryzinski, Phys. Rev. A138:305 (1965).

Problem 2. Surface sensitivity.

1. The experimental set-up of XPS is schematically shown on fig.1 in which A_0 is the area of the entrance slit of the analyser. The X-ray photoelectron intensity collected dI^A, relative to atoms of species A having N_A atoms (per unit volume) distributed between z and (z + dz), is given by[1]:

$$dI^A = \Phi_0 \left[N_A(z) \frac{A_0}{\sin\theta} dz \; \sigma^A(h\nu_0) \right] \exp(-z/\lambda\sin\theta) \cdot T$$

where θ is indicated in fig.1 and T is the collection efficiency of the analyser. Φ_0 represents the number of characteristic photons per second irradiating the effective area of the specimen $A_0/\sin\theta$.

a. Find the theoretical angular dependent curves for a clean flat surface of (for instance) pure silicon.

b. Same question for the overlayer signal intensity $I^O(\theta)$ and the substrate signal intensity, $I^S(\theta)$, in the case of a uniformly thin flat overlayer of thickness t.

118

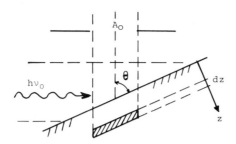

Fig. 1 (Problem 2). Experimental set-up of XPS.

c. Show the corresponding curves $I^O(\theta)$, $I^S(\theta)$ and $R(\theta) = I^O/I^S$. [For the sake of simplicity choose the case of SiO_2/Si and λ (Si in SiO_2) = λ (Si in bulk Si) and t = $\lambda/2$: the chemical shift effect allows one to distinguish between the two Si (2p level) lines].

2. In the Gallon's model[2], the sample is divided into slices of equal thickness a. The signals issued from the 1st slice (at the surface) are not attenuated; the signals issued from the jth slice [at the depth z such that: $(j - 1)a < z < ja$] are attenuated by the factor k^{j-1} where is k = $\exp(-a/\lambda\sin\theta)$. Assuming that the epitaxy of Al on GaAs follows the Franck, Van der Merwe, mode (a layer-by-layer epitaxy) give the intensity change of Al and As Auger signals (as a function of time) from 0 to 30 minutes when a monolayer is assumed to be deposited in 10 minutes.

3. Figure 2 shows the experimental change of the Auger signal

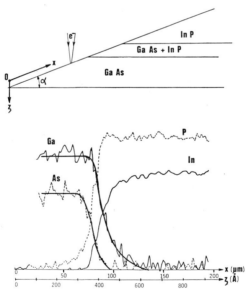

Fig. 2 (Problem 2). (Top). Schematic representation of the bevel of the InP/GaAs structure showing how a surface analysis along the bevel is converted into depth analysis. For the theoretical analysis, the interface region is divided into slices of thickness a and numbered 1,2,...,n. (Bottom). Corresponding Auger profiles relative to the four components: P (LMM line; E_K = 120 eV); As (LMM line; E_K = 1230 eV); In (MNN line; E_K = 410 eV); Ga (LMM line; E_K = 1070 eV).

intensities relative to the elements P, As, In, Ga when the incident electron beam is scanned along a chemical bevel cutting a InP/GaAs structure (see top of fig.2). Starting from this experimental data[3], find the concentration profiles of the four elements of interest by using Gallon's model and the expression of Sjazman and Leckey[4] for λ [with Ep = 14.8 eV and $\sin(\pi/2 - \theta) = 0.74$]. See the caption for the Auger kinetic energies.

1. C.S. Fadley, Electron Spectroscopy Vol.2. (G.R. Brundle and A.D. Baker eds) Academic Press, New York (1978), p.69.
2. T.E. Gallon, Surf. Sci. 17:486 (1969).
3. J. Cazaux et al Appl. Phys. 59:3598 (1986).
4. See ref.2, Probl.1.

Problem 3. Signal intensity, minimum detectable concentration and doses.

In AES, a pure elemental sample is irradiated by an incident beam of intensity $I_p \simeq 16$ nA. The quantum yield β for the Auger process under consideration is $\beta \simeq 10^{-4}$ ($\beta = N_A \cdot \sigma^A \cdot \lambda_A$) and the (angular and energetic) collection efficiency of the analyser is $T \simeq 5\%$ (of 4π sterad).

a. Evaluate the number of counts/s $I^0(S)$ that can be expected for the signal.

b. The same element is diluted in a matrix, and its spectrum is obtained in 10 minutes by exploring 20 channels in the sequential mode of acquisition (in order to obtain the signal and the background surrounding it).
. Find the minimum detectable concentration of this element if the background intensity, $I(B)$ is assumed to be $I(B) \simeq 10 \cdot I^0(S)$.
. Find the dose D (e^-/cm^2) received by the sample for incident spot sizes, d_0, of 10 μm and 100 Å.

c. The sample is now investigated by XPS. β and T values are assumed to have the above values (cf. Probl.1). The number of incident photons (per second) is $\phi \simeq 10^9$ X/s.
. Find the minimum detectable concentration of the element of interest by assuming now the use of a parallel recording system during 1 hour and a background intensity $I(B) \simeq I^0(S)/10$.
. Find the dose (X/cm^2) received by the sample for an irradiated area of 1 cm^2 and 1 μm^2.
. In the case of $\phi_0 = 10^9$ X/s and $d_0 = 1$ cm, a pinhole (d = 1 μm) is now inserted between the X-ray source and the sample in order to improve the lateral resolution of XPS: find the minimum time needed to identify an element of 10% atomic concentration (all the other parameters being unchanged).

Problem 4. Lateral resolution in SAM.

In SAM at normal incidence, the radial distribution function of the Auger electron created directly by the incident electron beam is given by[1]:

$$J_A = (I_A/\pi\sigma^2) \cdot \exp(-r^2/\sigma^2)$$

while the backscattering part of the Auger response function takes the form:

$$j(bs) = (R \cdot I_A/\pi\sigma'^2)\exp(-r^2/\sigma'^2).$$

a. Give the shape of the Point Spread Function [i.e. the sum $j_A + j(bs)$] for R = 0.8 and $\sigma' = 10\sigma$ by using the reduced variable r/σ. Find the resolving power given by the Rayleigh criterion[2], d_R, [two points can be distinguished if there is a 26.5% dip in intensity at the mid-point between them]. Find the resolving power given by the Sparrow criterion: d_S(FWHM).

b. This incident electron beam is scanned across a chemical edge. Give the shape of the obtained profile (Edge Spread Function). Evaluate the lateral resolution deduced from the criterion δ (10%-90%) [the distance between the two points giving 10% and 90% of the maximun Auger signal of an edge profile]. Evaluate the lateral resolution deduced from the criterion Δ (25%-75%).

c. The incident electron beam bombards a chemical edge (silver chemical edge on a tungsten surface) at oblique incidence ($\alpha = 50°$), the incidence plane being perpendicular to the chemical edge. In this mode, there is a shift "s" between the central point of the incident contribution and the central point of the backscattered contribution. Find the profile obtained at 20 keV when $\sigma = 30$ nm, $\sigma' = 0.29$ μm; R = 0.5; s = 0.077 μm. Compare this calculated profile to the experimental results obtained by Janssen and Venables[3] at 20 keV and the Monte-Carlo simulation performed by El Gomati et al.[4]

1. J. Cazaux, Surf. Sci. 125:335 (1983).
2. J. Cazaux, J. of Microscopy 145:257 (1987).
3. A.P. Janssen and J.A. Venables, Surf. Sci. 55:351 (1978).
4. M.M. El Gomati et al., Surf. Sci. 85:309 (1979).

ANSWERS TO PROBLEMS

Problem 1. Elemental identification.

a. B_1 → F (1s or K level) - $E_B \simeq 685$ eV

   A_1 → Sr (3d or $M_{4,5}$) $\simeq 135$ eV

   A_2 → Sr (3p or $M_{2,3}$) $\simeq 270$ eV

   A_3 → Sr (3s or M_1) $\simeq 357$ eV

   D → O (1s or K) $\simeq 530$ eV

   C → (X-ray induced) Auger line : F(KLL) : $E_K = 650$ eV.
Its measured pseudo binding energy is $E_B \simeq 600$ eV but its correct kinetic energy is $E_K = h\nu - E_B$.

b. $E_K = h\nu - E_B$; $E_K(A_1) \simeq 1120$ eV; $\lambda_{Sr}(A_1) \simeq 17.4$ Å

$$E_K(B_1) \simeq 570 \text{ eV}; \quad \lambda_F(B_1) \simeq 10.5 \text{ Å}.$$

c. $I(A) = \Phi_0 \cdot \beta \cdot T$ with $\beta \simeq N_A \cdot \sigma^X(A) \cdot \lambda(A)$.

$\dfrac{I(Sr)}{I(F)} \simeq \dfrac{\text{peak area}}{\text{ratio}} \simeq \text{(peak height)} \times \text{(FWHM)} = \dfrac{4}{7} \times 1.5 \simeq 0.86$

$\dfrac{I(Sr)}{I(F)} \simeq \dfrac{C(Sr)}{C(F)} \times \dfrac{\lambda(Sr)}{\lambda(F)} \times \dfrac{\sigma(Sr)}{\sigma(F)}$; $\dfrac{C(Sr)}{C(F)} \simeq 0.42$.

Proposed chemical composition: F_2 Sr.

d. Sr(LMM) at 1717, 1649, 1517, 1380 eV (E_K)

Sr(MNN) at 110 and 65 eV;   F(KLL) at 647 eV (E_K);   O(KLL) at 510 eV (E_K) (the sample is an insulator; caution: charging effects).

$$E_{n\ell}(F) = 685 \text{ eV}; \quad U(F) = 7.3; \quad Z_{n\ell} = 2$$

$$E_{n\ell}(O) = 530 \text{ N}; \quad U(O) = 9.43; \quad Z_{n\ell} = 2.$$

For F ; $g(U) = 0.183$; σ^e (F^{1s} , 5 keV) $\simeq 5 \cdot 10^{-20}$ cm^2 = $5 \cdot 10^4$ barn/at.

For O ; $g(U) = 0.16$; σ' (O^{1s} , 5 keV) $\simeq 7.5 \cdot 10^{-20}$ cm^2 = $7.5 \cdot 10^4$ barn/at.

e. The positions (in energy) of the core losses depends on the primary beam energy. Their kinetic energies are ($E_K = E_p - E_B$): $\simeq 685$ eV, 730 eV and 643 eV for the Sr series; $\simeq 315$ eV for F_{1s} loss and $\simeq 470$ eV for O_{1s} loss. The positions overlap that of the Auger lines. When changing the primary beam energy E_p, C.L. signals follow this change while Auger lines remain fixed.

Problem 2. Surface sensitivity.

Part 1. (See fig.3).

  a. $I^A = \int_0^\infty dI^A = \Phi_0 \cdot N_A \cdot \sigma^A(h\nu_0) \cdot \lambda \cdot T = I^A(\infty)$, independent of θ.

  b. $I^S = \int_0^\infty dI^A = I^A(\infty) \cdot \exp(-t/\lambda^A \sin\theta)$

  $I^O = \int_0^t dI^O = I^O(\infty) \cdot [1- \exp(-t/\lambda^O \sin\theta)$.

  c. $R = \dfrac{I^O(\infty)}{I^A(\infty)} \cdot [1- \exp(-t/\lambda^O \sin\theta)] \cdot \exp(t/\lambda^A \sin\theta)$

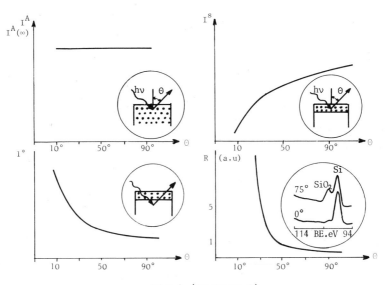

Fig.3 (Problem 2).

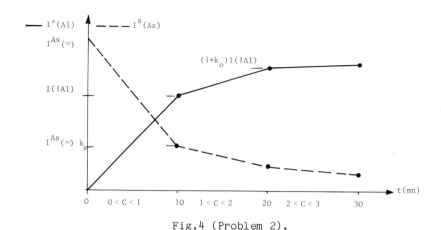

Fig.4 (Problem 2).

R is strongly dependent on θ (fig.3). In general $\lambda^O \neq \lambda^A$ because $E_K^O \neq E_K^A$.

This type of experiment allows one to distinguish between two situations. i). Homogeneous mixture of Si + SiO_2 and, ii). SiO_2/Si. In the latter case, it is possible to estimate t, even if its absolute value depends on the accuracy on λ (+ topographic + crystalline effects). This accuracy may be rather poor ($\pm 50\%$), but allows one to obtain an order of magnitude: either t = 4 Å, or t = 8 Å, or t = 20 Å.

Part 2. (See fig.4). When the coverage θ (C in fig.4) is $0 < \theta < 1$, the increase of the overlayer signal is a linear funtion of θ: $I^{Al} = \theta \cdot I$ (1 Al), where I(1 Al) is the contribution of one uncovered monolayer.

When $1 < \theta < 2$, the overlayer signal intensity is $I^{Al} = (1 + \theta \cdot k_O) \cdot I(1 \; Al)$. And so on. For a very thick overlayer, one obtains:

$$I^{Al}(\infty) = [1 + k_O + k_O^2 + \cdots + k_O^j + \cdots]I(1 \; Al) = I(1 \; Al)/(1 - k_O), \text{ where}$$

$$k_O = \exp(-a/\lambda_O \sin\theta).$$

When the coverage θ is $0 < \theta < 1$, the decrease of the substrate signal is also a linear function of θ:

$$I^{As} = (1 - \theta) \cdot I^{As}(\infty) + k_S \cdot \theta \cdot I^{As}(\infty)$$

$$I^{As} = [1 - (1 - k_S)\theta] \cdot I(1 \; Al)/(1 - k_S), \text{ and so on.}$$

The most striking result is that the increase of the overlayer signal (the decrease of the substrate signal) is given by broken lines instead of the curves given in the solution of Problem 2, Part 1 (fig.3 : I^S and I^O). In principle, this kind of experiment allows one to evaluate θ, λ_S, λ_O and to distinguish between the various epitaxial growth modes [see, for instance, Van Delft et al., Surf. Sci. 152/153:270 (1985)].

3. See ref.3 for details and answers. For the concentration profile alone see the fig.10 of this article.

Problem 3. Signal intensity, minimum detectable concentration and doses.

a. $I^0(S) = I_p \cdot \beta \cdot T \simeq 5 \cdot 10^5$ e/s.

b. The time per channel is $\tau = 30$ s. The number of counts for the signal and the background are $S = I^0(S)\tau$ and $B = I(B)\tau$ respectively.

. $x_m \simeq 3\sqrt{B}/S = 3[I(B)\tau]^{1/2}/[I^0(S)\tau] = 0.25 \cdot 10^{-2}$ at/at.

. $D = (I_p t)/(\pi d_0^2/4) = 1.2 \cdot 10^{22}$ e/cm^2 for $d_0 = 10$ μm.

and $1.2 \cdot 10^{28}$ e/cm2 for $d_0 \simeq 100$ Å

c. $I^0(S) = 5 \cdot 10^3$ e/s. $x_m \simeq 0.22 \cdot 10^{-3}$ at/at.

. $D \simeq 4.5 \cdot 10^{12}$ X/cm^2 for $d_0 = 1$ cm.

. $D \simeq 4.5 \cdot 10^{20}$ X/cm^2 for $d_0 = 1$ μm.

. Only 10 photons per second are striking the sample and

$$I^0(S) \simeq 5 \cdot 10^{-5} \text{ e/s ; } t_m = 9I(B)/[I(S)x_m]^2 = 1.8 \cdot 10^6 \text{ s ; } t_m > 20 \text{ days.}$$

Remarks. The aim of the present exercise is to give reasonnable orders of magnitude for the signal intensities, minimum detectable concentrations and doses in AES and XPS (Nevertheless the background values are, sometimes, slightly under estimated for XPS and over estimated for AES). This exercise illustrates the advantage of using a parallel detecting system. It shows also that, when the incident spot size is reduced, the advantage of improving the lateral resolution is paid for by an increase of the dose received by the sample (and radiation damages may occur) either in XPS or in AES. The last numerical application indicates that the improvement of the lateral resolution of XPS is not only an optical problem but also a X-ray source problem (for details see J. Cazaux, Appl. Surf. Sci. 20:457 (1985) and references therein).

Problem 4. Lateral resolution in SAM.

a. Concerning the resolving power in Scanning Auger Microscopy, the main question is to decide if it is incident spot size dependent or if it is governed by the lateral extend of the backscattered electron contribution. The answer depends on the criterion being used. The aim of the present exercise is to illustrate this debate by assuming that the radial distribution of the incident electron beam and the backscattered contribution have Gaussian forms (approximation). When the point spread function is considered, and when σ'(bs) is larger than σ ($\sigma' > 2.5\ \sigma$; this is always the case in SAM) the radial distribution of the Auger electron created by the backscattered effects is a faint halo (see fig.1; dotted line x 10) on which the central spike (due to the Auger electron directly created by the incident electron beam) is superimposed. The lateral resolution is only governed by the incident spot size (central spike). Numerical estimate of the resolving power leads to d (Rayleigh) $\simeq 2\sigma$ and d (Sparrow) $\simeq 1.6\sigma$.

b. When the incident electron beam is scanned along the r'Or axis and when the element being detected covers the half-plane of the sample surface defined by $-\infty < r < r_0$, each contribution (incident and backscattered) to the edge spread function corresponds to the convolution of a Gaussian excitation and the unit step function, leading to:

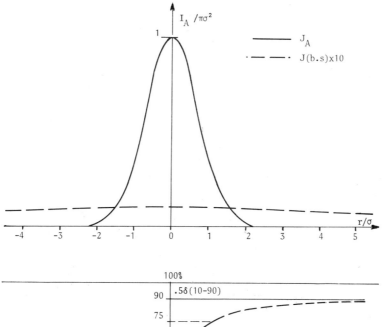

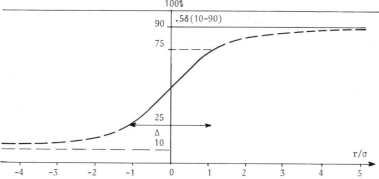

Figs 1 (top) and 2 (bottom) referring to the solution of Problem 4.

$$\frac{I(-\infty, r_0)}{I_A(1 + R)} = \frac{1}{2} + \frac{1}{2(1 + R)} \; \text{erf} \; t_0 + \frac{R}{2(1 + R)} \; \text{erf} \; t_0'$$

where erf t_0 (t$_0'$) is the error function of the reduced variable $t_0 = r/\sigma$ ($t_0' = r/\sigma'$). The result is shown in fig.2.

Next, when the most stringent criterion $\delta(10\% - 90\%)$ is applied to this profile, it is found that:
$$\delta(10\% - 90\%) \simeq 10\sigma \simeq \sigma'.$$

The lateral resolution is governed by the lateral dimension of the halo (of the backscatted contribution). When the less stringent criterion is used, $\Delta(25\% - 75\%)$, the result is $\Delta(25\% - 75\%) \simeq 2\sigma$. For details see Ref.1 of the text.

c. See fig.9a of ref.1 of the text.

125

REFLECTION ELECTRON MICROSCOPY

John M. Cowley

Department of Physics
Arizona State University
Tempe, AZ 85287, USA

INTRODUCTION

Reflection electron microscopy (REM) is now well established as a technique for the study of the structure of surfaces of crystals. Resolutions of better than 1 nm are possible and clear images can be obtained of surface steps 0.1 to 0.2 nm high. The image intensities are influenced only by the top few layers of atoms, and surface layers consisting of a fraction of a monolayer of foreign atoms can give high contrast. The images may be correlated with data from diffraction patterns from the surface and with microanalysis of the surface performed by use of electron energy loss spectroscopy (EELS) and, sometimes, Auger electron spectroscopy (AES). When the images are obtained in the scanning mode, with an electron beam of small diameter focussed on the specimen, the diffraction and microanalysis data may be obtained from very small regions of the surface. In this chapter, we give accounts of the current status of the techniques involved and of some recent applications.

The essential feature of the REM technique, as it is now practised, is the use of diffracted beams formed at relatively flat surfaces of crystals in the glancing-incidence forward-scattering mode. Most earlier imaging in the reflection mode relied on the detection of diffuse scattering, either because high-angle scattering was used or else because the specimens were covered with amorphous or poorly crystalline layers. The resolution was then relatively poor, and the images usually had poor contrast and depth of focus. The use of diffracted beams from clean crystal surfaces has allowed important improvements in all aspects of the imaging. Much of the treatment, therefore, must deal with the diffraction of electrons by perfect and imperfect crystals.

A wide range of electron beam energies has been used in reflection imaging. We will be concerned, to a limited extent, with the imaging of electrons in the "medium-energy" range from 2 to 20 keV. Most of our discussion will relate to imaging with 100 keV electrons, but the same treatments and discussions may be applied with little change, for electron energies from about 30 keV to 300 keV or above.

For electron energies of about 100 keV, the diffraction patterns used are reflection high energy electron diffraction (RHEED) patterns, obtained from flat crystal surfaces with the incident beam and the diffracted beams making angles of 10^{-2} to 10^{-1} rad with the surface (see chapter by Dobson, this volume). The images are then severely foreshortened in one direction, almost parallel to the incident beam, but at right angles to this direction much the same resolution can be obtained as in transmission electron microscopy.

When medium energy electrons are used, the electron wavelengths are greater (12.2 pm at 10 keV, as against 3.7 pm for 100 keV), and the angles of incidence are greater. The foreshortening of the images is less severe, but the resolution attainable is much poorer (10 nm or more). One big advantage of medium energy REM is that it is more amenable for use in ultra-high vacuum systems and so can be integrated readily with surface research instrumentation.

The higher energy REM imaging has been performed in TEM instruments. While TEM instruments provide highly desirable resolution capabilities, they traditionally have relatively poor vacuum systems with pressures in the specimen regions usually higher than 10^{-7} Torr. Since the REM technique is sensitive to fractions of a monolayer coverage by foreign atoms, there are relatively few specimens for which the capabilities of the REM technique can be utilized in standard, commercial TEM instruments. Special ultra-high vacuum TEM instruments have been built and are now being produced in increasing numbers and these will allow the REM and associated techniques to be used to full advantage.

EXPERIMENTAL TECHNIQUES

REM in TEM instruments

The flat face of a crystal which may be up to 1 or 2 mm in diameter, is mounted in a specially adapted specimen holder at the normal specimen position of a TEM instrument, within the high magnetic field of the objective lens. The face to be imaged is made almost parallel to the microscope axis. The incident beam is tilted, as in the usual configuration for dark-field TEM imaging, so that a chosen diffracted beam leaves the specimen in the direction parallel to the objective lens axis (see fig.1). In this way the effects of lens aberrations are minimized and the objective lens and subsequent lenses form the high-magnification image with the best possible resolution.

It should, ideally, be possible to tilt the specimen by rotation around the line in the surface perpendicular to the incident beam in order to vary the angles of incidence and diffraction, and also by rotation around the surface normal to allow a suitable azimuthal adjustment. Most tilting cartridges provided for conventional TEM instruments can be applied to approximate, or simulate, these motions.

A number of methods have been devised for mounting the millimeter-size specimens in place of the usual TEM grids. Hsu[1] describes several arrangements by which specimens of a variety of shapes can be attached to holders made from thin plates or grids. For the noble metals, Au and Pt, use of a near-spherical crystal, formed by melting the end of a thin wire, is particularly effective[2]. A short length of the unmelted wire, wrapped around a support, serves to hold the ball in place. Cleavage of a single crystal, previously shaped and mounted in the specimen holder, has also provided suitable crystals faces[3].

If one of the strong, low-angle reflections is used to form the REM image, the image intensities can be relatively strong, comparable to those for TEM dark-field images. However, with the diffraction angles of about 10^{-2} radians common for such reflections at 100-200 keV, the foreshortening of the images is a serious limitation. If high angle reflections are used, the foreshortening factors may be as small as 15 to 20, but the image intensities are an order of magnitude smaller. Hence some compromises must usually be made, and images are usually observed and recorded at relatively low magnifications of 20,000 to 100,000.

It is possible to obtain good REM pictures of some surfaces in conventional electron microscopes in spite of the presence of thin layers of absorbed gas, contamination or oxidation products (fig.2). If the image is formed by selecting a strong diffracted beam by use of a small objective aperture, the contribution to the image from thin surface layers may be very small. Diffraction from these layers gives a broad diffuse background or diffraction maxima in other parts of the diffraction pattern, and the fraction of this scattering which enters the objective aperture is very small.

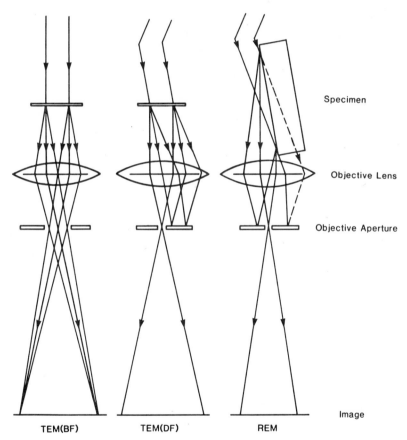

Fig. 1. Ray diagrams illustrating the formation of an image by the objective lens for: TEM bright field, with interference between transmitted and diffracted beams; TEM dark field, with tilted incident beam and one diffracted beam selected by the objective aperture; and REM, with the same geometry as for TEM (DF) but with beams diffracted from the extended face of a bulk specimen. [Courtesy, T. Hsu].

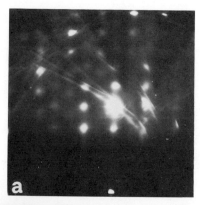

Fig. 2. Diffraction and reflection image from a Pt (111) surface: (a) RHEED
pattern with (555) beam strongly excited: (b) REM image showing
single-atom surface steps. The arrows indicate the stepdown
direction. The strain-field of a dislocation, D is seen at the end
of a surface step. [Courtesy, T. Hsu].

However, the presence of unwanted surface layers of imperfectly-known
form and content is a major disadvantage when surface reactions or
modifications are to be investigated, especially when elevated temperatures
are to be used. It is clearly desirable to have specimens prepared and
maintained under ultra-high vacuum conditions.

So far, it has not been found possible to build TEM instruments with
complete ultra-high vacuum (UHV) systems. The complexity of the microscope
column and the requirements for compactness, stability and precision of the
construction are not compatible with current UHV practice. Instead, TEM
instruments have been modified to provide, as far as possible, an UHV
environment for the specimen by use of liquid-helium cooled cryoshields[4].
In this way the partial pressures of condensable gases can be kept below
10^{-10} Torr and evidence suggests that atomically cleaned surfaces may be
mantained in a clean condition for considerable periods. Devices for in
situ cleaning, heating and treatment of surfaces have been built into the
specimen chambers[5] so that observations have been made of surface reactions
at temperatures of up to 1000 °C.

The observation and recording of RHEED patterns is an essential part
of REM experiments. If an analytical TEM instrument is used, further
information concerning the crystal surfaces may be available. Convergent
beam diffraction patterns may be obtained with beams of diameter 10 nm or
so[6].

The early attempts to obtain EELS signals from surfaces with the REM geometry were not encouraging[7]. However Wang and Cowley[8] have shown that if the EELS signals are collected over relatively long times it is possible to detect signals from surface layers which are sufficient to allow useful, if approximate, analysis of the composition of surface layers. Also, in favourable cases, the fine structure on the inner shell edges may be recorded sufficiently well to be analysed in terms of the energy states and environments of surface atoms (from ELNES and EXELFS data, respectively).

SREM in STEM instruments

In its imaging modes, the scanning transmission electron microscopy (STEM) instrument is known to give results similar to those for TEM. The principle of reciprocity suggests that, in principle, for the same electron-optical conditions, the image contrast is the same for the two types of instrument. In practice, the STEM image is formed by scanning a focussed beam over the specimen and various signals may be collected as a function of time during the scan, in addition to those signals from the transmitted and diffracted beams. The scanning mode has some disadvantages in terms of image quality but the variety of image signals provides some distinct advantages. From each pixel of the image in turn, it is possible to detect the microdiffraction pattern, the EELS signals, and the intensities of emitted characteristic X-rays or Auger electrons, and also the secondary electrons.

The same characteristics of the scanning systems are of value also in reflection imaging. The scanning reflection electron microscopy (SREM) images are similar in principle to those of REM. In practice, the image quality is sometimes poorer, but this defect is compensated by the possibility of obtaining diffraction or microanalysis data from very small portions of the imaged area. Resolutions of better than 1 nm have been achieved[9] and microdiffraction patterns from regions 1 nm in diameter may be recorded readily (fig.3). So far, little has been done with the microanalysis signals. Some results have been obtained with EELS only in the low-loss region, for energy losses of 0-30 eV, and also with secondary electron imaging. In these cases the signals detected are relatively large, being within an order of magnitude or two of the elastically scattered beam signals. For processes involving the excitation of inner shell electron energy levels of the sample, the signals are much lower and problems arise in getting adequate signal-to-noise levels.

In a STEM instrument, the electron beam is focussed on the specimen by the objective lens. The incident beam is thus convergent and the diffraction pattern produced is a convergent beam (CBED) pattern in reflection, as in transmission. The characteristics of convergent beam RHEED patterns have been explored and described by Shannon et al.[6] and by Lehmpful and Dowell[10]. With a large angle of convergence, the patterns are dominated by the lines and areas produced by resonance diffraction effects. For smaller angles of convergence, the separate diffraction spots appear as in RHEED patterns, except that the spots are not sharp but enlarged into finite disks, usually showing some fine structure within them. The detector collects some chosen portion of the diffraction pattern.

Application of the reciprocity relationship reminds us that the very obvious influence on the image intensity of the detector position in the SREM mode must reflect an equally strong influence of the incident beam direction on the image intensity in the REM mode.

In order to obtain an image signal sufficiently strong for high-resolution SREM, even with a strong diffracted beam, it is necessary to use a very bright source. Only field emission guns give sufficient intensity in small probes. The necessity for observing the reflection diffraction pattern with a small probe requires that an efficient two-dimensional detector system be added to the dedicated STEM instrument. Such a system has been described by Cowley[11,12].

The vacuum levels in dedicated STEM instruments are usually considerably better than for commercial TEM instruments. Pressures of 10^{-8}

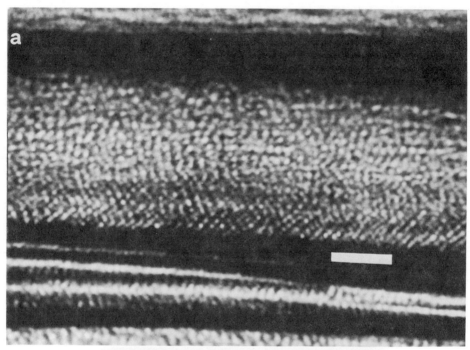

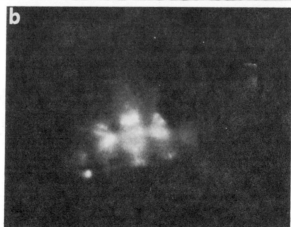

Fig. 3. (a) SREM image of small Pd crystal aligned along steps, about 2.5 nm apart, on a MgO(100) cleavage face after electron irradiation. Marker is 15 nm. (b) Convergent beam reflection electron diffraction pattern of Ni layer on MgO cleavage face with a beam diameter at the specimen of about 0.5 nm. [Courtesy, H.J. Ou].

to 10^{-9} Torr in the specimen chamber have been standard. For some newer instruments, pressures of 10^{-10} Torr are approached. Also some recent instruments are incorporating devices for <u>in situ</u> specimen treatment or for specimen preparation and treatment in an attached UHV chamber.

Medium energy SREM

The imaging of surfaces with the REM technique using electron beams in the medium energy (2-20 keV) range has been developed largely as an extension of the UHV secondary electron microscopy (SEM) systems, constructed for surface research (see, for example, Venables et al.[13]). The probe-forming lens has a working distance of several centimeters to allow access to the specimen for other surface research devices, and detectors such as the cylindrical mirror analyser (CMA) commonly used by Auger electron spectroscopy and other electron spectroscopies.

The medium energy reflection electron diffraction pattern is observed by use of a suitable fluorescent screen or channel-plate electron multiplier system and a suitable electron-multiplier detector is used to collect particular diffracted beams as the incident beam is scanned over the specimen.

In one experimental device of this sort[14] image resolutions of about 10 nm have been obtained (fig.4). In other instruments which have incorporated more complete surface analysis capabilities[15,16] resolutions not much poorer than this have proved effective in revealing details of surface structure including surface steps, dislocation strain fields and the domains of surface superstructures.

IMAGE FORMATION

Because REM images are produced by detecting diffracted beams from surfaces, the image intensities are affected, as in dark field TEM, by any factors which influence the amplitudes or intensities of the diffraction.

Fig. 4. Medium energy (18 keV) SREM image of a Si(111) cleavage face, formed in UHV, showing the distribution of domains of the 2×1 super-lattice stucture. Marker is 0.8 μm. [Courtesy, H.J. Ou].

Thus the images reveal changes of structure amplitude, due to changes in composition or crystal structure, and changes in crystal lattice orientation. Also they are sensitive to variations of phase of the electron waves and show fine morphological features by phase-contrast imaging.

The glancing angle-geometry used for REM introduces a number of important differences from the TEM case. For flat surfaces, the foreshortening factor may be 20 to 60, so that the angular relationships between directions in the surface are grossly distorted. Only one line across the surface, perpendicular to the diffracted beam being imaged (in the REM case), is in focus. The out-of-focus distance varies linearly with distance from this line and may be as high as 10 μm or more in relatively low magnification images.

Small projections from a surface are imaged without foreshortening. Undulations on a surface can have an exaggerated effect in the image, depending on their orientation relative to the diffracted beam. For appreciable surface roughness the interpretation of the image may become very complicated. In all cases, azimuthal rotation of the specimen can produce very large changes in the appearance of the image.

Geometry of images

The foreshortening factor for a flat surface is $(\sin\theta)^{-1}$ where θ is the angle between the surface and an axial diffracted beam used for imaging in the REM case. For either REM or SREM it may be said that θ is the angle between the surface and the axis of the objective lens (before the specimen for SREM; after, for REM). The in-focus line on the surface (OF in fig.5) is usually recognisable. The line at right angles to this (OA in fig.5) is the projection of the objective axis direction on the surface and may serve

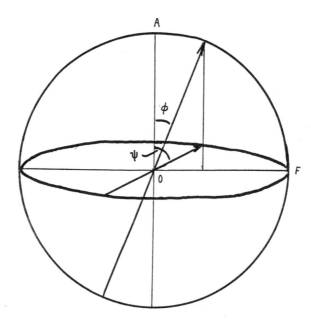

Fig. 5. The geometry of foreshortened REM images. A circle in the surface becomes an ellipse in the image. A line on the surface making an angle φ with the line OA in the surface, in the plane containing the objective lens axis and perpendicular to the surface, appears to make an angle ψ with OA.

as a reference direction, the zero for the azimuthal angle φ. A straight line on the surface making an angle φ with this direction appears in the image to make an angle ψ such that

$$\tan\psi = \tan\phi/\sin\theta. \qquad (1)$$

Thus, small angles between the surface line and the reference direction will appear magnified by the foreshortening factor. If the objective aperture is made large and includes two strong diffracted beams (or, in the case of SREM, two incident beam directions which diffract electrons into the detector) a line on the surface gives two lines in the image at different angles, ψ, intersecting at the in-focus position[17]. Problem 1 deals with some of the aspects of REM geometry.

A small projection from a flat surface is seen doubled in the REM image. One image of it is formed because it intercepts the diffracted beams from the surface; another, "mirror image", is formed because it intercepts the incident beam before it reaches the surface and is diffracted. This mirror image is directly below the direct image only if the incident and diffracted beam lie in the same plane perpendicular to the surface.

The images of large steps on the surface can also be interpreted in terms of simple geometric constructions. The appearance of steps up and steps down is very different, as suggested by fig.6. A step up, as seen from the objective lens side, appears to be twice its actual size because of the addition of the mirror image. A step down will not be visible except for some diffraction and phase contrast effects at the edge.

Phase contrast

For small steps on the surface, of height close to the resolution limit of microscope or less, the main contribution to the image contrast comes from the phase difference between the electrons diffracted from the top and bottom of the step (cf. fig.6). As in the case of TEM imaging of the edge of a thin object, a phase step gives little contrast in focus, but out-of-focus a strong black-white contrast develops, reversing in sign from under-focus to over-focus. In REM the large range of focus values present ensures that the out-of-focus black-white contrast will be a dominant effect, as is seen in fig.2. Similar phase contrast effects also become prominent in the out-of-focus regions for surface defects such as

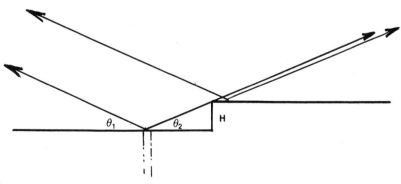

Fig. 6. REM imaging of a step of height H. The beam travels from left to right for a step down (for REM) and from right to left for a step up, when viewed from the objective lens.

dislocation strain fields which appear narrow in the incident beam direction because of the foreshortening effect.

For small surface steps the in-focus contrast is usually not zero. Although on occasions it can be unobservably small, it is often possible to observe a thin dark line at the in-focus position, and this contrast may differ for up- and down- steps. The in-focus contrast can also be strongly influenced by the diffraction conditions, becoming strong when the surface-channelling resonance is strong. There are important questions to be answered as to whether the surface step contrast is purely phase contrast or whether it includes a component of diffraction contrast resulting from a relaxation the crystal lattice around the step. The evidence on this question will be reviewed in the section on Lattice relaxation. The theoretical basis for calculating the image contrast to be expected for any assumed model of a step will be discussed in sections on Image theory and Applications of REM.

Diffraction contrast

The intensities in REM images depend on all those factors which affect the intensity of the chosen RHEED diffraction spot. From the point of view of kinematical diffraction, the diffraction intensities depend on the structure amplitude for the reflection, determined by the nature and positions of the atoms in the unit cell. The intensities are also strongly dependent on the orientation of the crystal, with large changes for small fractions of a degree tilt. Thus one expects to see contrast from changes of surface structure or composition and from the local variations of crystal orientation associated with the strain fields of defects, as in TEM.

In REM the diffraction conditions are always strongly dynamical. The diffraction intensities are determined by the coherent interaction of a number of diffracted beams, and in the REM geometry these interactions may be strongly influenced by the surface geometry through refraction effects and surface channelling phenomena. Thus the dynamical diffraction effects are prominent and in general are more complicated than those familiar in TEM studies of crystals and their defects. A further complication is that at surfaces there may be appreciable modifications of the diffraction conditions by surface relaxations and the presence of surface layers which introduce non-periodic components superimposed on the periodic crystal substrate. The sensitivity of REM and RHEED to the surface perturbations provides much of the interest in the use of these techniques for surface studies, but the complications of the diffraction effects emphasize the need for an adequate theoretical treatment and an adequate control of the experiment parameters.

The importance of surface structure in producing diffraction contrast in REM images is well illustrated by the almost black-white contrast between domains of surface super-structures involving a monolayer or less of surface atoms, as in the observations of Au on Si(111) by Osakabe et al.[18] or the observations of the Si(111) 7×7 structure by Osakabe et al.[19,20]. The various types of contrast exhibited by the strain fields of dislocations intersecting a crystal surface are illustrated by Peng and Cowley[21].

In high resolution REM the image contrast produced by changes in diffraction conditions will be complicated by phase contrast effects, especially in out-of-focus regions of the image since any change in diffraction amplitudes across a sharp boundary will inevitably be accompanied by a discontinuity in the phase of the diffracted beam. The

clearest images of diffraction contrast effects are thus to be obtained
close to the in-focus positions. Problem 2 deals with questions of
foreshortening, depth of focus and resolution under different imaging
conditions.

REM images of periodic surface structures

With resolutions of better than 1 nm possible, it is to be expected
that images should be obtained showing the larger periodicities present in
some crystal structures. The 2.6 nm periodicity of the Si(111) 7×7 surface
superstructure has been seen clearly and used effectively by Tanishiro et
al.[22], (see fig.7). Artificially periodic structures formed by molecular
beam epitaxy have been imaged by Yamamoto and Muto[23] and Hsu[24]. The
particular aspects of the imaging of periodic structures in REM which
deserve mention are those associated with the glancing-incidence geometry
and image foreshortening. Clearly periodicities which are comparable with
the resolution limit will be visible only when the periodicity is in a
direction almost exactly perpendicular to the incident beam. Thus the wide
range of defocus values in an image could introduce perturbations of the
appearance of the periodic structure through the Fourier image phenomenon.

Fig. 7. REM image showing the 2.3 nm fringes due to the Si(111) 7×7
       superstructure crossing surface steps (dark bands), with a lateral
       shift of the fringes at some steps. [Courtesy, Dr K. Yagi: see
       Tanishiro et al. 1986].

The form of the image of a periodic object changes periodically with defocus (see Chapter on: Fundamentals of HREM, by Smith). The same image contrast should reappear for differences of defocus equal to $2a^2/\lambda$ where a is the periodicity. At intervals of a^2/λ in the defocus values, the image will be reproduced but with a shift sideways of $a/2$. At intermediate defocus values other forms of the image will appear, including forms in which the periodicity may appear to be halved.

For example, for a 1 nm periodicity, and 100 keV electrons, the repeat distance, with lateral translation of $a/2$, is a^2/λ = 270 nm which, with a foreshortening factor of 40, becomes 6.7 nm in the image. For a 0.5 nm periodicity the repeat distance would appear to be about 1.7 nm. Thus a false appearance of a two-dimensionally periodic structure could be generated, or false conclusions could be drawn of the presence out-of-phase boundaries.

Experimental considerations

The experimental realities of reflection imaging limit the accomplishments to some extent. In the TEM instruments the incident beams are not plane parallel waves. The finite beam convergence and the rotational aberrations of the condenser lenses produce complications in that the incident beam direction is a function of position on the specimen surface, leading to contrast variations other than those due to defocus.

For the scanning reflection modes, there is always a problem in getting sufficient signal strength to produce images of reasonable quality at high resolution. There is a tendency to use objective and detector apertures as large as possible in order to increase the signal strength, but an increase of the aperture sizes leads to restrictions on the depth of focus and a reduction of phase-contrast effects.

The reflection imaging techniques, however, can produce high resolution imaging information concerning the surfaces of bulk crystals and equivalent information is not available from any other technique. The image information can be obtained in parallel with diffraction data, and with the microanalysis of surface layers by use of EELS techniques and complementary imaging methods such as SEM. As the techniques are further developed and used in instruments designed with UHV specimen environments and other desirable facilities, REM and SREM will inevitably become major tools for surface research.

IMAGE THEORY

General

The introductory discussion of REM and SREM imaging in the previous sections suggests that, while the image contrast for the reflection case has features in common with that for TEM, there are essential differences which make it desirable to discuss the situation in more detail. For the most part, our discussion on the theoretical basis will deal with idealized cases. For the REM case, a plane parallel incident wave will be assumed and correspondingly, a very small detector aperture will be assumed for SREM. The effects of finite beam convergences for REM and finite detector sizes for SREM can usually be introduced by the incoherent averaging of intensities. Unless otherwise indicated, the results for REM can be considered, on the basis of the reciprocity principle, to apply equally for SREM if the electron optics are equivalent[25].

Most of the obvious differences from the TEM case arise from the use of glancing angles of incidence and diffraction, the consequent foreshortening of the images and the large ranges of defocus values present in most images. The less obvious, but highly significant differences, include those dependent on the dynamical diffraction effects since dynamical diffraction processes at surfaces are markedly different from those in transmission through thin films.

Refraction effects at surfaces become important at glancing angles. The refractive index for electrons is given approximately by

$$n - 1 = \phi(r)/2E, \tag{2}$$

where $\phi(r)$ is the potential distribution in the solid and E is the accelerating potential of the incident electron beam. Since ϕ is of the order of 10-20 eV, n-1 is approximately 10^{-4} for 100 keV electrons. For beams making angles of less than 10^{-3} with the surface, the refraction deflection at the surface may be comparable with the Bragg angles.

It is usually assumed for convenience in theoretical treatments that the boundary of a crystal is sharp, with a planar discontinuity between the zero potential of space and the average inner potential of the crystal. This assumption may well be inadequate to describe the situation in many cases. Until more is known of the actual potential distributions of surfaces, the perturbations of the potential distribution by various surface layers and defects, calculations based on this assumption must be treated with caution.

Much of the theory of RHEED, has been done, as in the LEED case, by ignoring the refraction effects at the surface until the final determination is made of the diffracted beam directions in vacuum. In this way some important effects are overlooked. A strong channelling of the electron beams parallel to the surface can occur as a direct consequence of the refractive index change which acts as a potential barrier to electrons leaving the surface. As will be seen, this so-called surface resonance effect can have marked influence on reflection image contrast.

Surface step contrast

The very strong contrast effects given in REM images by surface steps, one atom or less in height, are prominent features which have been attributed to phase contrast imaging. On the assumption that no relaxation of the crystal lattice takes place around the step, the lattice orientation and crystal structure are taken to be unchanged by the step and the phases and amplitudes of the diffracted beams leaving the surface are unaffected by the presence of the step. The contrast then arises from the phase difference due to the differences in path length for electrons diffracted from the top and bottom of the step. From fig.6 it seems that for an incident beam making a glancing angle θ_1 with the surface, a diffracted beam making an angle θ_2 and a step height H, the path difference is

$$H[1 - \cos(\theta_1 + \theta_2)]/\sin\theta_2 . \tag{3}$$

For θ_1 and θ_2 equal to the Bragg angle, θ_B, for the atomic planes parallel to the surface, of spacing d_n, the path difference becomes equal to $2d\sin\theta_B = \lambda$. The phase differences is then 2π and no contrast is produced. Relatively large phase differences, of π or more, can be produced if θ_1 and θ_2 differ from θ_B or if the step height differs from d_h.

The simple theory for the imaging of a step then follows the TEM theory for imaging the edge of a thin specimen. When the jump in phase is small (<<1), the familiar weak phase object approximation (WPOA) can be made, giving the result that the in-focus contrast is zero in the absence of spherical aberration and strong black-white line pairs appear out of focus, with contrast and line separation increasing with defocus. The contrast reverses from underfocus to overfocus and with change of sign of the phase step. Osakabe et al.[19,20] have given an analysis of the variation of image contrast with defocus and the sense of the step (up or down) for this case. The effects of spherical aberration are usually not important because small objective aperture sizes are used to isolate the diffraction spots forming the image. For high resolution conditions, with large objective apertures, the spherical aberration will have the effect of giving a minimum, but non-zero, contrast for a slight underfocus, as in TEM.

Since, however, the phase difference at a step may not be small, a somewhat more complicated treatment, based on the phase object approximation (POA) rather than the WPOA, is needed. This has been described by Cowley and Peng[26]. A typical set of results of this treatment is shown in fig.8, calculated for the case of a phase difference of 1 rad. The contrast is seen to vary strongly with the defocus, with the sign of the phase change and also with the quantity B which measures the displacement of the center of the objective aperture from the diffracted

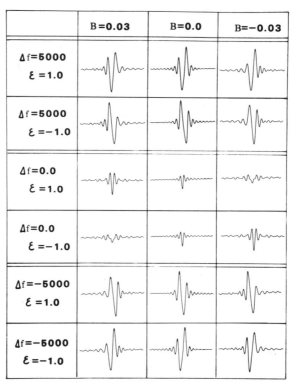

Fig. 8. Calculated intensity profiles based on the phase object approximation, for surface steps in REM with a phase shift $\varepsilon = \pm 1$ rad, for various values of the defocus Δf in Å and the deviation B in Å^{-1} of the objective aperture center from the center of the diffraction spot. [from Cowley and Peng, 1985].

beam position. Far out of focus, the step image is dominated by the black-white defocus contrast effect. In focus, for B = 0, a narrow dark line is produced instead of the zero contrast for the WPOA. For B non-zero, an asymmetric component is added to the in-focus image.

If the displacement of the objective aperture in a direction perpendicular to the surface is made so large that the diffracted beam is excluded, the step image will appear as a sharp white line on a black (zero intensity) background (see, for example, Osakabe et al.[18]). This follows because the diffraction spot from the perfect crystal surface will be a sharply-defined reflection, but a discontinuity such as a step gives a streak in the diffraction pattern, perpendicular to the surface and so will scatter into the displaced aperture.

The above description of the image contrast of a step in terms of phase contrast imaging, relies on a simplifying assumption which, by analogy with the TEM case, we call a column approximation. The amplitude of a diffracted wave at a point on the surface is assumed to depend on the amplitude of the incident wave at that point and on the content of a column of material centered on the point, extending into the crystal at right angles to the surface and having cross-section dimensions smaller than the resolution of the imaging process (see fig.6). The question of the validity of this approximation can be resolved only when an adequate theoretical description of the interaction of the electron beams with a crystal surface is available (see section on Dynamical calculations). In practice, the approximation appears adequate for most medium-energy imaging. For the 100 keV energy range, it can fail significantly in particular cases, partly because the resolution attainable is relatively good and partly because the lower glancing angles of incidence can enhance the surface resonance effects which involve the channelling of electrons along surface atomic layers for relatively large distances.

Strain field contrast

The local variations of crystal lattice orientation associated with the strain field around an extended defect, such as a dislocation, modify the amplitudes of the diffracted beams and hence produce strain-field contrast in REM images, as in dark-field TEM. For REM, as in TEM, the column approximation provides a good first approximation for the image contrast. For TEM with crystal thicknesses of less than about 100 nm, the columns chosen are parallel to the electron beam direction and the column widths may be no more than 0.3 - 0.5 nm. For REM, columns chosen perpendicular to the surface are almost perpendicular to the electron beam and it must be expected that the column approximation will be less effective.

Given this approximation, however, the image intensities are calculated by combining information on the variation of local crystal lattice orientation with position on the surface, and theoretical results on the variation of diffracted beam intensity with orientation. The theory used is the dynamical diffraction theory for RHEED for a perfect crystal. Only in special cases is the relatively simple two-beam dynamical theory for the reflection case relevant. For general cases and especially near zone-axis orientations, the many-beam dynamical diffraction theory is needed. This theory has been formulated for the reflection case on the basis of the original Bethe theory, by Colella[27] and further developed by Moon[28] and others. The Bloch-wave solution of the Schrödinger equation for electrons in an infinite crystal is matched at a planar surface with the vacuum wave consisting of the incident and diffracted beams.

The reflection theory is more complicated than that for transmission because waves going both ways, in and out, with respect to the entrance face must be considered.

In order to treat the strain fields due to a dislocation, for example, the intensity of the diffracted beam needs to be calculated as a function of angle of incidence (angle of the incident beam with the surface) and the azimuthal angle. For a pure screw dislocation perpendicular to the surface only the variation with angle of incident is relevant. For a pure edge dislocation perpendicular to the surface, only the azimuthal angle has an effect. In general both variations are involved. The strain field at a free surface is not the same as in bulk because the surface must relax so that the net force acting on a surface atom is zero. Expressions for the strain fields of dislocations in bulk and at a free surface are given, for example, by Hirth and Lothe[29].

The REM images of strain fields for screw and edge dislocations at surfaces were calculated by Shuman[30] using the column approximation and not taking surface relaxation into account. The results for screw dislocations suggested black-white streaks, as observed in images such as fig.9.

More recently, the effects of surface relaxation have been included. It has been shown[21] that, even for the simplest case of a screw dislocation perpendicular to a surface in an isotropic crystal, the images can be more complicated. Fine detail near the dislocation core is visible under reasonably high resolution imaging conditions. For edge-dislocations, or mixed-dislocations, the images may show considerable complexity with a strong dependence on the diffraction conditions (see Problem 3).

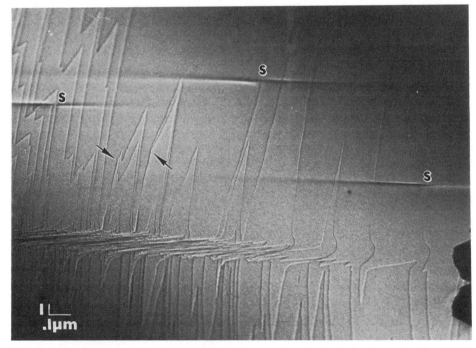

Fig. 9. REM image of a GaAs(110) cleavage face showing the strain fields from dislocations, S, at the terminations of surface steps and places (arrowed) where steps two-atoms high split into two smaller steps. [from Hsu et al., 1984].

Dynamical calculations for imperfect crystals

The model of a perfectly periodic crystal structure terminated on a geometric plane has served as a starting point for considering dynamical scattering effects in the reflection mode. For a useful theoretical treatment of REM image contrast, two further stages of complication are necessary. Firstly, any treatment of diffraction from surfaces for RHEED or REM must include the effect of modifications of the structure and periodicity of the surface planes. The topmost surface atom layers may be relaxed so that the interplanar separations are different from those in bulk. Also there may be layers of physically absorbed or chemically bonded foreign atoms on the surface. Secondly, since the objective of REM is usually the study of the local variations of structure on the surface, it is necessary to have a theory, and a means for calculation, of the diffraction effects associated with changes of structure within the surface planes.

The first requirement may be met in several ways. Moon[28] suggested that the change of lattice plane spacings at the surface can be treated formally by assuming an artificial, large unit cell periodicity perpendicular to the surface such that the penetration of the electron wave into the crystal is less than one unit cell. Britze and Meyer-Ehmsen[31] considered that the crystal could be described in terms of slices taken parallel to the surface having different structure. For each slice the Bloch wave solution could be obtained as for a three-dimensional crystal and the solutions could be matched at the boundary planes.

A somewhat different approach by Maksym and Beeby[32] assumes the crystal to be periodic in the two dimensions parallel to the surface but non-periodic in the direction perpendicular to the surface. The dependence of the wave function in the direction perpendicular to the surface is obtained from a system of differential equations. Ichimiya[33] used a multi-slice formulation, with slices again chosen parallel to the crystal surface, defining a transfer matrix to relate the diffraction beam amplitudes between slices for the transition of the waves in and out of the crystal.

An extension of the Maksym-Beeby approach has been made by Kawamura et al.[34] for example, for perturbations of the surface layers, by surface steps, assuming a periodic array of steps. The use of a large unit cell in the surface layer periodicities, however, involves a large number of Fourier coefficients and large computations, even when only a one-dimensional long period is considered.

For the treatment of an isolated defect on a crystal surface, it is possible, in principle, to apply the above methods, assuming an artificial large periodic unit cell (the "periodic continuation" method). This method has been used successfully for equivalent TEM cases and is now a standard procedure in multislice calculations. The dimensions of the artificial unit cell must be appreciably larger than the dimensions of the defect. Even after taking account the spread of the electron wave in the crystal and the spread due to defocus effects, there must be no appreciable overlapping of the effects on the wave field due to neighbouring defects of the periodically repeated set. In the REM case, the complications due to the glancing angle of incidence make this procedure more difficult. It has been shown[35] that for electron energies in the 20-100 keV range, the perturbation of the wave field due to a surface step may extend along the crystal surface in the beam direction for a distance of 20-50 nm. Thus if an artificial periodicity is to be imposed for slices of crystal taken parallel to the surface, at least one of the unit cell dimensions must be 50 nm or more, the calculations then become impossibly cumbersome.

The difficulties involved with such an approach, taking slices of crystal parallel to the crystal surface, motivated the alternative approach of using a multi-slice formulation of the dynamical diffraction problem with the slices taken perpendicular to the crystal face and almost perpendicular to the incident beam. This approach was suggested and briefly explored by Cowley and Warburton[36]. More recently it has been extended and applied to the calculation of RHEED patterns and the REM contrast of surface steps by Peng and Cowley[35].

The slices perpendicular to the crystal surface are basically non-periodic since they include the electron wave function both inside the crystal and in the vacuum. The periodic continuation assumption involves the assumption of a periodic array of crystals as in fig.10. The artificial periodicity must be large enough to include a depth of crystal greater than the effective penetration of the electron beam, plus a width of the vacuum great enough to contain the amount of wave field necessary to produce sufficiently well-defined diffracted beams. A periodicity of about 5 nm may be sufficient in some cases, but 10 nm is more satisfactory.

The plane wave incident on the object is usually masked so that the waves which will strike the crystal face of interest are transmitted and the waves which would enter the edges of the crystal blocks are excluded. The wave is then propagated from one slice to the next as in the multislice calculations for TEM. The potential distributions in the slices are chosen to incorporate any desired model of the surface structure, including any relaxation effects and the appropriate variation of potential from the crystal to the vacuum.

As the wave is propagated from slice to slice, the wave-field is seen to build up in the crystal and in the vacuum. An equilibrium state is achieved after propagation through 50 to 100 nm (fig.11). The wave field in the vacuum may then be selected and Fourier transformed to produce the RHEED pattern amplitudes.

Once the equilibrium state has been established, the effect of any

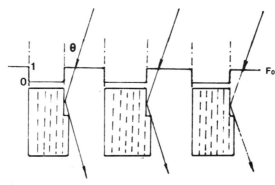

Fig.10. The scheme used for calculating RHEED patterns and REM images. The crystal is assumed to be repeated periodically. The incident beam makes a small angle, θ, with the surface of interest. The wave field is propagated from slice to slice with slices taken perpendicular to the surface. A modulating function, F_0, on the incident beam, is zero for the crystal faces nearly perpendicular to the incident beam, so that the beam will strike only the crystal face of interest. [Peng and Cowley, 1986].

defect on the wavefield can be found by making appropriate changes in a
few, or all, of the remaining slices and continuing the propagation through
the further slices until a new equilibrium state has been achieved.
Fourier transform of the vacuum wave function then shows the effect of the
defect on the RHEED pattern. Operation on the wave function by the
transfer function of the objective lens allows the REM image intensities to
be calculated (fig.12).

Calculations made in this way are expensive of computer time but not
to the extent of being impracticable. For two-dimensional periodicities of
1 × 11.3 nm, the calculations for propagation through a distance of about
200 nm takes about 10 hours on a VAX 750 computer. Considerable reductions

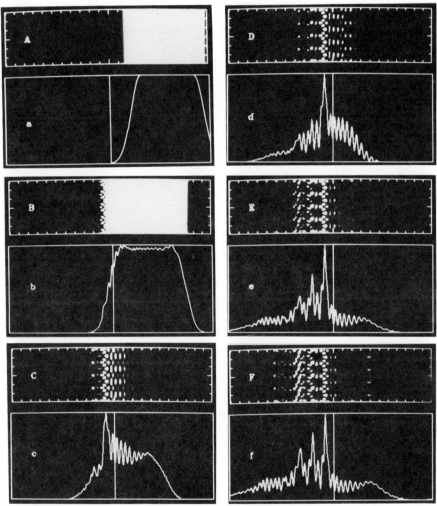

Fig.11. Propagation of the electron beam along the crystal surface,
calculated for Pt(111) with θ = 54.1 mrad, 19 keV. Figures A,
B,...,F show the two-dimensional distribution of intensity in
slices at distances 2.9, 87, 174, 290, 364 and 406 Å from the
entrance face. Figures a, b,...,f show the intensity profiles,
averaged across the slice for the same slices [from Peng and
Cowley, 1986].

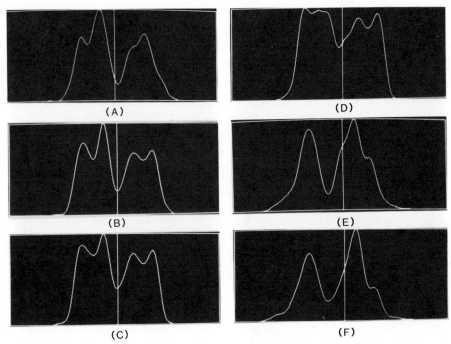

Fig.12. Calculated intensity profiles in a REM image of a single-atom strip
on a Au(111) face for θ = 28 mrad, 40 keV for defocus values:
(A) -2800;   (B) -1800;   (C) -1200;   (D) -600;   (E) +2000,   and
(F) +2600 Å. [Peng and Cowley, 1986].

of this computing time may well be possible. A separate calculation must
be made for each direction of incidence and for each model of the crystal
surface. However, once the form of the equilibrium wave field for a given
angle of incidence has been established, this wave field can be used as the
input wave for the calculation of the effect of any defect.

Surface channelling and resonance effects

Early RHEED observations showed that a specularly reflected beam may
be produced whether or not the electron beam is at the Bragg angle for
reflection from the crystal lattice planes. Geometrically the specular
beam direction is given as if by mirror reflection from the surface plane.
The specular beam is especially strong when it lies on a Kikuchi line,
which may be parallel to, or inclined to, the surface. It was shown by
Miyake et al.[37] that this condition corresponds to the condition that a
diffraction beam should be produced in a direction almost parallel to the
crystal surface. The geometric conditions under which this effect arises
have been described by Peng and Cowley[38].

When a strong diffracted beam lies almost parallel to the crystal
surface, a convenient classical description suggests that the electrons do
not have a sufficient momentum perpendicular to the surface to allow them
to escape over the potential barrier of the inner potential and hence are
reflected back and are trapped in the surface layer, travelling relatively
large distances parallel to the surface.

The resulting build-up of intensity in the surface layers produces an enhancement of all diffracted beams, and especially of the specular beam, related to the surface diffracted wave by a strong Bragg reflection condition. In terms of the wave picture it may be said that a surface resonance condition is created, which channelling of the waves along the surface layers.

These simple pictures are consistent with results of the dynamical diffraction calculations such as those of Marten and Meyer-Ehmsen[39] and Peng and Cowley[35]. For particular angles of incidence the wave field in the crystal is seen to be strongly concentrated along particular atom rows, the penetration of the wave into the crystal is reduced to one or a few atomic layers and the effects of a perturbation such as a surface step is propagated for a greater distance along the surface.

The effects of the surface resonance on the REM image contrast can be striking. Hsu and Peng[40] recognized three categories of specular reflections as giving characteristic step images; those at a Bragg angle with no resonance effect (BNR), those with a resonance effect but not at a Bragg angle (RNB) and those at a Bragg angle with a resonance effect (BR). When neither a Bragg reflection nor a resonance effect is present the image obtained with the specular reflection is usually too weak to observe.

The characteristic variation of step contrast with defocus, as described by the phase object approximation and illustrated in fig.8 is given by the BR and BNR cases, although when the resonance condition is present the in-focus image is modified: the contrast may be appreciable in-focus and may differ for steps-up and steps-down.

For the RNB condition, the contrast near to the in-focus position is different. Both the steps-up and the steps-down give a strong dark line. The asymmetric black-white contrast, reversing with the sign of defocus, does not become obvious until the defocus values are large. A loose argument, giving a qualitative explanation for this observation, is that a surface step disrupts the surface resonant wave field, decreasing the intensity of the specular reflection for the considerable distance along the surface required to re-establish the resonance condition.

It is possible that surface resonant conditions may have equally strong effects on the images of the strain fields associated with dislocations and other defects. The changes of crystal lattice orientation in the strain field may well increase or decrease the magnitude of the resonance effect. As yet such effects have not been calculated or recognized in experimental observations.

Discussion

As yet the theoretical basis for the interpretation of REM images is in an unsatisfactory state. Simple ideas based on geometric optics, phase contrast and column approximations can give qualitative descriptions of the contrast effects which appear to be satisfactory, at least as a first approximation, for imaging with moderate resolution. For the quantitative interpretation of image contrast, especially under high-resolution conditions, a proper dynamical diffraction treatment is essential. The dynamical diffraction theories for perfect crystals and those for samples with perfect two-dimensional periodicity parallel to the surface, are in reasonably good shape. However it is difficult to apply these treatments to the cases of importance in REM where the main objects of study are usually the local structural and compositional perturbations.

The multislice computing method with slices perpendicular to the surface has been applied to a few special cases, but many more calculations are needed before an adequate description is possible of the effects on the image of the various parameters of the diffraction and imaging geometry. It is desirable to test the effects of various assumptions that can be made regarding the structure of the surface, including perturbations of the potential distribution at the surface, surface relaxations, models for atom displacements at defects and the effects of chemically different atoms in surface layers or concentrated at defects. It is possible that in time, a sufficient number of calculations may be accumulated to provide a basis for detailed deductions concerning these effects: but a more satisfactory basis for image interpretation is clearly desirable.

APPLICATIONS OF REM

In an increasing number of laboratories, the techniques of reflection imaging are being applied to a variety of problems. Some of the experiences gained are discussed here, to serve both as illustrations of the nature of the information which may be obtainable, and also as indications of the unresolved problems on which further work is desirable. The discussion is undoubtedly biased towards the interest of the author and his associates, but the topics may be of sufficient general interest to give them wider significance.

Surfaces steps and surface reaction

The clear visibility of steps in REM images introduce many possibilities for the investigation of the role of surface steps in determining the physical and chemical properties of surface. For simple metals and monoatomic semiconductors, the nature of the step seen on a surface is rarely in doubt. Growth steps on the principle faces and the slip steps, formed by the passage of partial dislocations intersecting the surface, are usually of a form which is well defined by observations or inferences from other techniques. The fact that the steps most commonly seen are of one atom height has been confirmed in several ways. Osakabe et al.[18] observed the movements of steps on silicon heated in vacuum as the silicon atoms evaporated from the surface and showed that the assumption of single-atom height gave agreement with known evaporation rates. These authors and Hsu et al.[41] observed the termination of steps at the intercepts on screw dislocations on surfaces. The height of such steps is given by the Burgers vector of the dislocations. In the case of a GaAs(110) surface, Hsu et al.[41] showed that the step heights could correspond to either one or two layers of atoms, depending on the nature of the dislocations.

The role of surface steps is fundamental for surface reactions, surface phase transformation and crystal growth processes. Inferences have been drawn from various analytical and diffraction techniques as to the way in which surface steps influence these processes and a limited amount of evidence is available from direct TEM imaging. REM however, offers the possibility for high resolution observation of the configurations of surface steps and their changes in the course of the reactions.

An outstanding example of the use of REM for such purposes is provided by the observations, at temperature, by Osakabe et al.[19,20] of the transformation from the Si(111) 1×1 to the 7×7 surface superstructure. The transformation was shown to be initiated at the surface steps, with the 7×7 structure growing on the top side of the steps. Detailed studies[42] later gave insight into the nature of the transformation. More recently[22], high resolution REM studies have shown clear images of the 2.3 nm periodicity of the 7×7 lattice and have shown the dynamics of the interaction of the surface steps with this periodic modification of the surface layers of atoms (fig.8).

Many observations have been made by Hsu and collaborators[1,2] of steps on small, almost spherical, crystals of Pt and Au formed by melting the ends of thin wires in air. The growth steps on near-planar 111 facets of these crystals are often indented, presumably by the annihilation of vacancies which diffuse to the surface and then to the steps during the cooling process. Vacancies reaching the surface well away from surface steps nucleate to form circular depressions in the surface, one atom deep (fig.13).

During the cooling of this crystals many dislocations travel through the crystal and leave traces on the surface in the form of steps of height equal to the projection of the Burgers vector on the surface normal. These traces usually take the form of straight lines in principal crystallographic directions. If these steps are formed after the crystal is cool they cut across the growth steps with no apparent interaction. However the point of intersection is one of high energy. If the temperature is

Fig.13. REM images of Pt(111) surface taken with the (555) reflection showing growth steps, slip traces (ON, O'N'; OP, OP') and a sub-surface dislocation line. Figures (a) and (b) show the same area but with a difference in azimuth of approximately 25°. In (b) the arrowed step appears to be doubled. The dark band H' in (b), due to a sub-surface dislocation, would be expected at about M in (a) but is not visible. Loops between growth steps are mostly atom-deep depressions formed by the aggregation of vacancies. In (b), differences in intensity across single-atom steps are evident. [Hsu and Cowley, 1983].

sufficiently high to allow diffusion of atoms over the surface there will be diffusion away from these points. The step contours become rounded off, leaving characteristic configurations of bent steps as in fig.13. From the observation of the remnants of the step traces, it is possible to make deductions concerning the history of the events that have taken place on the surface. Observations of the diffusion processes at temperature in an appropriate ultra-high vacuum or a controlled atmosphere, should allow the derivation of diffusion coefficients of step energies under a variety of conditions.

Lattice relaxation at surface steps

Elementary arguments may be made to suggest that the energy associated with a step on a crystal surface may be minimized if the crystal lattice around the step is allowed to relax. The resulting strain field around the step should then influence the intensity of diffraction from the surface planes. As a consequence, the REM images of surface steps should show some component due to the strain field in addition to the phase contrast considered earlier on.

Osakabe et al.[20] used an ingenious technique to estimate the magnitude of this strain field. They noted that the in-focus contrast of steps is not zero, as predicted by the weak phase object approximation, but usually a weak, dark line is observed. They attributed this line to strain contrast. This line was seen to disappear, however, when the step crossed the streak of strain-field contrast due to an emergent screw dislocation at a particular value of the distance from the dislocation. The strain field at the step may then be considered as equal and opposite to that due to the dislocation at that position.

The uniqueness of this interpretation was called into question by Cowley and Peng[26] who argued that the weak phase object approximation does not usually apply for step contrast considerations. If the phase change at the step is not very small, the prediction based on the phase object approximation is that a dark line should always appear at the in-focus position, even without any strain field (fig.8).

Further evidence suggesting the existence of a strain field has been described by Peng and Cowley[21,38]. These authors noted that in high resolution REM images, the images of steps on Pt and other materials often appear to be split into two parallel lines (fig.13). They suggested that the variation of the lattice plane inclination with respect to the incident beam could take the diffraction angle through the Bragg angle once and then back through the Bragg angle to the original angle, giving a double peak of diffracted intensity. This could account for the appearance of a doubled bright line sometimes observed, but less easily for the doubled dark line, also observed.

An alternative explanation for the doubling of step lines is provided by Lehmpful and Uchida[43] who suggested that the dynamical diffraction condition is perturbed by the step, resulting in an oscillation of the wave field with the periodicity of the extinction distance. Further experiments are clearly needed to establish whether this or another explanation is appropriate.

In general it would seem that, although the REM technique has, in principle, sufficient sensitivity to detect even small strain fields at steps, the strong variations of image contrast with diffraction conditions will require that the experimental parameters be determined with considerable accuracy and extensive dynamical diffraction calculations are required to establish the basis for the interpretation of the images.

Sub-surface defects

Evidence has been found in several cases that faults or other defects in atomic layers below the surface can influence image contrast. In images of platinum crystals, large neighbouring surface regions, 100 nm or more in extent, separated by a single surface step, show very different intensities in REM images (fig.13). Dr A. Howie (private communication) pointed out this peculiarity and suggested that it may arise because of the presence of stacking faults in the surface layers. The occurrence of surface layer stacking faults of this type has been suggested on the basis of various forms of indirect evidence. Also the high resolution electron microscopy of the edges of thin crystals[44], the profile imaging technique, has shown direct evidence in the case of gold.

In the generally accepted notation, the sequence of stacking of the hexagonal close-packed layers of atoms in (111) planes is denoted as ABCABC... for face-centered cubic metals. For a stacking fault in the top layer of atoms on a (111) face, the sequence would be ABCABA. That is, the top layer is placed directly above the second previous layer, instead of above the third previous one, giving a local hexagonal close-packed sequence. A single surface step could then terminate the anomalously stacked top layer, leaving a normal face-centered cubic stacking sequence in the layers of atoms.

Under kinematical diffraction conditions, the top few layers would give clearly different diffraction patterns, according to whether the faulted top atomic layer was present or not. In the presence of the prevailing strong many-beam dynamical diffraction conditions, however, the conditions for observing the intensity differences are less clearly defined, and detailed calculations are required for accurately determined experimental conditions in order to provide unambiguous interpretations of the observation.

In the case of cleaved graphite crystals, it is known from transmission electron microscopy that stacking fault ribbons may occur parallel to the cleavage faces. It has been reported by Hsu and Cowley[45] that clear evidence for such ribbons is seen in REM pictures of graphite cleavage faces. The image contrast varies greatly with the choice of the diffraction beam used to form the image. The depth of the ribbons below the surface can not be determined, but the evidence of computations suggests that the penetration of the electrons into a crystal is of the order of only a few atomic layers, so that the depth is probably no more than about 1 nm.

In many images of metal crystal surfaces, lines appear which resemble, in some respects, those due to surface steps, but are much more diffuse (fig.13). These lines may be straight or may be curved into broad arcs[21,38]. Their contrast is usually a minimum when they are parallel to the incident beam direction. These lines have been attributed to the strain fields of dislocations or, more probably, partial dislocations lying sufficiently far below the surface to ensure that the main perturbation of the surface atomic planes is small and is spread over distances of 10 nm or more. Rough estimates suggest that the depth of the dislocation line below the surface may be 2-5 nm. The conclusion is that such defects extend over distances of several micrometers at a depth below the surface which remains almost constant.

Surface treatments and surface reactions

The number of investigations of surface treatments and reactions carried out using surface imaging and associated methods is still rather small, partly because there are few microscopes which can provide suitable means for treating surfaces under the ultra-high vacuum conditions required to ensure that the results are meaningful. A few representative investigations will be mentioned here as a guide to the nature of the projects which are now considered feasible.

Studies have been made by Ogawa et al.[46] and also by Claverie et al.[47] of the effects of ion bombardment on surface structure. It was shown that a low dose of Ar ions is sufficient to remove all step contrast from REM images of Si surfaces, but after subsequent annealing at 600°C, the steps reappear in exactly the same positions as before. Heavier doses of ion bombardment however, render the surface amorphous to a considerable depth so that, on annealing, the surface layers are recrystallized with no apparent memory of their earlier forms. Studies of the oxidation of Si(111) surfaces have been made by Shimizu et al.[48]

The SREM technique has been applied in a series of studies of the deposition of metals on magnesium oxide surfaces. Small gold crystals, approximately 2 nm in diameter evaporated on MgO smoke crystal faces, were shown to form in exact parallel orientation on the MgO(100) faces[49]. The foreshortening of the SREM images revealed, very clearly, the tendency for the gold particles to be aligned, apparently on steps on the MgO surfaces. The angles between pairs of lines of particles were measured. Removal of the foreshortening factor then revealed a strong tendency for the lines of particles to be inclined to each other, and to the [100] direction, by 2 or 3°. A straight-line step at such a small angle to the [100] direction can arise only if there is a periodic array of kinks along the step, about 6 nm apart. Similar small angles between extended straight steps have been observed, using REM, on cleavage faces of GaAs crystals (fig.9)[41].

Observations have been made of thin films of various metals, including Cu and Pd, on MgO crystal cleavage faces, in the vacuum of 10^{-8} Torr in the specimen preparation chamber of the HB5 STEM instrument[3]. The specimens were transferred to the observation position without exposure to any higher pressure. The metal films, as deposited, appear to consist mostly of small crystals growing epitaxially, in parallel orientation with the MgO, along the MgO surface steps. Under irradiation by the electron beam, however, the metal films are seen to change. In the case of Pd, small crystals, less than 1 nm in diameter are seen to be distributed along parallel lines, arranged periodically about 3 nm apart[50] (see fig.3). These parallel lines appear to correspond to periodic arrays of steps on the MgO surface since similar, but less extensive sets of parallel steps, 3 nm apart, have been observed on freshly cleaved MgO surfaces. After irradiation, the Pd is in (110) orientation on the MgO(100) faces. Sets of steps, decorated with Pd, are observed to make angles of 1-2° with each other.

Associated techniques

The scanning mode used for SREM imaging, is similar to STEM in that it lends itself to the acquisition of information through parallel modes of data collection. Microdiffraction patterns can be obtained from areas of diameter roughly equal to the resolution limit of the images. This possibility has been used extensively, for example, in the determination of the phases present and the epitaxial arrangements of 1-2 nm particles of metals deposited on oxide substrates[51]. In principle, microanalysis may be performed of regions of comparable size when the beam is stopped on the specimen and the characteristic emitted X-rays or energy losses of the incident beam are observed. For the SREM case, however, the signal levels are too low to be useful for these analytical modes.

The secondary electron signals are useful. In the grazing-incident SREM configuration, the SEM signals obtained are relatively strong but have a rather poor resolution of 2 nm or more. For crystal faces making greater angles to the incident beam or for small particles, resolutions of 1 nm or better may be obtained. The secondary electron signal gives information on the specimen which is complementary to that from STEM or SREM. It is sensitive to the morphology and so may be very useful in allowing an appreciation of the three-dimensional form of bulk specimens or large crystals, so that faces may more easily be aligned for orientations appropriate for SREM. The intensity of the secondary electron signal is dependent on the work function of the emitting surface and so may provide information on changes of surface composition on a monolayer scale.

In the REM mode, the diffraction pattern obtained using the selected area electron diffraction technique represents the average of the structure over the illuminated area of the specimen. It has high angular resolution and can provide accurate data in crystal faces which are seen, from the REM image, to be free of gross faults.

From the large illuminated region of a REM image, strong diffraction beams may be obtained and may be used to provide EELS signals of useful strength. The first attempts to make use of EELS signals for microanalysis were not encouraging. Krivanek et al.[7] found that with the glancing angle incidence in the REM mode, the surface plasmon modes are strongly excited. Multiple scattering by these plasmons produces a high background signal for larger energy losses so that the signal-to-noise ratio is poor for the detection of inner-shell edges. More recent work[8] has suggested, however, that it is possible to derive useful information from the EELS spectra provided that the signals are collected for a sufficiently long period (fig.14).

From the surfaces of heavy-metal crystals it is possible to obtain well-defined M edges and also K or L edges from thin layers of lighter elements on the surfaces. From cleavage faces of MgO, the oxygen K-edge may be recorded in sufficient detail to allow a comparison of the near-edge fine structure with that obtained in transmission. The EXELFS analysis of the oxygen K-edge structure reveals systematic differences from the data obtained in transmission through thin MgO film. The differences are consistent with a model in which oxygen atoms are absorbed on the surface above the Mg ions of the terminated bulk structure.

The data for this determination was obtained from a MgO crystal set to give the 400 reflection with a strong resonance contribution from 200 and 220-type reflections. Under these conditions the dynamical diffraction calculations show that the electron wave field in the crystal is almost entirely confined to the top one or two planes of atoms. As compared with the transmission case, the ratio of the strengths of the O and Mg K-edges is enhanced, suggesting that an excess of oxygen is present in the surface layers. This result is consistent with the analysis of the EXELFS data.

It thus appears that EELS analysis, performed in conjunction with REM imaging, with an appropriate adjustment of the diffraction conditions, as deduced from the SAED pattern, can be an effective means for the chemical analysis of surface layers and even for a study of surface modifications of the bonding distributions.

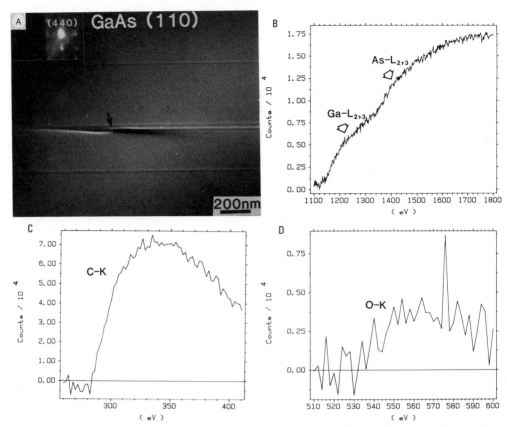

Fig.14. REM and EELS from a GaAs(110) surface. (A) REM image from the (440) reflection (insert) showing a very smooth face: (B) the L-edges for Ga and As resolved in the EELS spectrum from the (440) spot; (C) the K-edge from carbon present in a thin contamination layer on the surface; (D) the K-edge from oxygen absorbed on the GaAs surface or contained in the contamination layer. [Wang and Cowley].

CONCLUSION

The techniques of REM and SREM have been proved as useful methods for the study of surface structures. The images suffer from severe foreshortening, but for reasonably flat surfaces it is possible to get images showing fine detail of areas many micrometers in diameter. The resolution obtained in the one direction in the surface perpendicular to the incident beam can be better than 1 nm and several contrast-producing mechanisms can be involved, giving sensitivity to the micromorphology of the surface (steps one-atom high) and variations of composition and structure of the surface layers. The sensitivity to surface structure is comparable to that of LEED: fractions of a monolayer of added atoms can give large changes of signal strength.

The surface imaging technique may be used in conjunction with microdiffraction, giving the crystal structures of small surface regions, or microanalysis using EELS to show variations in surface composition for very small surface areas. The combination of these techniques can provide a very powerful approach to problems in surface science.

In comparison with the profile imaging method (D.J. Smith, this volume) the resolution attainable with REM is not so good and the information given on structural variation as a function of distance below the surface is less direct. However the possibility of obtaining information on the two-dimensional variation of structures over large surface areas is an important advantage.

The STM technique applied to surface studies in vacuum (N. Garcia, this volume) can provide better resolution than REM in some cases and does not suffer from the foreshortening effect or from radiation damage effects. However, at least in the current stage of its development, the STM is limited with respect to REM in that the images are of much smaller areas and the interpretation of image contrast is more difficult. Both imaging methods can be associated with a range of analytical techniques but the information gained is quite different in the two cases. It may well be that STM and REM will be combined in the future to allow a full range of complementary techniques to be applied to appropriate specimens. Instruments can be envisaged that will allow images to be obtained simultaneously by the two methods.

Relative to other surface analysis techniques, the reflection imaging modes suffer from the fact that the necessary instrumentation is mechanically complicated and has been developed with provision for only moderately high vacuum. The ultra-high vacuum versions of the electron microscopy instruments, now being produced, are necessarily cumbersome, inconvenient and expensive: but they are necessary to allow the surface imaging and associated techniques to be used to full advantage for providing high spacial resolution in surface science.

ACKNOWLEDGEMENTS

The author is grateful to his associates, Tung Hsu, H.J. Ou, L.M. Peng, and Z.-L. Wang for useful discussion and the provision of illustrations. The work of this laboratory in the areas described has been supported by NSF grants DMR 8302415 and DMR 8510059 and U.S. Department of Energy grant DE-FG02-86ER45228, and has made use of the resources of the ASU Facility for High Resolution Electron Microscopy supported by NSF grant DMR 8611609.

REFERENCES

1. T. Hsu, Norelco Reporter no 1 EM, 31:1 (1984).
2. T. Hsu and J.M. Cowley, Ultramicroscopy, 11:239 (1983).
3. H.-J. Ou and J.M. Cowley, in: "Electron Microscopy 1986", T. Imura, S. Maruse and T. Suzuki eds, The Japanese Soc. Electron Microscopy, Tokyo, Vol. II p.1361 (1986).
4. K. Takayanagi, K. Kobayashi, K. Yagi and G. Honjo, J. Phys. E., 11:441 (1978).
5. Y. Tanishiro, K. Takayanagi, K. Kobayashi and K. Yagi, Acta Cryst., A37:C300 (1981).
6. M.D. Shannon, J.A. Eades, M.E. Meichle and P.S. Turner, Ultramicroscopy, 16:175 (1985).
7. O.L. Krivanek, Y. Tanishiro, K. Takayanagi and K. Yagi, Ultramicroscopy, 11:215 (1983).
8. Z.-L. Wang and J.M. Cowley, Ultramicroscopy, 21:77 (1987).

9. J.M. Cowley, in: "Microbeam Analysis 1980", D.B. Wittry, ed., San Francisco Press, San Francisco, p.33.
10. G. Lehmpful and W.C.T. Dowell, Acta Cryst., A42:569 (1986).
11. J.M. Cowley, in: "Proc. 37th Annual Meeting EMSA", G.W. Bailey ed., Claitor's Publ. Divis. Baton Rouge, p.472 (1979).
12. J.M. Cowley, J. Electron Micr. Techniques (1988) In press
13. J.A. Venables, A.P. Janssen, P. Akhter, J. Derrien and C.J. Harland, J. Microscopy, 118:351 (1980).
14. E.S. Elibol, H.-J. Ou, G.G. Hembree and J.M. Cowley, Rev. Sci. Instrum., 56:1215 (1985).
15. T. Ichinokawa, Ultramicroscopy, 15:193 (1984).
16. M. Ichikawa, T. Doi, M. Ichihashi and K. Hayakawa, Japan. J. Appl. Phys., 23:913 (1984).
17. J.M. Cowley, R. Glaisher, J.A. Lin and H.-J. Ou, in: Proc. 44th Annual Meeting EMSA, G.W. Bailey ed., San Francisco Press, San Francisco, p.684 (1986).
18. N. Osakabe, Y. Tanishiro, K. Yagi and G. Honjo, Surface Sci., 97:393 (1980).
19. N. Osakabe, Y. Tanishiro, K. Yagi and G. Honjo, Surface Sci., 102:424 (1981a).
20. N. Osakabe, Y. Tanishiro, K. Yagi and G. Honjo, Surface Sci., 109:353 (1981b).
21. L.-M. Peng and J.M. Cowley, Micron and Microsc. Acta, 18:171 (1987a).
22. Y. Tanishiro, K. Takayanagi and K. Yagi, J. Microscopy, 142:211 (1986).
23. N. Yamamoto and S. Muto, Japan. J. Appl. Phys., 23:L806 (1984).
24. T. Hsu, J. Vacuum Sci. Technol., B3:1035 (1985).
25. J.M. Cowley, Appl. Phys. Letters, 15:58 (1969).
26. J.M. Cowley and L.M. Peng, Ultramicroscopy, 16:59 (1985).
27. R. Colella, Acta Cryst., A28:1 (1972).
28. A.R. Moon, Z. Naturforsch., A27:390 (1972).
29. J.P. Hirth and J. Lothe, "Theory of Dislocations", John Wiley, New York (1982).
30. H. Shuman, Ultramicroscopy, 2:261 (1977).
31. K. Britze and G. Meyer-Ehmsen, Surface Sci., 77:131 (1978).
32. P.A. Maksym and J.L. Beeby, Surface Sci., 110:423 (1981).
33. A. Ichimiya, Japan. J. Appl. Phys., 22:76 (1983).
34. T. Kawamura, P.A. Maksym and T. Iijima, Surface Sci., 148:L671 (1984).
35. L.-M. Peng and J.M. Cowley, Acta Cryst., A42:552 (1986).
36. J.M. Cowley and P. Warburton in: "The Structure and Chemistry of Solid Surfaces", G.A. Somorjai, ed., J. Wiley and Sons, New York, (1969).
37. S. Miyake, K. Kohra and M. Takagi, Acta Cryst., 7:393 (1954).
38. L.-M. Peng and J.M. Cowley, J. Electron Microscopy Tech., 6:43 (1987b).
39. H. Marten and G. Meyer-Ehmsen, Surface Sci., 151:570 (1985).
40. T. Hsu and L.-M. Peng, Ultramicroscopy, 22:217 (1987).
41. T. Hsu, S. Iijima and J.M. Cowley, Surface Sci., 137:551 (1984).
42. Y. Tanishiro, K. Takayanagi and K. Yagi, Ultramicroscopy, 11:95 (1983).
43. G. Lehmpfuhl and Y. Uchida, in: Proc. 44th Annual Meeting EMSA, G.W. Bailey, ed. San Francisco Press, San Francisco, p.376 (1986).
44. D.J. Smith and L.D. Marks, Ultramicroscopy, 16:101 (1985).
45. T. Hsu and J.M. Cowley, in: "The Structure of Surfaces", M.A. Van Hove and S.T. Tong, eds, Springer-Verlag, Berlin, p.55 (1984).
46. S. Ogawa, Y. Tanishiro, K. Kobayashi, K. Takayanagi and K. Yagi in: "Electron Microscopy 1986", T. Imura, S. Maruse and T. Suzuki, eds, The Japanese Soc. Electron Microscopy, Tokyo, Vol.II, p.1349 (1986).
47. A. Claverie, J. Faure, C. Vieu, J. Beauvillain and B. Jouffrey in: "Electron Microscopy 1986", T. Imura, S. Maruse and T. Suzuki, eds, The Japanese Soc. Electron Microscopy, Tokyo, Vol.II, p.1357 (1986).

48. N. Shimizu, Y. Tanishiro, K. Kobayashi, K. Takayanagi and K. Yagi, Ultramicroscopy, 18:453 (1985).
49. J.M. Cowley and K.D. Neumann, Surface Sci., 145:301 (1984).
50. H.-J. Ou and J.M. Cowley, Ultramicroscopy, 22:207 (1987).
51. J.M. Cowley, in: "Catalyst Characterization Science", M.L. Deviney and J.L. Gland, eds, American Chemical Society, Washington, p.329 (1985).

PROBLEMS

1. A REM image is formed of the (111) surface of a gold crystal ($a_0 = 0.407$ nm). The plane of incidence makes an angle of 10^{-2} radians with the [01$\bar{1}$] direction in the surface plane. A stacking fault on a ($\bar{1}$11) plane intersects the surface creating a step along the [011$\bar{1}$] direction. Find the orientation of the image of the step in the REM image formed with: (a) the (666) reflection and, (b) the (577) reflection. What lateral displacement of the image of this step is caused by a large step, 2 nm high, which is perpendicular to the incident beam in the case of: (1), a step up and, (2), a step down?

2. How are
(a) the foreshortening factor,
(b) the depth of focus and
(c) the resolution
affected by
(1) the angle of convergence of the incident beam,
(2) the angular range of the diffracted beams which are detected,
(3) the angle made by the incident beam (axial ray) with the surface,
(4) the angle made by the diffracted beam with the surface, and
(5) the refraction effect at the crystal surface,
in cases of REM (objective lens after the specimen) and SREM (objective lens before the specimen).

3. An edge dislocation in a crystal having isotropic elastic properties is perpendicular to the crystal surface. A REM image is formed with the incident beam parallel to the terminating plane of atoms. Sketch, roughly, the REM image formed with a high-order reflection from the lattice planes parallel to the surface under the assumption of: (a), 2-beam diffraction conditions, (b), many-beam diffraction conditions when the variation of diffracted intensity with azimuth angle is as in the diagram below.

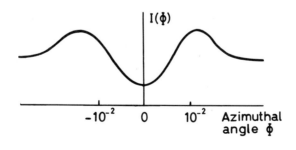

HINTS ON PROBLEM SOLUTIONS

1. The important factor affecting the image geometry is that the surface is to be viewed along the axis from the objective lens. The reflection used for REM imaging is assumed to be directed along this axis, towards the objective lens. A line on the surface appears in the image to be in a direction determined by the azimuthal angle, ϕ, and the angle of the diffracted beam with the surface, θ, as in equation (1).

2. The foreshortening factor depends only on the angle between the objective lens axis and the specimen surface.

The depth of focus depends on the objective aperture size or the angular width of the diffracted beam, whichever is smaller. The angular width of the diffracted beam may depend on the incident beam convergence (in REM) and the crystallinity of the specimen.

The resolution limit is determined by the objective aperture size.

3. (a) Under two-beam conditions $\underline{g} \cdot \underline{b} = 0$ where $\underline{g}$ is the diffraction plane vector and $\underline{b}$ is the Burgers vector. Hence the contrast is zero.

(b) Under the many-beam diffraction conditions the contrast is non-zero because the strain field of the dislocation gives an azimuthal variation of the diffracted intensity. It is necessary to map the strain field in terms of azimuthal angle change on the surface, convert the variation of azimuthal angle to variation of intensity using the given diagram and then apply the foreshortening factor to deduce the form of the image. Note that the strain field will have symmetry about the plane of incidence through the dislocation core in this case and hence the image intensity will have a plane of symmetry.

AN INTRODUCTION TO REFLECTION HIGH ENERGY ELECTRON DIFFRACTION

Peter J. Dobson

Philips Research Laboratories
Cross Oak Lane, Redhill, Surrey RH1 5HA, U.K.

INTRODUCTION

Reflection high energy electron diffraction (RHEED) uses electrons of energy from a few keV to ~100 keV directed at low angles of incidence (0.01 to 0.1 rad) with respect to a surface. RHEED is a fairly old technique. There are several examples[1,2] of its use in the 1930's and it was extensively employed for corrosion studies and epitaxial growth in the 1950's and 1960's [3-6]. However the enthusiasm which existed around 1970 for performing a complete surface structure determination using the low energy electron diffraction technique deflected interest away from RHEED. One of the reasons for this was the sensitivity of RHEED to topography, which the surface scientist took to be a serious disadvantage! The recent revival of interest in RHEED stems from its wide use in monitoring the films and layers grown by molecular beam epitaxy (MBE) particularly of the III-V semiconductors[7-9]. The technology which has developed for MBE of the III-Vs has now been applied to the growth of II-VIs, group IV, various metals and even oxides. It now seems very likely that the molecule-by-molecule or atom-by-atom approach which is central to the MBE concept can be applied to grow any natural occuring material and to grow completely new "artificial" crystal structures. This has already been achieved for atomic layer superlattices of the III-V systems and the group IV system.

Reflection high energy electron diffraction RHEED is ideally suited to any thin film deposition technique because none of the equipment intrudes into the experimental growth region (see fig.1). The electron gun and viewing screen are positioned at low angles (~5°) to the substrate surface. In this chapter we will adopt a fairly pragmatic approach, and after a brief description of the vacuum requirements, sample mounting constraints and details of a typical RHEED system, we will see how the diffraction patterns can be interpreted. The diffraction aspects of RHEED will be kept to a fairly simple level but I shall indicate where we need to adopt a more rigorous approach.

VACUUM, SAMPLE MOUNTING CONSTRAINTS

I shall assume that we wish to monitor the growth of some layers in a vacuum chamber. These layers could be made by vacuum deposition from material that is evaporated from small ovens or cells, or they could be grown by some gas reaction on a surface. The use of RHEED does place a constraint on the ambient atmosphere within the growth chamber. The vacuum must be high enough that the electrons are not scattered by the background gas molecules. In practice, for electrons of 10 keV this means a vacuum of better than 10^{-5} mbar is needed, therefore placing a constraint on the thin film deposition techniques that can be <u>continuously</u> monitored using RHEED. In practice it is only realistic to use the technique to monitor deposition in high vacuum.

Therefore, the high pressure chemical vapour phase and plasma deposition methods are not readily combined with RHEED. It has become customary to refer to high vacuum deposition as molecular beam epitaxy (MBE) although this term is often a misnomer, with many materials being deposited from atomic beams and the thin films not always being epitaxial. The term MBE gained acceptance through its wide use to grow epitaxial III-V layers from molecular beams (dimer or tetramer) of group V elements. There has recently been a trend towards supplying the group III component in the form of metallo-organic molecules and the group V component as the hydride (e.g. arsenic as arsine AsH_3). This development is known variously as metallo-organic molecular beam epitaxy (MOMBE) or chemical beam epitaxy (CBE)[10,11].

All high vacuum deposition techniques require certain pre-requisites to be met:
a) The vacuum must be the highest that can be achieved in order that molecules from the residual atmosphere are not incorporated into the thin films.
b) All sources of gas molecules from internal surfaces should be minimised.
c) The material used for construction of the vacuum chamber, substrate holders, furnaces, shutters, etc. must be chosen carefully to avoid contamination of the growing layers by diffusion or vapour transport.

In practice, it has become customary to construct the vacuum chambers from stainless steel which has been subjected to special cleaning procedures after fabrication. These procedures involve chemical or electrochemical removal of machining and welding debris and often proprietary methods to reduce the internal surface outgassing rate (e.g. glass bead shot peaning). Non-porous materials are used for internal components and care exercised in the choice of refractory metals for use in hot regions, e.g. evaporator furnaces, substrate heaters. In short, all of the requirements of good ultra high vacuum practice must be employed. The

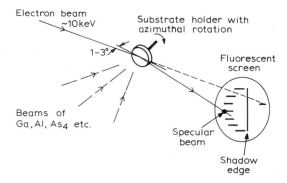

Fig. 1. Schematic diagrams of the application of RHEED to MBE or any thin film growth technique.

choice of the type of vacuum pump depends on the potential gas loading. Ion pumps and cryopumps can be used in situations where the gas load is small, but liquid nitrogen trapped diffusion pumps or turbomolecular pumps are needed when the gas load is likely to be high. It is now customary to surround the deposition chamber with liquid nitrogen cooled surfaces to trap gas molecules and minimise the possibility of them reaching the substrate.

The need for the highest possible vacuum can be seen by recalling that from the kinetic theory of gases, the arrival rate of molecules at a surface is [see problem 1a)]:

$$\nu = 0.25 \ n \ v_a \tag{1}$$

where n is the number density of molecules at pressure p and temperature T

$$n = p/kT \tag{2}$$

and v_a is the average velocity of the molecules of mass M

$$v_a = 2\sqrt{(2kT/\pi M)} \tag{3}$$

with k = Boltzmann's constant. Thus:

$$\nu = p/\sqrt{(2\pi MkT)} \tag{4}$$

For example, if we substitute numbers for nitrogen (M = $28 \cdot 1.66 \cdot 10^{-27}$ kg) at 300 K and a pressure of 10^{-6} torr = $1.33 \cdot 10^{-6}$ mbar = $1.33 \cdot 10^{-4}$ Nm^{-2}, we find from eq. (4) that $\nu = 3.8 \cdot 10^{18}$ m^{-2} s^{-1}. This corresponds to approximately half a monolayer a second. These numbers are representative of many typical demountable glass bell jar vacuum deposition systems and it is immediately apparent that there is a potentially high risk of incorporation of gas molecules in a growing thin film. Even an improvement in the pressure by four orders of magnitude to 10^{-10} mbar will give a gas bombardment rate of 10^{-4} monolayer per second. Incorporation of these molecules into the growing layer depends on other factors such as the sticking coefficient and reactivity of the gas molecules with the material in the deposit.

A few words are necessary regarding techniques for cleaning the surface of materials to be studied or the substrates which are used for epitaxial growth. Obviously, care should be taken to avoid unnecessary contamination prior to insertion in the vacuum chamber. Subsequently, there is a wide choice of techniques available and the method adopted depends on the actual material and purpose behind the experiment or film deposition. For small scale surface science experiments on certain crystal planes of brittle or easily cleaned material, cleavage in situ in the vacuum chamber is convenient. For large area substrates such as those encountered in the MBE of semiconductors a carefully chosen chemical pre-treatment designed to leave a thin uniform layer of protective oxide is employed. The substrate is then heated in vacuum to remove the oxide.

In the case of III-V material the heating is done in a beam of arsenic molecules in order to maintain a stoichiometric composition. In certain cases, particularly for metals, it is necessary to ion bombard with neon or argon to remove oxide or carbonaceous products. Argon ion beams of 500-5000 eV are employed with current densities in the range 1 to 200 $\mu A \cdot cm^{-2}$. Crystallographic damage can be minimised by employing low angles of incidence and by choosing the ion mass to be approximately equal to the mass of the atom to be removed. Damage is annealed out by thermal treatment, taking care not to overheat the sample with the attendant out-diffusion of trace impurities. Sometimes it may be necessary to resort to in situ oxidation/reduction cycles by heating the material alternately in a low pressure ambient of oxygen or hydrogen.

Modern analysis and growth equipment is fitted with a vacuum sample load-lock stage which enables specimens or substrates to be inserted into the UHV chamber via a small bakeable intermediate vacuum system. The purpose of this is to eliminate the need to expose all of the walls of the UHV chamber to atmospheric contamination each time a sample is changed. This greatly reduces the pumping time and enables a continuous UHV processing of epitaxial material to be achieved. Many of the design principles of vacuum systems and surface cleaning procedures are covered extensively in the literature[12-15].

THE SAMPLING DEPTH IN RHEED

Electrons are strongly scattered by matter and we should distinguish between elastic and inelastic scattering. There are many ways of describing the magnitudes of elastic and inelastic scattering but conceptually it is useful to approach the problem by considering the mean free paths involved. The mean free path Λ_i is related to the atomic scattering cross section σ_i by

$$\Lambda_i = 1/(N\sigma_i) \tag{5}$$

where N is the number density of atoms in unit volume of the solid and the suffix i is used to distinguish different scattering processes. Some books also introduce other scattering cross sections, for example Heidenreich[16] uses both the cross section per unit mass S and the volume cross section Q. These are related to σ via

$$S_i = N\sigma_i/\rho \tag{6}$$

and

$$Q_i = N\sigma_i \tag{7}$$

where ρ is the density of the material. Now, the elastic scattering cross section σ_e is related to the atomic scattering factor f(s) and the scattering angle β by

$$\sigma_e = \int_0^\pi |f(s)|^2 \sin\beta \, d\beta \tag{8}$$

where the scattering parameter $s = (4\pi/\lambda)\cdot\sin(\beta/2)$ and λ is the de Broglie wavelength of the electrons. Various approaches have been made to evaluate eq. (8) and these are summarized by Heidenreich[16]. A useful approximation is given by Humphreys[17]

$$\sigma_e = A\cdot\lambda^2\cdot Z^{4/3}/[\pi(1 - v^2/c^2)] \tag{9}$$

where v is the incident electron velocity, c is the velocity of light and Z is the atomic number. A is a fitting parameter which depends on the approximations to the atom wavefunctions which are used and it lies in the range from 1 to 2 with 1.8 being appropriate for a Thomas-Fermi model. If we take aluminium at an energy of 10 keV this will give a cross section of $9\cdot10^{-17}$ cm^2 which is equivalent to a total elastic scattering mean free path of 200 Å. In fig.2, values of the mean free path derived from equations (5) and (9) are compared with the quasi-elastic, thermal diffuse scattering (or phonon) mean free path Λ_{TDS} derived for Al by Howie[18]. Λ_{TDS} does not include the coherent (Bragg) contribution to the total elastic scattering.

The inelastic mean free path results from a combination of scattering processes. The total inelastic cross section σ_{inel} is made up from contributions from single electron scattering σ_{se}, plasmon scattering σ_{pl} and core level excitation σ_c.

$$\sigma_{inel} = \sigma_{pl} + \sigma_{se} + \sigma_c \qquad (10)$$

$$(\Lambda_{inel})^{-1} = (\Lambda_{pl})^{-1} + (\Lambda_{se})^{-1} + (\Lambda_c)^{-1} \qquad (11)$$

The most important contribution is that due to plasmon scattering. The problem has been discussed thoroughly by Ritchie and Howie[19] (see also the chapter by Howie on Localised Surface Imaging and Spectroscopy in the STEM).

Seah and Dench[20] have compiled lists of inelastic mean free path data for application to escape depths in different electron spectroscopies. They suggest that for medium/high energies there is a reasonable fit to the empirical relation

$$\Lambda_{inel} = 0.41 \cdot a^{3/2} \cdot E^{1/2} \qquad (12)$$

where Λ is in nm, E is the primary electron energy in eV and a is related to the atomic size (in nm) given by

$$a = (\rho/A \cdot M_n)^{1/3} \qquad (13)$$

where ρ is the density, A is the atomic weight and M_n is the nucleon mass. This gives rather low estimates of inelastic mean free paths, e.g.: for Al at 10 keV it gives $\Lambda_{inel} \simeq 55$ Å. In fig.2 we show the theoretical prediction of a model due to Penn[21] which also fits quite well with the compilation of data by Seah and Dench. This model predicts Λ_{inel} for Al to be 100 Å at 10 keV whereas Howie's[18] estimate for inelastic scattering is

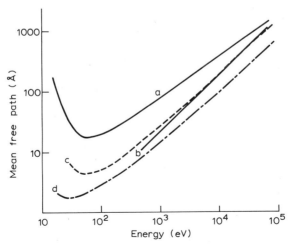

Fig. 2. Examples of various mean free paths for Aluminium. a) The quasi-
elastic or thermal diffuse scattering mean free path taken from
ref.18. b) The elastic mean free path derived from equations 5 and
9 in the text. c) The inelastic mean free path taken from ref.18.
d) Theoretical prediction of Penn for inelastic mean free path.

≈200 Å. Most of the discrepancy here probably arises because Howie does not include the inelastic scattering at small angles which would still be accepted by a typical electron microscope aperture or contribute to the intensity measured in a RHEED experiment where energy filtering is not employed.

How then, does this affect the depth sampled by RHEED? If we treat the problem simply as one of the passage of electrons through material at a low incident angle, then for an angle of 20 mrad inside the solid, after traversing 200 Å the beam will have reduced to 1/e of its initial intensity and this would occur at 3.4 Å depth. Clearly most of the (elastic) diffracted intensity will come from the first atomic layer or two. The situation is, of course, much more complicated since the amplitude scattered by the top layer of atoms does depend on multiple scattering from the other layers and these simple concepts of mean free paths should not be taken too precisely. Furthermore, in RHEED, excitation of surface plasmons from outside the crystal will significantly reduce the incident beam amplitude and that has not been allowed for in the above treatment.

However, the main points to remember are that the mean free paths at energies used in RHEED are relatively small and they have a dependence on energy as shown in fig.2. A more rigorous multiple scattering dynamical treatment would give similar conclusions and Britze and Meyer-Ehmsen[22] have shown that the effective penetration depth varies with azimuthal angle in silicon, giving a range of values 5 to 10 Å at 10 keV and 30 mrad incidence angle. We can now use this idea of very limited beam penetration to develop a simple model of the Ewald sphere and reciprocal lattice construction to explain many RHEED patterns.

LIMITED PENETRATION MODEL OF RHEED

If we start from a familiar description of the scattering from a small regular crystal we can see the effect of limited penetration by artificially restricting one dimension of the crystal. Taking a crystal with an orthorhombic shape and with dimensions $N_a a$ x $N_b b$ x $N_c c$ where the crystal cell is defined by a x b x c and N_a, N_b, N_c are the numbers of units along the orthogonal directions x, y, z, the structure factor can be written (replacing the usual sum by continuous integral):

$$F = f \int_{crystal} \exp[2\pi i(K_x x + K_y y + K_z z)]dv \qquad (14)$$

where K_x, K_y, K_z are the magnitudes of the components of the wavevector K along the x, y and z directions.

$$\underline{K} = \underline{k}_s - \underline{k}_i \qquad (15)$$

and $\underline{k}_s$ is the scattered wavevector and $\underline{k}_i$ is the incident wavevector.

We can re-write equation (14)

$$F = f \int_{-aN_a/2}^{aN_a/2} \exp(2\pi i K_x x) \ dx \int_{-bN_b/2}^{bN_b/2} \exp(2\pi i K_y y) \ dy \int_{-cN_c/2}^{cN_c/2} \exp(2\pi i K_z z) \ dz \qquad (16)$$

Now each integral term has a maximum if

$$K_x a = h \ ; \quad K_y b = k \ ; \quad K_z c = l \qquad (17)$$

These are the Laue conditions for strong Bragg diffraction to occur, i.e.

when $\underline{K}$ coincides with the reciprocal lattice vector $\underline{g}$

$$\underline{K} = \underline{g} = h\underline{a}^* + k\underline{b}^* + l\underline{c}^* \tag{18}$$

where the asteriscs denote reciprocal lattice vectors. The intensity of the diffraction is given by $I = FF^*$ where F^* denotes the complex conjugate of F and then,

$$I = |f|^2 \left[\frac{\sin^2 N_a K_x(a/2)}{\sin^2 K_x(a/2)} \cdot \frac{\sin^2 N_b K_y(b/2)}{\sin^2 K_y(b/2)} \cdot \frac{\sin^2 N_c K_z(c/2)}{\sin^2 K_z(c/2)} \right] \tag{19}$$

Strictly speaking this result only follows when F is evaluated by exact summation of the atomic contributions rather than by the integral approximation of eq.(14). Each term here is the familiar interference function and they give the effective "size" and "shape" of each reciprocal lattice point. If any of the N's is reduced, the interference function is broadened in that particular direction. Now we can see the effect of limited penetration by saying that we effectively only have a few crystal planes parallel to the surface which can contribute to diffraction. In the limit of $N_c \rightarrow 1$ then the set of reciprocal lattice "points" is replaced by a set of reciprocal lattice "rods" as shown in the fig.3. This approximation can be thought of as relaxation of the third Laue condition. If we were to consider the diffraction by a long linear chain of atoms on a surface we would relax the second Laue condition for the direction normal to the chain lying in the surface and thus generate a reciprocal lattice "sheet".

This representation is simply that of taking the Fourier shape transform of our diffracting object.

METHODS FOR CALCULATING THE "SIMPLE" RECIPROCAL LATTICE

There are several different but self-consistent ways of estimating the quasi two-dimensional reciprocal lattice and each one has its merits for different circumstances. We can either take the bulk 3-D reciprocal lattice and relax the appropriate Laue condition, which is rather like smearing out the reciprocal lattice points into rods along the direction normal to the surface, or we can estimate the two dimensional reciprocal lattice rod arrangement from a model of the 2-D lattice. If the atoms are all identical this procedure is very simple and can merely be taken as an extension of the 3-D case, i.e.:

Inter-rod spacing = (Inter-row atom spacing)$^{-1}$.

The equivalence of the two approaches is illustrated by estimating the

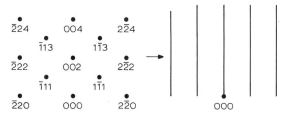

Fig. 3. The replacement of reciprocal lattice "points" by "rods" in going from three dimensions to two dimensions. The indexing is given for the [110] azimuth of an (001) surface of a face-centred cubic lattice

reciprocal lattice appropriate for a face-centred (001) surface. If we take a model of the surface as in fig.4a, we identify some characteristic inter-row spacings for well defined low index azimuths. We can then use these to construct a plan "end on" view of the reciprocal lattice rods as in fig.4b. We generate exactly the same inter rod spacings by taking the 3-D reciprocal lattice sections for the approximate azimuths and relaxing the third Laue condition. Sections given in the appendix of Hirsch et al[23] are particularly useful. Here we can note some merits of each approach. In the first instance it is easy to work backwards from a diffraction pattern to a hard ball model of a surface, although one will seldom reach an unambiguous model. In the second case, some three dimensionality is retained and relationships with the bulk crystal directions can be recognized.

This is especially important if some degree of penetration of the beam occurs such as on a rough or undulating surface, or if thermal diffuse scattering effects enhance the original hkl reciprocal lattice features. An example of this occured in a recent study of the $LiNbO_3$ surface[24]. However for cases where more than one atom type is present, we have to estimate 2-D structure factor for the surface lattice. Again we use a very similar approach to the 3-D case but only use two Miller indices

$$F = \sum_{\substack{\text{surface cell}}} f_n \exp[2\pi i(Hx_n + Ky_n)] \tag{20}$$

where x_n, y_n define the atom positions in the unit cell, H, K are the 2-D Miller indices defined according to Wood's[25] notation, and f_n are the

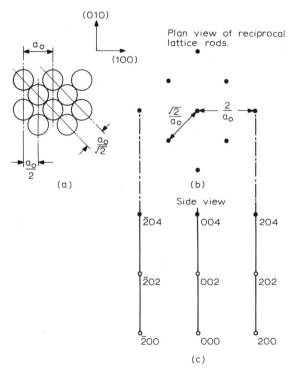

(a)

(b)

(c)

Fig. 4. a) Shows the plan view of the top layer of a face-centred cubic (001) lattice. b) Shows the plan view of the reciprocal lattice rods from this surface. c) Shows the side view with the bulk or three-dimensional reciprocal lattice points indicated on the "rods".

atomic scattering factors. A warning is necesssary at this point. H and K are not to be equated with the hkl used in the 3-D notation. The reader should consult Wood[25] for details of the 2-D notation, here it suffices to say that a cell defined by m x n has sides which are m and n times respectively the size of those of the primitive surface unit cell. We can now illlustrate the use of a 2-D structure factor calculation, taking as an example the (001) surface of a sodium chloride type of lattice with atomic scattering factors f_A and f_B and define our primitive cell as in the fig.5a. Using equation (20):

$$F_{HK} = f_A\{1 + \exp[2\pi i(H \cdot 1 + K \cdot 1)]\} +$$

$$+ f_B\{\exp[2\pi i(H \cdot 1 + K \cdot 0) + \exp[2\pi i(H \cdot 0 + K \cdot 1)]\}$$

Now for the integral order beams:

$$F_{01} = F_{10} = F_{11} = f_A(1 + 1) + f_B(1 + 1) = 2f_A + 2f_B.$$

For the half order beams:

$$F_{0\ 1/2} = F_{1/2\ 0} = f_A(1 - 1) + f_B(1 + 1) = 0$$
$$F_{1/2\ 1/2} = f_A(1 + 1) + f_B(-1 - 1) = 2f_A - 2f_B$$

and generally all beams with $F_{m/2\ n/2}$ where m, n are odd integers will have this structure factor. The resultant reciprocal lattice is in fig.5b with these m/2,n/2 points shown. For practising, see Problems 2a), b), c).

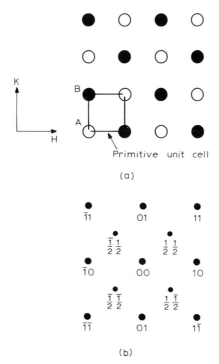

Primitive unit cell

(a)

(b)

Fig. 5. a) Schematic diagram of a NaCl type lattice with atoms A, B. The two-dimensional indices H, K defined as shown. b) The reciprocal lattice of this surface. The larger "points" indicate integral order beams with structure factor $2(f_A + f_B)$ and the smaller points are for the fractional order beams with structure factor $2(f_A - f_B)$.

Having established the rules for deriving the reciprocal lattice appropriate for RHEED in the approximation of limited penetration we now proceed to consider the kinematic Ewald construction. This is shown in fig.6. It will be seen that we expect to see <u>points</u> of intersection of the Ewald sphere with reciprocal lattice rods and these will lie on an arc known as a Laue zone. At first sight this seems contrary to experience, since many RHEED patterns display streaks and many authors have stated that long streaks were indicative of "well ordered flat surfaces". This is not true! The streaks that are observed in practice have several origins which we will discuss, but here we emphasise that a perfectly ordered flat surface will show spots lying on an arc. There are several departures from this ideal situation. Firstly, atoms are in motion, i.e. lattice vibrations or phonons have an effect on the diffraction condition. Secondly, the electron beam is not necessarily perfectly parallel or monoenergetic. Thirdly, the crystal may depart from perfection in being bent or mosaic structured or the surface may show particular types of disorder. Finally we should also consider what patterns we expect from steps and highly disordered layers.

Scattering by Phonons

In all diffraction problems the scattering by phonons plays an important role. As the temperature is increased, atoms vibrate more strongly about the equilibrium lattice sites and the net intensity of diffracted radiation (electrons, X-rays or neutrons) in a particular Bragg reflection is reduced by an amount given by the Debye-Waller factor $\exp(-2M)$, where:

$$M = [8\pi^2 \langle u^2 \rangle \sin^2(\beta/2)]/\lambda^2 \qquad (21)$$

and the mean square atomic displacement $\langle u^2 \rangle$

$$\langle u^2 \rangle = (4.364/A\theta)[\Phi(x)/x + 1/4] \qquad (22)$$

where A is the atomic weight, θ is the Debye temperature and $\Phi(x)$ is the Debye function defined by

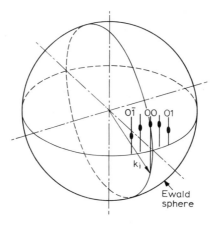

Fig. 6. Ewald construction showing intersection of the sphere of radius $k = 1/\lambda$ with reciprocal lattice rods. Note that these points of intersection lie on a Laue zone. Only the lowest order Laue zone is shown.

$$\Phi(x) = (1/x) \int_0^X [y/(\exp y - 1)] dy \quad \text{and} \quad x = \theta/T$$

(see James[26] and Guinier[27] for more details). Now, the "missing" intensity from the Bragg reflections is not uniformly scattered into the background, but is concentrated close to the reciprocal lattice points. This is known as thermal diffuse scattering and we can think of this as resulting from the participation of a phonon to conserve momentum

$$\underline{k}_s = \underline{k}_i \pm \underline{q} + \underline{g} \tag{23}$$

where $\underline{k}_s$ is the scattered wavevector, $\underline{k}_i$ is the incident wavevector, $\underline{g}$ is a reciprocal lattice vector and $\underline{q}$ is a phonon wavevector. This is fully described for X-ray diffraction by James and simple treatments for electron diffraction can be found in books by Heidenreich[16] and Hirsch et al[23]. The Ewald construction for the three-dimensional case is shown in fig.7a. In the quasi two-dimensional situation of RHEED the situation is less clear since the vector $\underline{g}$ is no longer precisely defined. In this case many more possibilities exist to conserve momentum and energy as illustrated in fig.7b. This is probably consistent with the explanation of streaking in RHEED which was offered by Beeby[28] and Holloway[29]. From fig.7b we can see that a wide spread of scattered beams in the plane of incidence may result.

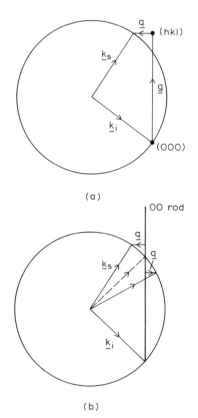

(a)

(b)

Fig. 7. Ewald constructions to indicate scattering by phonons.   a) The three-dimensional case associated with the condition close to the Bragg reflection (hkl).   b) The two-dimensional case with the phonon wavevector $\underline{q}$ providing momentum gain or loss. The dashed line represents the scattered wavevector for the non-phonon loss specular beam. The angles have been exaggerated for clarity.

The Ewald construction can be used to estimate the geometric extent of streak extension in RHEED. Crystallographic effects can be incorporated into the reciprocal lattice and electron beam energy spread and angular divergence or convergence can be incorporated in the incident and scattered wavevectors. Energy spread of the electron beam will not affect the angular spread of the true specular beam as is obvious from fig.8a. The effect of beam convergence or divergence can be estimated from fig.8b and the equivalent diagrams for non-specular beams. Generally we can say that for the specular beam the angular spread of the streaking should reflect the angular spread of the incident beam and with care this should not exceed 2 mrad and can be neglected.

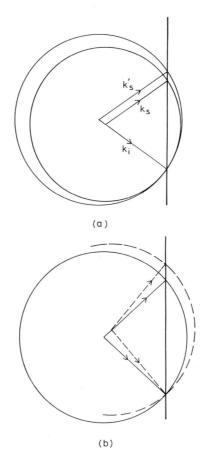

(a)

(b)

Fig. 8. a) The effect of energy spread can be estimated from this construction, where

$$k_s = \frac{(2mE)^{1/2}}{h} \quad \text{and} \quad k'_s = \frac{[2m(E + \Delta E)]^{1/2}}{h}$$

Note that k_s and k'_s are parallel, i.e. there is no angular spread for the specular beam. b) The effect of divergence or convergence can be seen from such a construction. For the specular beam, the spread in scattered angles is of course equal to the spread of incident angle.

STREAKS FROM CRYSTALLINE FACTORS

There are two major causes of streaking and they are easily recognized
and distinguished. The first is essentially a three dimensional effect,
and results from having a physically bent or mosaic crystal. The Ewald
construction is shown in fig.9a. It is often the case that the mosaic is
isotropic, resulting in a diffuse "fan" of reciprocal lattice rods and the
streaking shows a broadening at higher scattering angles.

In some cases, such as when thin samples are used, e.g. mica, the
bending or deformation is predominant only in one direction and this can
usually be recognised in the RHEED patterns. The second major
crystallographic cause of streaking is the existence of one dimensional
disorder in the surface. Any one-dimensional feature such as a row of
atoms, crystallographic step or anti-phase boundary will give a sheet in
reciprocal space. The Ewald construction is illustrated in fig.10 along
with an example. This has been well documented in the literature. The
origin of such one-dimensional disorder has received very little attention
and it is likely that more determined assessment along the lines suggested
by Welberry[30] for X-ray diffraction in bulk material will be worthwhile.
Here we should note that Welberry gives several two-dimensional models and
Fourier transforms of one-dimensional disorder which are directly
applicable to RHEED from surfaces.

The effect of steps and facets

This is best dealt with along the lines already well established for
LEED, in particular by Henzler[31]. In fig.11 we show the effects of steps
on the reciprocal lattice construction. Good examples of their effects can
be seen in the literature[32-34]. The idea underlying the explanation of

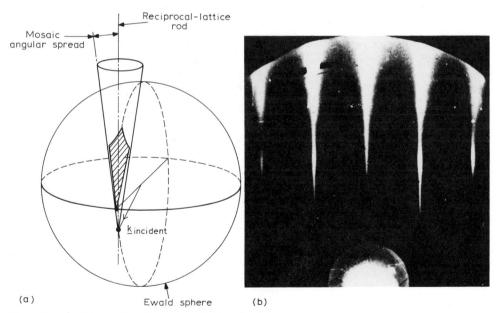

(a)

(b)

Fig. 9. a) Schematic diagram of the Ewald construction for a bent or mosaic
crystal with variation of angles ± α about the normal. The shaded
region represents all possible 00 reciprocal lattice rods. b) An
example of fan shaped streaks from an epitaxial (111) layer of gold
grown on a mosaic crystal of tungsten (110).

171

these features is that facets have associated with them, their own set of reciprocal lattice rods which are normal to the facet surface. Similarly, for steps, one should think of the reciprocal lattice as being a convolution of the rod system associated with the (low index) terrace with the rod system of the real geometric surface.

REFRACTION IN RHEED

When an electron beam encounters the surface it experiences an abrupt change in potential. This results in a change in the de Broglie wavelength of the electron when it enters the solid, and consequently there is a refraction effect. This effect is quite small but significant at low angles of incidence. Essentially, this is Snell's law applied to electron beams and it can be dealt with conveniently using the Ewald construction as illustrated in fig.12. The radius of the Ewald sphere outside the surface is given by:

$$|\underline{k}_o| = (2mE)^{1/2}/h$$

where m is the free electron mass, h is Planck's constant and E is the energy of the electron with respect to the vacuum level (see fig.14). Inside the solid the energy is now $E - V_o$ where V_o is the inner potential and the radius of the Ewald sphere is:

$$|\underline{k_i}| = [2m(E - V_o)]^{1/2}/h \qquad (24)$$

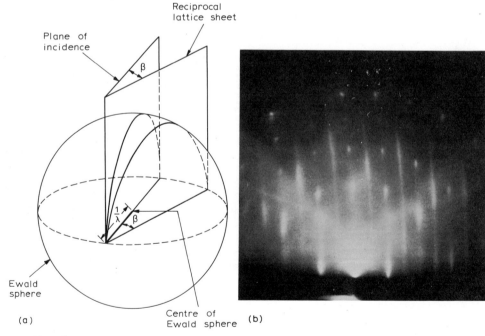

Fig.10. a) Reciprocal lattice – Ewald sphere construction showing the intersection of a reciprocal lattice sheet at an angle of β with respect to the incident plane. This will give rise to a curved streak in the RHEED pattern. b) An example of curved streaks from the [010] azimuth of a GaAs (001) 2×4 reconstructed surface.

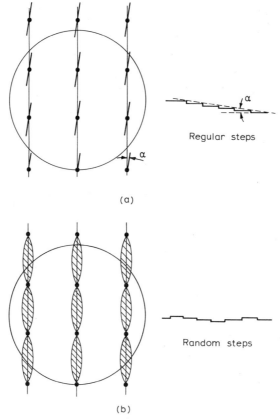

Regular steps

(a)

Random steps

(b)

Fig.11. These diagrams show the projection of the Ewald sphere on a section
of reciprocal space, i.e. the Laue circle for different step
configurations. a) Regular steps. Each reciprocal lattice point
has associated with it a rod inclined at an angle α representing
the angle between the surface and the crystal plane or terrace.
b) Random steps. There is now a continuum of possible rods through
each point, represented by the shaded regions.

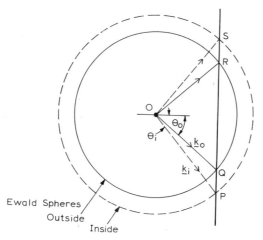

Fig.12. Schematic diagram of the Ewald sphere construction to show the
effect of refraction.

From fig.13 it will be seen that V_0 is negative and consequently

$$|\underline{k_i}| > |\underline{k_o}|.$$

The refraction effect can be estimated by wave-matching via the "rod" PQRS. If we allow the beam to penetrate the surface we are shifting our origin of reciprocal space from Q to P and this latter point is the origin of three-dimensional reciprocal space. The specular beam is OS inside the solid and is refracted to OR outside. From this construction we can see that

$$|\underline{k_o}|\cos\theta_O = |\underline{k_i}|\cos\theta_i \qquad (25)$$

$$\cos\theta_O = (1 - V_0/E)^{1/2} \cdot \cos\theta_i \ . \qquad (26)$$

This can be seen to be very important in RHEED [see Problem 1c)] if we estimate the angle of incidence of a 10 keV beam to excite a Bragg reflection which occurs <u>inside</u> a crystal. If we take as an example the case of GaAs (001), to excite the 004 beam, the angle inside the crystal is θ_i where

$$\theta_i = \sin^{-1}(g_{004}/2k_i), \quad g_{004} = (0^2 + 0^2 + 4^2)^{1/2}/a_0 \quad \text{and} \quad a_0 = 5.65 \ \text{Å};$$

for 10 keV electrons

$$|\underline{k_i}| = [2m(10\ 014.5)]^{1/2}/h$$

having assumed a value of 14.5 eV for the inner potential. Hence $\theta_i = 2.47°$. Now, using the relationship (26) between θ_O and θ_i, $\theta_O = 1.16°$. Note that this is a large effect! This estimation of refraction is therefore very important for the calculation of three-dimensional bulk diffraction conditions in the specular and non-specular beams. Use can be made of this effect to estimate the inner potential, and an example of this for LiNbO$_3$ was recently published by Petrucci et al[24]. The inner potential V_0 is the mean potential, i.e. V_{000} which is obtained from a Fourier expansion of the real crystal potential

$$V(r) = \sum_{g} V_g \exp(2\pi i\underline{g}\cdot\underline{r}) \qquad (27)$$

where $\underline{g}$ is a reciprocal lattice vector. It can be obtained from a knowledge of the crystal cell and the atomic form factors. A particularly simple and apparently reliable method for estimation of V_0 is outlined by Vainshtein[35].

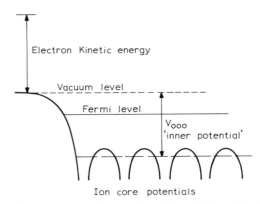

Fig.13. Schematic diagram to illustrate the change in energy an electron experiences when it enters a solid.

The simple view of refraction given here is not strictly correct, especially when more than one diffracted beam is excited. Under these conditions we have to take dynamical interactions between diffracted beams into account and introduce other Fourier coefficients into the problem. This was discussed in the context of LEED by Stern and Gervais[36] and their Ewald construction/Dispersion surface diagrams can be applied directly to RHEED. An early account of inner potential was given by Miyake[37].

KIKUCHI LINES

In the background of RHEED patterns from well ordered crystals there is a regular pattern of lines of intensity. These are known as Kikuchi lines, and they result from the diffraction of electrons which have been incoherently scattered on entering the solid. This lack of coherence can be due to loss of energy to phonon, plasmon or interband excitations, or scattering from steps, etc. on the surface. The "incoherent" electrons are diffracted by the crystal planes and give a pattern consisting of lines. The geometry of such patterns has been treated in detail in the literature (see for example Hirsch et al[23]), and here we give a simple account based on the Ewald construction. If we start by ignoring refraction effects and treat the electrons as an inner source of isotropically distributed electrons within the crystal, then for a particular Bragg condition (hkl) to be satisfied the Ewald sphere must always touch (000) and (hkl). The locus of the centre of the Ewald sphere will therefore project onto a line which is the perpendicular bisector of the vector defining hkl.

For any particular azimuthal direction the procedure to be adopted is to first of all plot the reciprocal lattice points hkl in that projection (these are conveniently given in the Appendix of Hirsch et al[23]), and then construct lines which are the perpendicular bisectors of these points. This gives the Kikuchi line geometry. Refraction effects can be allowed for by application of equation (26). An example of this is given for the [110] direction of GaAs (001) in fig.14 which shows the unrefracted and refracted lines for a beam of 12.5 keV and inner potential of 14.5 eV.

There are some occasions when steps are present on a surface such that both the refracted and unrefracted Kikuchi lines can be seen. This is illustrated schematically in fig.15. Strictly speaking, refraction must occur for both the beams emerging via the step and the terrace, however the angle of deviation for the former is negligible. An example of this for the horizontal (001) Kikuchi lines from an (001) cleavage surface of MgO is shown in fig.16.

SURFACE RESONANCES

This term can be rather confusing, since it has been used to describe different phenomena in surface science. Here we use the term to describe beams which travel parallel to the surface and we distinguish two cases. The reader is referred to an excellent account by Miyake and Hayakawa[38] for more detail.

Beam threshold resonance

Consider the situation where the angle of incidence is adjusted to just excite a particular HK surface beam, i.e. the Ewald sphere is tangential to the HK reciprocal lattice rods, fig.17. When this condition is approached there will be a marked change in the intensity of the specular beam since dynamic coupling between the new HK beams and the 00 beams will be "switched on". The occurrence of this condition is also marked by an effect analogous to Kikuchi lines. Since the beam threshold condition is when the Ewald sphere is tangential to a particular HK rod, and it passes through the origin of reciprocal space, the locus of the sphere describes a parabola (see Ichimiya et al[39]). These parabolae are a common feature in the background and their origin is similar to that of Kikuchi lines, viz: they result from incoherently scattered electrons which satisfy this beam threshold surface resonance condition.

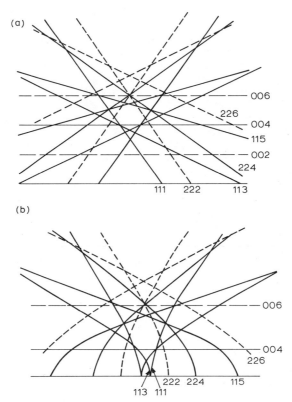

Fig.14. a) Kikuchi pattern for the <110> azimuth of GaAs (001) with no refraction correction. The dashed lines refer to weakly allowed Bragg reflections. b) The same Kikuchi pattern but with refraction estimated for a 12.5 keV energy and inner potential of 14.5 eV.

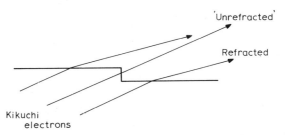

Fig.15. Schematic diagram showing that Kikuchi electrons emerging via a terrace are refracted strongly and the angle of emergence is greatly reduced in comparison with those which emerge via steps.

Trapped beam resonance

The strong refractive effect of the inner potential has important implications for any Bragg beams which are excited at low angles to the surface. It becomes quite likely that such beams are internally reflected and cannot emerge. The only external manifestation of their existence is via dynamic coupling between these beams and other beams which do emerge. One example of this might be for the simultaneous excitation of the $22\bar{4}$ and $22\bar{2}$ beams along the [110] azimuth on a GaAs (001) surface. In fig.18 we show the projection along the beam direction for this condition. The circle is the projection of the Ewald sphere <u>inside</u> the solid. Now, the $22\bar{4}$ beams only make an angle of 1.1° (at $12.\overline{5 \text{ keV}}$) with respect to the

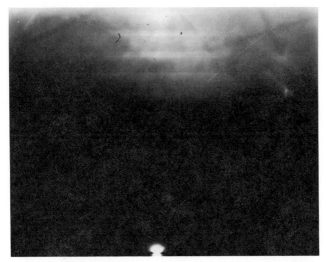

Fig.16. Horizontal [00ℓ] Kikuchi lines from a (001) cleavage surface of MgO. Note that the lines occur in pairs. The upper line of the pair has emerged via a step and the lower, more refracted line has emerged via a terrace.

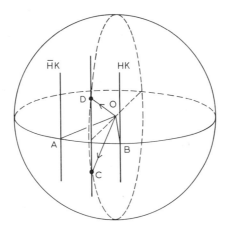

Fig.17. Schematic diagram of the Ewald sphere - reciprocal lattice construction for a beam emergence surface resonance condition. The Ewald sphere is just tangential to the reciprocal lattice rods HK at A and B. The incident beam is along OC, the specular is OD and the surface beams are OA and OB.

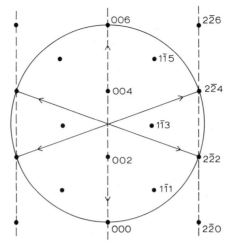

Fig.18. The trapped beam resonance condition for the <110> directions on the GaAs (001) surface. The circle is the projection of the Ewald sphere on the lowest order Laue zone. Note that six beams are excited.

surface and therefore cannot escape. This is a particularly unusual condition since for the GaAs structure the 222 and 006 which are simultaneously excited are all very weak beams because of the structure factor. The angle of incidence for this condition is given by

$$\sin\theta_i = g_{006}/2|\underline{k}_i| \qquad \text{so that} \qquad \theta_i = 3.31° \qquad \text{for } E_p = 12.5 \text{ keV.}$$

If we now use equation (26) to account for refraction we obtain $\theta_0 = 2.77°$ for $V_0 = 14.5$ eV. The specular beam does show a maximum at around this angle (see Larsen et al[40]). However, with so many beams excited, one should be cautious about using a simple refraction correction involving V_{000}, the first term in the Fourier expansion of the crystal potential.

THEORETICAL INTENSITY CALCULATIONS

In comparison with LEED there are very few RHEED intensity calculations. The early RHEED calculations were based on Bethe's n beam dynamical diffraction theory[22,41]. Recent calculations have been based on 2-dimensional Fourier expansion methods[42,46]. Maksym has recently summarised the ideas behind this approach, so they will not be dealt with here and the reader is referred to Maksym's recent review[47] for more details.

FINAL REMARKS

It is not possible in a short introductory article to do full justice to any subject. The intentions here were to provide a users' introductory guide to the subject and inevitably much has been omitted. Amongst the more important omissions are any detailed discussion of inelastic scattering, and the use of RHEED for microscopic image formation. The former subject has not been covered in a systematic way in any review articles or, to author's knowledge in any recent publications. Suffice to say here that almost all RHEED studies report the diffraction patterns or intensities which <u>include</u> inelastically scattered electrons. Determination of the significance of this, especially for detailed surface structure work remains a problem for the future. The subject of reflection high energy electron microscope imaging has taken giant steps forward in recent years and the reader is referred to the chapter by Cowley in this volume for details.

REFERENCES

1. G.P. Thomson and W. Cochrane, "Theory and Practice of Electron Diffraction", Macmillan, London (1939).
2. G.I. Finch and H. Wilman, Ergebn. Exakt. Naturw., 16:354 (1936).
3. H. Raether, Ergebn. Exakt. Naturw. 24:54 (1951).
4. E. Bauer, "Elektronenbeugung - Theorie, Praxis und Industrielle Anwendungen", Verlag Moderne Industrie, Munchen (1958).
5. Z.G. Pinsker, "Electron Diffraction", Butterworths, London (1953).
6. E. Bauer, "Techniques of Metals Research", Ed. R.F. Bunshah, Interscience - Wiley, New York (1969), Vol.2, Chapter 15.
7. A.Y. Cho and J.R. Arthur, Progr. Solid State Chem. 10:157 (1975).
8. R. Ludeke, IBM J. Research and Development, 22:304 (1978).
9. B.A. Joyce, J.H. Neave, P.J. Dobson and P.K. Larsen, Phys. Rev. B29:814 (1984).
10. W.T. Tsang, Appl. Phys. Lett. 45:1234 (1984).
11. W.T. Tsang, Journal Electronic Materials, 15:235 (1986).
12. J.C. Rivière, "Practical Surface Analysis", Eds D. Briggs and M.P. Seah, Wiley, London (1983), p.17.
13. R.W. Roberts and T.A. Vanderslice, "Ultrahigh Vacuum and its Applications", Prentice-Hall, New Jersey (1963).
14. W. Espe, "Materials of High Vacuum Technology", Pergammon Press, Oxford (1966).
15. P.A. Redhead, J.P. Hobson and E.V. Kornelsen, "The Physical Basis of Ultrahigh Vacuum", Chapman and Hall, London (1968).
16. R.D. Heidenreich, "Fundamentals of Transmission Electron Microscopy", Interscience, New York (1964).
17. C.J. Humphreys, Rep. Prog. Phys. 42:1825 (1979).
18. A. Howie, in: "Electron Diffraction (1927-1977)", Eds P.J. Dobson, J.B. Pendry and C.J. Humphreys, I.O.P. Conf. Series No.41 (1978), p.1.
19. R.H. Ritchie and A. Howie, Phil. Mag. 36:463 (1977).
20. M.P. Seah and W.A. Dench, Surface and Interface Analysis 1:2 (1979).
21. D.R. Penn, Phys. Rev. B13:5248 (1976).
22. K. Britze and G. Meyer-Ehmsen, Surface Sci. 77:131 (1978).
23. P.B. Hirsch, A. Howie, R.B. Nicholson, D.W. Pashley and M.J. Whelan, "Electron Microscopy of Thin Crystals", Krieger, New York (1977).
24. M. Petrucci, C.W. Pitt and P.J. Dobson, Electronics Lett. 22:954 (1986).
25. E.A. Wood, J. Appl. Phys. 35:1306 (1964).
26. R.W. James, "The Optical Principles of the Diffraction of X-Rays", Bell, London (1962).
27. A. Guinier, "X-ray Diffraction in Crystals, Imperfect Crystals and Amorphous Bodies ", Freeman, San Francisco (1963).
28. J.L. Beeby, Surface Science 80:56 (1979).
29. S. Holloway, Surface Science 80:62 (1979).
30. T.R. Welberry, Rep. Prog. Phys. 48:1543 (1985).
31. M. Henzler, Appl. Phys. 9:11 (1976).
32. G.W. Simmons, D.F. Mitchell and K.R. Lawless, Surface Science 8:130 (1967).
33. F. Hottier, J.B. Theeten, A. Masson and J.L. Domange, Surface Science 65:563 (1977).

34. R.H. Milne, Surface Science 122:474 (1982).
35. B.K. Vainshtein, "Structure Analysis by Electron Diffraction", Pergamon Press - MacMillan, New York (1964), p.242.
36. R.M. Stern and A. Gervais, Surface Science 17:273 (1969).
37. S. Miyake, Proc. Phys. - Math. Soc. Japan 22:666 (1940).
38. S. Miyake and K. Hayakawa, Acta Cryst. A26:60 (1970).
39. A. Ichimiya, K. Kambe and G. Lehmpfuhl, J. Phys. Soc. Japan 49:684 (1980).
40. P.K. Larsen, P.J. Dobson, J.H. Neave, B.A. Joyce, B. Bolger and J. Zhang, Surface Science 169:176 (1986).
41. R. Collela and J.F. Menadue, Acta Cryst. A28:16 (1972).
42. P.A. Maksym and J.L. Beeby, Surface Science 110:423 (1981).
43. P.A. Maksym and J.L. Beeby, Surface Science 140:77 (1984).
44. P.A. Maksym, Surface Science 149:157 (1985).
45. A. Ichimiya, Jap. J. Appl. Phys. 22:176 (1983).
46. H. Marten and G. Meyer-Ehmsen, Surface Science 151:570 (1985).
47. P.A. Maksym, NATO Advanced Research Workshop on "Thin Film Growth Techniques for Low Dimensional Structures", Eds R.F.C. Farrow, S.S.P. Parkin, P.J. Dobson, J.H. Neave and A.S. Arrott, Plenum Press, New York (1987), p.95.

PROBLEMS

Question 1

a). In an MBE chamber the total pressure is 5×10^{-11} torr at the sample. Estimate the bombardment rate of gas molecules and hence the worst possible contamination (in atoms cm^{-3}) that will occur for a growth rate of 3 Å s^{-1}. (Assume atom layers 3 Å apart and 6×10^{14} cm^{-2} density and take a gas molecular weight = 28). Boltzmann's constant $k = 1.38 \times 10^{-23}$ JK^{-1}, mass of nucleon = 1.66×10^{-27} kg.

b). A GaAs (001) crystal has been cut off orientation by 1.5° such that the surface consists of regular terraces. If the step height is monomolecular, estimate the terrace length (a_0 = 5.65 Å).

c). For an electron beam of 10 keV directed down the terraces, calculate the angle with respect to the terrace for excitation of the 008 beam (assuming V_0 = 15 eV). Use the approximation

$$\lambda_e = \left[\frac{150.4}{E} \right]^{1/2} \text{ Å for E in eV.}$$

The sample is 300 mm from a fluorescent screen. Sketch the position of the straight through beam, the shadow edge, the 008 beam from the terrace and the 004 Kikuchi lines.

Question 2

a). Distinguish between the diffraction features you would expect for a Si (001) 2x1 and 1x2 surface reconstruction. If both domains are present, show how this may be distinguished from a (2x2) structure.

b). Sketch the expected pattern for the first two Laue Zones for the [110] and [010] directions. Estimate the incident angle for the 0, 1/2 beams to emerge for an energy of 10 keV (Si : a_0 = 5.43 Å).

c). Show that if atoms are arranged in a (2x4) unit cell as shown below, the (0, 1/4) (0, 1/2) and (0, 3/4) beams will be forbidden.

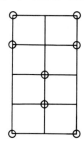

Question 3

Discuss how the intensity of the following features changes when epitaxial growth is initiated under conditions of layer by layer growth:

a) The specular beam at a strong primary diffraction condition.

b) A fractional order beam.

c) The background.

How would the observations be changed if layer growth changed to island growth (i.e. Stranski - Krastanov mode).

ANSWERS

Question 1

a). Pressure of 5×10^{-11} torr = 6.5×10^{-11} mbar
$$= 6.5 \times 10^{-9} \ \text{Nm}^{-2}.$$

The rate of bombardment $\nu = [\frac{P}{2\pi m_N \ MKT}]^{1/2}$

$$= \frac{6.5 \times 10^{-9}}{(2\pi \times 28 \times 1.66 \times 10^{-27} \times 1.38 \times 10^{-23} \times 300)^{1/2}}$$

$$= 1.87 \times 10^{14} \ \text{m}^{-2}$$

$$= 1.87 \times 10^{10} \ \text{cm}^{-2}.$$

This is ~ 3×10^{-5} of a molecular layer.

In 1 cm^3 we have 1.3×10^{-8} layers = 3.3×10^7.

Hence we have incorporated:

$$1.87 \times 10^{10} \times 3.3 \times 10^7 \ \text{molecules cm}^{-3}$$

$$= 6.23 \times 10^{17} \ \text{cm}^{-3}.$$

Note this is a large number in terms of carrier densities. If these molecules were incorporated in an electrically active state they could give carrier or trap densities ≈ 10^{18} to 10^{17} cm^{-3} which is very significant.

For a GaAs (001) crystal which is 1.5° off orientation, the step height is monomolecular and of height $a_0/2 = 2.825$ Å.

The terrace length is L = 2.825/tan 1.5°

$$= 108 \text{ Å}.$$

c). If θ_i is the angle inside the crystal, for the Bragg condition for the 008 beam $\sin\theta_i = g_{008}/2K_i$

where $g_{008} = 8/a_0 = 1.416$ Å^{-1}

$$K_i = 1/\lambda = (E/150.4)^{1/2} = 8.15 \text{ Å}^{-1}.$$

Hence $\sin\theta_i = 1.416/2\times8.15$

$$\theta_i = 4.98°.$$

The angle outside the crystal θ_0, with respect to the terrace is given by

$$\cos\theta_0 = (1 + V_0/E)^{1/2} \cos\theta_i$$

and substitution gives $\theta_0 = 4.46°$.

The figure Q1(a) shows the construction to identify the shadow edge and the main features are indicated in figure Q1(b).

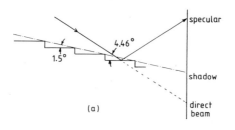

(a)

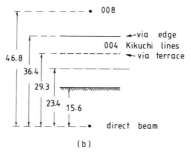

(b)

The 004 Kikuchi line occurs at an angle of θ_i^k inside the crystal with respect to the terrace, where

$$\theta_i^k = \sin^{-1} (g_{004}/2K);$$

$$\theta_i^k = \sin^{-1} (0.708/2\times8.15)$$

$$= 2.49°.$$

This is the angle which the Kikuchi line makes with respect to the projection of the terrace if it emerges, unrefracted, via the step edge. This corresponds to a distance of 13 mm from the projection of the terrace, i.e. 36.4 mm from the direct beam.

For the Kikuchi line which is refracted as it emerges via the terrace, the angle is given by θ_i^k, where:

$$\cos\theta_O^k = (1 + V_0/E)^{1/2}.\cos\theta_i^k,$$

hence:

$$\theta_O^k = 1.13°$$

which corresponds to a distance of 5.9 mm from the projection of the terrace, or 29.3 mm from the direct beam.

Question 2

a). The Si (001) surface shows (2x1) and (1x2) reconstructions which arise from adjacent terraces which are separated in height by $a_0/4$.

The accompanying figure Q2(a) shows the situation for the (2x1), (1x2) surface and the (2x2) reconstruction.

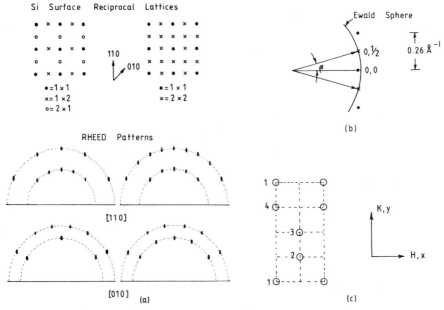

b). For the (0, 1/2) beams to emerge, the angle of incidence must be such that the Ewald sphere is just tangential to the 1/2 order rods as shown in the plan view of Fig.Q2(b). This occurs when

$$\sin\theta_i = \sin\beta \quad \text{and} \quad \beta = \sin^{-1}(0.13/K_i)$$

i.e. $\theta_i = 0.913°$.

c). If we define atom positions and axes as shown in the figure Q2(c) we can construct the structure factor $F = \Sigma\, f_n \exp[2\pi i(Hx_n + Ky_n)]$ and we assume that the atomic scattering factors are all equal.

For general surface indices H,K we have for each term in the summation:

atom (A) $4f$ exp $2\pi i$(H.0 + K.1)
      (B) $4f$ exp $2\pi i$(H.1 + K.1)
      (C) $4f$ exp $2\pi i$(H.1 + K.2)
      (D) $4f$ exp $2\pi i$(H.0 + K.3)

If H,K = 0, 1/4, the sum of these terms is

$$f\left[4 + 4\exp(\pi i/2) - 4 + 4\exp(3\pi i/2)\right]$$

$$f(4 + 4i - 4 - 4i) = 0$$

For H, K = 0, 1/2, the sum is

$$f(4 - 4 + 4 - 4) = 0$$

For H, K = 0, 3/4, the sum is

$$f(4 - 4i - 4 + 4i) = 0$$

Hence these 1/4 order beams are forbidden. This means that models for the GaAs (001) 2×4 surface which are based on an array of arsenic dimers arranged in the manner are inconsistent with the sharp, well defined, 1/4 order features seen from such surfaces.

Question 3

a). If the specular beam is at a strong primary diffraction condition, any change to the surface topography will increase the diffuse scattering at the expense of the specular intensity. The specular intensity will therefore fall.

b). A fractional order beam may show a marked reduction of intensity if the growth changes the reconstruction such as might occur for III-V semi-conductors where the ratio of group III to V surface atom population changes when growth starts. On the other hand, if the fractional order beam is not sharp, but elongated and diffuse, an increase might be seen since the diffuse scattering is increased by the 2-D growth centres.

c). Generally the background intensity will increase when growth is initiated because diffuse scattering is increased. If layer growth changes to island growth, the oscillations of intensity cease with a general change of background intensity. Strong modulation along streaks, and the emergence of broad diffraction spots will be seen.

INTENSITY OSCILLATIONS IN REFLECTION HIGH ENERGY

ELECTRON DIFFRACTION DURING EPITAXIAL GROWTH

P. J. Dobson[1], B. A. Joyce[1], J. H. Neave[1] and J. Zhang[2]

[1] Philips Research Laboratories, Redhill, Surrey, U.K.
[2] Physics Dept., Imperial College, London SW7 2BZ, U.K.

INTRODUCTION

The technique of reflection high energy electron diffraction (RHEED) has been used for many years in studies of thin film growth. In the 1950's and 1960's, many studies of the epitaxial growth of metal films relied on RHEED to determine whether the growth occurred via island formation, or two-dimensional layersand whether or not defects such as stacking faults or twin boundaries were present[1,2]. With the extensive use of molecular beam epitaxy (MBE) for the growth of semiconductor layers[3-5] RHEED has been routinely used to monitor the growth. It provides a simple qualitative check on the morphology and crystalline order of the growing layer. In 1981 it was noted that when growth is initiated the intensity of the RHEED pattern oscillates[6-8]. These oscillations have been studied extensively[9-24] and can be used to provide a fairly precise indication of the growth rate in MBE. The variation of diffracted features with time can also give much more information about the surface processes which occur during and after deposition of atoms or molecules. This subject has been extensively reviewed by the authors[9,16,17,20,23,24] and others[10-15,21] recently, so only a brief summary of the main aspects will be given here.

DIFFRACTION CONDITIONS AND RHEED INTENSITY OSCILLATIONS

One of the main problems in the area has been that for a given set of growth conditions, oscillations of the diffracted intensity can be readily observed over a wide range of diffraction conditions, and many published accounts have failed to report the exact angle of observation or incidence angle and energy of the electrons. Furthermore, the detected intensity can often be such that it represents the sum of two features, for example,the specular beam and some diffuse background feature adjacent to the specular beam. We have addressed this problem in particular in a recent publication[22].

A typical set of RHEED oscillation results from a specular beam from GaAs (001) is shown in fig.1[23]. These measurements were made under identical growth conditions. Note in particular how the sense and "phase" of oscillations changes in going from an incident angle of 1.09° to 1.69°. In fig.1b we show the specular beam intensity for the GaAs (001) 2x4 surface under steady state conditions. Note that the changes to the nature of the RHEED intensity oscillation follow the specular intensity variation. In general we see an initial reduction (as for $\theta = 1.09°$) when the specular intensity is high and an initial increase ($\theta = 1.69°$) for angles where the specular intensity is low. This observation, which has also been made on the [110], [010] and [130] azimuths has led us to propose a simple step edge scattering model to explain the RHEED intensity oscillations[9,16,17,22-44].

The bases for the model are as follows. Firstly, we treat the <u>change</u> in specular scattering as an optical problem and note that a step edge of ~3 Å will have a marked effect on electrons of de Broglie wavelength

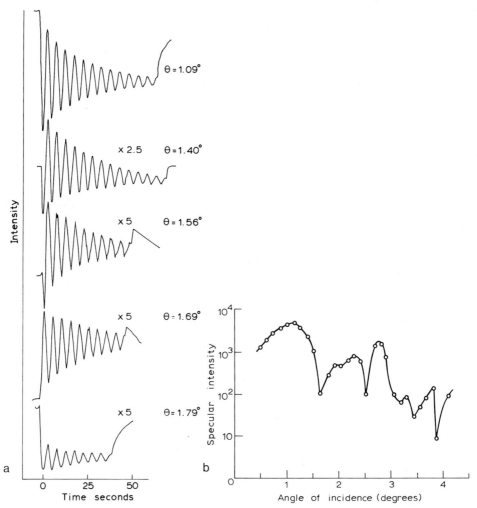

Fig. 1. (a). RHEED intensity oscillations in the specular beam observed at different angles of incidence for the [110] azimuth of a GaAs (001) 2x4 surface. Beam energy = 12.5 keV. (b). The variation of RHEED intensity as a function of angle of incidence for the [110] azimuth of a GaAs (001) 2x4 surface under static, i.e. equilibrium, conditions. Beam energy = 12.5 keV. This result is referred to as a "rocking curve".

~0.1 Å. Secondly, we accept that dynamical scattering effects must be important and we use experimental rocking curves (specular intensity as a function of incident angle) as a guide to these effects. In this way, strong coupling between beams in the crystal which give a strong maximum in specular intensity can be empirically allowed for in a description of the RHEED. Thirdly, the presence of steps on the surface can lead to increases in intensity of either the diffuse background, or of Kikuchi fatures. The latter manifests itself via a refraction effect and occurs because the refraction of a Kikuchi beam is almost negligible for beams emerging via the step edge in contrast to those which emerge via a terrace.

The main consequences for our step edge scattering model can be seen by referring to fig.2 in which we show the effect on the specular beam intensity of the changing step density which results from a regular two-dimensional nucleation and growth. The reduction of the specular intensity is matched by an increase in diffuse scattering and this may be observed by monitoring any feature in a diffraction pattern which owes its presence to diffuse scattering.

In contrast to this model, Lent and Cohen[11] treat the problem in terms of two level interference in which electrons scattered from one terrace interfere with those scattered from the adjacent terrace, either constructively (in-phase) or destructively (out-of-phase) depending on the angle of incidence and energy of the electrons. Whilst some of these effects may be present in some data, we believe that such a model does not fully account for changes in the scattering with changes in azimuthal direction and cannot account for an initial increase in the intensity as for fig.1 at θ = 1.69°. The simple step edge scattering model can explain

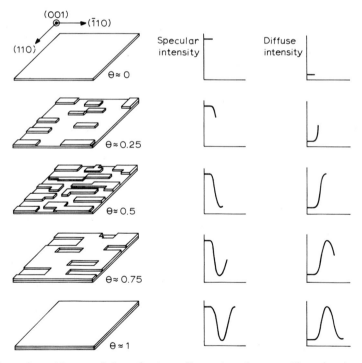

Fig. 2. A schematic model of two-dimensional growth showing how the specular beam intensity reaches a minimum when the step edge density is highest and the diffuse intensity reaches a maximum. θ refers to the fractional monolayer coverage.

these effects and, furthermore, is consistent with similar oscillatory phenomena in epitaxial growth, e.g. changes in electrical conductivity with thickness of metal films[25-27], changes in overpotential during electrolytic deposition[28,29] and the periodic variation in Auger peak shape during epitaxial growth[30,32].

RHEED OSCILLATIONS AND GROWTH EFFECTS

We will now make use of the simple step edge scattering concept, in combination with a more detailed description of the diffraction process where appropriate, to describe the origins of some of the main phenomena observed by RHEED during growth.

Damping of Oscillations.

One of the most characteristic features of the intensity variations is the damped nature of the oscillations. On the basis of our step edge scattering model this implies that the step density is periodically changing and reducing with time. When the oscillations are completely damped this presumably corresponds to a constant step density which would occur when the arriving adatoms migrate to the step edges rather than nucleate new two-dimensional growth centres on the terraces. Under these conditions, growth occurs by regular step movement. A similar conclusion was reached on the basis of electrical conductivity measurements for metal films[25-27].

Van Hove et al.[33] have also illustrated an example of damping of oscillations which also shows a beating of two oscillation frequencies. This results from a non-uniform flux over the substrate which is sampled by the electron beam. The locally different growth rates then produce cancellation (damping) and beating effects.

Increase of the Oscillation Amplitude.

Neave et al.[9] observed an increase in the amplitude of oscillation when another molecular or atomic beam was turned on. This implies an increase in the changes of step density and is consistent with the new atomic species providing new nucleation sites on terraces (i.e. an impurity nucleation effect) or changing the surface diffusion kinetics (i.e. as may be expected to occur for the growth of AℓGaAs instead of GaAs). Some care is needed in interpreting the results, especially for III-V growth, where the group III/V flux ratio and different group III elements have a strong influence on the nature of the reconstruction[34,35].

Absence of Oscillations.

An absence of oscillations may indicate, as discussed above, that the step density is constant and growth occurs by regular step movement. On the other hand, if three-dimensional nucleation is occurring there will be an absence of oscillations and this situation can usually be easily recognised by the overall appearance of a spotty pseudo-transmission RHEED pattern.

Initial Transient Effects.

There is an increasing number of observations which show something happening before regular oscillations appear. This initial behaviour may be in the form of an abrupt change in the intensity when the shutter is opened[17,36] or it may be in the form of a slow almost featureless variation before the regular oscillations[38]. Both effects probably have one thing in common. When the growth is initiated, by opening a group III shutter in the case of III-V growth, the surface adjusts to its new situation. In the case of GaAs, opening the Ga shutter will cause an abrupt change to the surface III/V ratio. The surface reconstruction of GaAs is known to be very dependent on the Ga/As ratio and one can expect a surface which was initially a well ordered (2x4) arsenic stabilized surface to show a tendency towards the (3x1) surface. We have shown[34] that very large differences in specular intensity occur between these reconstructed surfaces, so we would expect a sudden rapid transient change in the intensity when we change the Ga/As surface ratio. This effect is well illustrated in fig.3 under different diffraction conditions[17]. We note also that this explanation could account for similar observations by Briones et al.[36] in which the variation with arsenic flux was examined. Sugiura et al.[37] have shown that the time to grow the first monolayer differs from that to grow subsequent layers and this is probably a consequence of the different equilibrium step density on the initial surface.

In the case of the "transient" behaviour in the growth of Si (001) recently studied by Ichikawa and Doi[38] using RHEED/Reflection electron microscopy there was a slow change of intensity before regular oscillations of RHEED intensity were observed. A similar observation had been made by Sakamoto et al.[39]. Ichikawa and Doi tentatively associate this stage with the break up of the Si (111) 7x7 reconstruction. Their RHEED/Reflection electron microscopystudy is of great significance since it has provided the most direct evidence that RHEED intensity oscillations are associated with

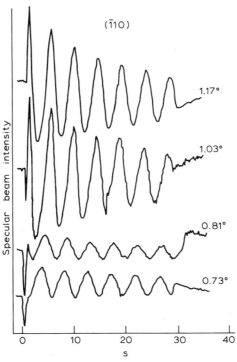

Fig. 3. Examples of the initial transient behaviour for the specular beam intensity along the [110] azimuth of a GaAs (001) 2x4 surface. Note the dependence of the behaviour on diffraction conditions (angle of incidence). Beam energy = 12.5 keV.

layer-by-layer 2-D nucleation. Furthermore, they provide good visual evidence of the importance of the "diffuse" and specular elastic components of scattering in RHEED.

Recovery Effects.

When growth is stopped, we might expect the reverse of the initial transient to occur, i.e. the Ga/As surface ratio changes, leading to a different reconstruction and hence a different specular diffracted intensity. The change in reconstruction is probably rapid, i.e. < 1 second. After this, there will be other changes due to surface diffusion, etc.. These changes are essentially thermodynamic in origin and they result from the surface attaining a minimum energy configuration, i.e. for a well oriented sample the steps and disorder boundaries will tend to be eliminated by surface diffusion processes. There will be a range of activation energies and rate constants for this stage and the changes are slow, i.e. of order seconds to several minutes. The recovery process has been examined by several groups[9,10,35,40,41] and generally most agree that there are fast and slow processes although there are significant differences regarding the interpretation. Lewis et al.[41], for example, associate the fast process with a smoothing of the growth front, whereas we interpret this as a surface atomic reconstruction effect.

The recovery effects have led to the concept of "growth interrupts" to improve the morphology of interfaces of quantum wells grown by MBE[42-44]. The idea behind this is that during growth, the limited adatom mobility leads to terrace lengths which are small, i.e. of the order of the effective migration length of the adatoms ~100 Å[23,45]. Therefore, when the shutters are operated to grow a quantum well of GaAs within AℓAs, these terraces of ~100 Å are "frozen in" to the interface. If growth is interrupted when changing material sources for ~1 to 2 minutes to allow for surface re-management, larger terraces and subsequently flatter interfaces may be achieved. The growth interruption is performed by closing the group III shutters, and leaving the arsenic flux on. This procedure has been demonstrated to lead to a large improvement of the quality of quantum wells as judged from low temperature photo-luminescence[42,44]. However, one should reserve some judgement in view of the possibility of contamination of the interface by residual gas molecules during the interrupt period. This is likely to be particularly severe for an aluminium alloy surface. A further possible complication which probably occurs for the interruption of an alloy surface during growth is that surface enrichment of one component is likely to occur.

Frequency Doubling and Phase Differences in RHEED Oscillations.

In fig.1 we showed how the apparent sense or phase of oscillations changes with the diffraction condition. The two extremes are shown at incident angles of 1.09° and 1.69°. In the latter case the intensity increases initially and we interpret this on the basis that we are seeing an increase in the diffuse scattering from the step edges, i.e. the specular beam intensity at this angle is low and the increase in diffuse scattering dominates the measured intensity. We illustrate this schematically in fig.4a and also show the case where changes in specular intensity are comparable to the changes in diffuse scattering. In this case we expect to see the appearance of "harmonics". The conditions for such a transition from dominant specular scattering to dominant diffuse scattering can be quite subtle and occur over a very small angular range as shown in fig.4b. This again highlights the need to be aware of the diffraction condition dependence of RHEED intensity charges.

There is an example in the literature of an extra period of oscillation which is different from the above description. Sakamoto et al.[18] observed that for the growth of Si on Si (001) they could obtain oscillations which were indicative of the completion of layers either $a_0/2$ or $a_0/4$ in depth, depending on their diffraction conditions. This was neatly resolved as being a consequence of the fact that the (2x1) reconstruction occurs on the terrace which is $a_0/4$ above or below the (1x2) reconstruction. By monitoring the features in the first fractional Laue zone which are unique to either the (2x1) or (1x2) structure they were able to show that the growth proceeded in a monolayer ($a_0/4$) fashion and that the specular beam intensity is strongly influenced by dynamical electron diffraction interactions.

We recently completed an analysis of the apparent phase relationship of the peaks in RHEED intensity oscillations for a wide range of diffraction conditions for the growth of GaAs on the GaAs (001) 2x4 surface[46]. We conclude that there are very few conditions under which maxima or minima correspond exactly to the completion of a layer. This is almost certainly a consequence of the combination of specular and diffuse scattering in the measured intensity.

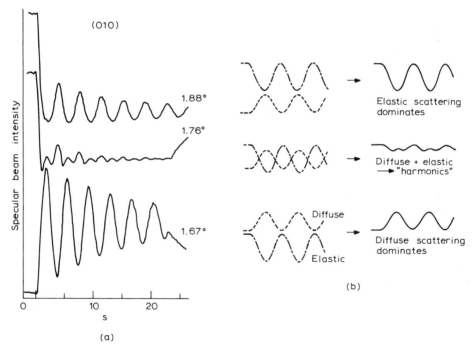

Fig. 4. (a). The change in the "phase" and appearance of harmonics in RHEED intensity oscillations at different angles of incidence. These results are for the specular beam along the [010] azimuth of a GaAs (001) 2x4 surface. Beam energy = 12 keV. (b) A qualitative explanation of the effect in terms of the relative contributions of the elastic specular scattering and diffuse scattering to the measured intensity. The angle of 1.67° corresponds to a minimum of intensity in the rocking curve under static, i.e. non-growing, conditions.

SUMMARY

We have given a brief summary of the phenomenon of RHEED intensity oscillations. They occur because of layer-by-layer growth via a two-dimensional nucleation process. It is important to realise that the diffraction conditions under which the observations are made are specified and well understood. Oscillations of the intensity of almost all diffraction features can be seen when growth is initiated and in general the elastically scattered specular intensity falls and the diffuse intensity increases when the step edge density increases. The behaviour of the RHEED intensity can be used to infer large amount of new information concerning epitaxial growth. It should be emphasised that whilst we have restricted most comments to the growth of III-V semiconductors and silicon there are numerous recent examples of RHEED intensity oscillations for the growth of II-VI semiconductors[47,48] and metals[49,50].

REFERENCES

1. D.W. Pashley, Advances in Physics 5:173 (1956).
2. D.W. Pashley, Advances in Physics 14:327 (1965).
3. A.Y. Cho and J.R. Arthur, Progr. Solid State Chem. 10:157 (1975).
4. R. Ludeke, IBM J. Research and Development 22:304 (1978).
5. B.A. Joyce, J.H. Neave, P.J. Dobson and P.K. Larsen, Phys. Rev. B29:814 (1984).
6. J.J. Harris, B.A. Joyce and P.J. Dobson, Surface Sci. 103:L90 (1981).
7. G.E.C. Wood, Surf. Sci. 108:L441 (1981).
8. J.J. Harris, B.A. Joyce and P.J. Dobson, Surface Sci. 108:L444 (1981).
9. J.H. Neave, B.A. Joyce, P.J. Dobson and N. Norton, Appl. Phys. A31:1 (1983).
10. J.M. van Hove, C.S. Lent, P.R. Pukite and P.I. Cohen, J. Vac. Sci. Technol. B1:741 (1983).
11. C.S. Lent and P.I. Cohen, Surface Sci. 139:121 (1984).
12. P.R. Pukite, C.S. Lent and P.I. Cohen, Surface Sci. 161:39 (1985).
13. B.F. Lewis, T.C. Lee, F.J. Grunthaner, A. Madhukar, R. Fernandez and J. Maserjian, J. Vac. Sci. Technol. B2:419 (1984).
14. M.Y. Yen, T.C. Lee, P. Chen and A. Madhukar, J. Vac. Sci. Technol. B4:590 (1986).
15. T.C. Lee, M.Y. Yen, P. Chen and A. Madhukar, J. Vac. Sci. Technol. A4:884 (1986).
16. B.A. Joyce, P.J. Dobson, J.H. Neave, K. Woodbridge, J. Zhang, P.K. Larsen and B. Bolger, Surface Sci. 168:423 (1986).
17. P.J. Dobson, B.A. Joyce, J.H. Neave and J. Zhang, J. Crystal Growth 81:1 (1987).
18. T. Sakamoto, T. Kawamura and G. Hashiguchi, Appl. Phys. Lett. 48:1612 (1986).
19. J. Aarts, W.M. Gerits and P.K. Larsen, Appl. Phys. Lett. 48:931 (1986).
20. B.A. Joyce, J.H. Neave, J. Zhang, P.J. Dobson, P. Dawson, K.J. Moore and C.T. Foxon, in: NATO Advanced Research Workshop on "Thin Film Growth Techniques for Low Dimensional Structures", R.F.C. Farrow, S.S.P. Parkin, P.J. Dobson, J.H. Neave and A.S. Arrot, eds, Plenum Press, New York (1987), p.19.
21. P.I. Cohen, P.R. Pukite and S. Batra, in: NATO Advanced Research Workshop on "Thin Film Growth Techniques for Low Dimensional Structures", R.F.C. Farrow, S.S.P. Parkin, P.J. Dobson, J.H. Neave and A.S. Arrott, Plenum Press, New York (1987), p.69.
22. J. Zhang, J.H. Neave, P.J. Dobson and B.A. Joyce, Appl. Phys. A42:317 (1987).

23. B.A. Joyce, P.J. Dobson, J.H. Neave and J. Zhang, Surface Sci. 174:1 (1986).
24. B.A. Joyce, P.J. Dobson, J.H. Neave and J. Zhang, Surface Sci. 178:110 (1986).
25. C. Pariset and J.P. Chauvineau, Surface Sci. 78:478 (1978).
26. J.P. Chauvineau, J. Crystal Growth 53:505 (1981).
27. D. Schumacher and D. Stark, Surface Sci. 123:384 (1982).
28. V. Bostanov, R. Roussinova and F. Budevski, J. Electrochem. Soc. 119:1346 (1972).
29. R.D. Armstrong and J.A. Harrison, J. Electrochem. Soc. 116:328 (1969).
30. Y. Namba and R.W. Vook, Thin Solid Films 82:165 (1981).
31. Y. Namba, R.W. Vook and S.S. Chao, Surface Sci. 109:320 (1981).
32. R.W. Vook and Y. Namba, Appl. Surface Sci. 11/12:400 (1982).
33. J.M. van Hove, P.R. Pukite and P.I. Cohen, J. Vac. Sci. Technol. B3:563 (1985).
34. P.K. Larsen, P.J. Dobson, J.H. Neave, B.A. Joyce, B. Bolger and J. Zhang, Surface Sci. 169:176 (1986).
35. J.H. Neave, B.A. Joyce and P.J. Dobson, Appl. Phys. A34:179 (1984).
36. F. Briones, D. Golmayo, L. Gonzalez and J.L. de Miguel, Japan J. Appl. Phys. 24:L478 (1985).
37. H. Sigiura, M. Kawashima and Y. Horikoshi, Japan J. Appl. Phys. 25:1847 (1986).
38. M. Ichikawa and T. Doi, Appl. Phys. Lett. 50:1141 (1987).
39. T. Sakamoto, N.J. Kawai, T. Nakagawa, K. Ohta and T. Kojima, Appl. Phys. Lett 47:617 (1985).
40. F-Y Jhang, P.K. Bhattacharya and J. Singh, Appl. Phys. Lett. 48:290 (1986).
41. B.F. Lewis, F.J. Grunthaner, A. Madhukar, T.C. Lee and R. Fernandez, J. Vac. Sci. Technol. B3:1317 (1985).
42. T. Hayakawa, T. Suyama, K. Takahashi, K. Kondo, S. Yamamoto, S. Yano and T. Hijikata, Appl. Phys. Lett. 47:952 (1985).
43. K. Sakaki, M. Tanaka and J. Yoshino, Japan J. Appl. Phys. 24:L427 (1985).
44. D. Bimberg, D. Mars, J.N. Miller, R. Bauer and D. Oertel, J. Vac. Sci. Technol. B4:1014 (1986).
45. J.H. Neave, P.J. Dobson, B.A. Joyce and J. Zhang, Appl. Phys. Lett. 47:100 (1985).
46. B.A. Joyce, J. Zhang, J.H. Neave and P.J. Dobson, accepted for publication in J. Appl. Phys.
47. T. Yao, H. Taneda and M. Funaki, Japan J. Appl. Phys. 25:L952 (1986).
48. R.L. Gunshor, L.A. Kolodziejski, M.R. Melloch, M. Vaziri, C. Choi and N. Otsuka, Appl. Phys. Lett. 50: 200 (1987).
49. C. Koziol, G. Lilienkamp and E. Bauer, Appl. Phys. Lett. 51: 901 (1987).
50. S.T. Purcell, B. Heinrich and A.S. Arrott, Phys. Rev. B35:6458 (1987).

EMISSION AND LOW ENERGY REFLECTION ELECTRON MICROSCOPY

Ernst Bauer and Wolfgang Telieps

Physikalisches Institut
Technische Universität Clausthal
and SFB 126 Göttingen-Clausthal
D-3392 Clausthal-Zellerfeld, FRG

INTRODUCTION

Surface electron microscopy requires that the electrons used for imaging either originate predominantly from the first few atomic layers or that their emission probability depends strongly upon surface structure and/or properties. With fast electrons this is achieved by illumination parallel to the surface (edge-on TEM), or nearly parallel to the surface (REM) or by imaging with the weak diffracted beams produced by surface features (superstructures or thickness differences: plane-view TEM). Imaging with slow electrons, on the other hand, relies either on the low escape depth of slow electrons and/or on their transmission probability through the surface.

There are many ways in which microscopy with slow electrons can be done. The most popular one at present is scanning microscopy[1] in which a finely focussed beam is scanned across the surface and the emitted secondary or Auger electrons are used for the formation of SEM or SAM images, respectively. In commercial systems electrons from several keV to several 10 keV are used in the illuminating beam and resolutions below 10 nm have been achieved in SEM with 30 keV electrons. Recently, the incident energy has been reduced so much that scanning imaging with slow reflected electrons (SLEEM) has become possible, with resolutions down to 50 nm at 500 eV[2,3].

Another way of imaging is projection: the electrons are guided along diverging electric or magnetic field lines from the point of emission on the surface to an image disc in the field emission electron microscope (FEM)[4] or the magnetically collimated photoelectron microscope[5]. In order to achieve better than 1 μm resolution with electrons of several eV energy, a magnetic field of about 10 Tesla is needed at the specimen, which requires a superconducting magnet. In contrast, FEM is very simple but limited to microscopic surfaces with a small radius of curvature ($r \approx 100$ nm) which is necessary in order to achieve the high fields required for field emission. This small radius ensures simultaneously a high magnification; the resolution is about 1 nm. FEM is very useful where surface processes - adsorption, diffusion, crystal nucleation, etc. - are to be studied on many surface orientations simultaneously, in particular when small surfaces are of interest as in heterogeneous catalysis.

However, quantitative image interpretation can be a problem due to variations of the electric field F across the field emitter tip which has an influence on the emission distribution j_{FE} similar to that of the work function $e\Phi$. j_{FE} is given by the Fowler-Nordheim equation

$$j_{FE} = \alpha(F/\Phi)\exp(-\beta\Phi^{3/2}/F)$$

where α, β are slowly varying functions of $F^{1/2}/\Phi$. Thus, changes of the local Φ, which is the quantity characterizing a given surface element, during a surface process may be accompanied by changes in local F. A second problem can be the competition between the various planes on the surface which makes the determination of the local coverage difficult. Finally, the surface process to be studied may be influenced by the strong field. This is even more important in the ion imaging analog to FEM, i.e. FIM[6]. Here, the fields needed for field ionization of the typical He or Ne imaging gas are much higher, usually several 10^{10} Vm^{-1} vs. several 10^7 Vm^{-1} in FEM. These high fields can cause field dissociation of two-dimensional structures on the surface, produce field-induced structures, e.g. reconstruction, and field-desorb many surface species, even at the low temperatures (usually $\leq$ 20 K) required for atomic resolution. This restricts the application of FIM even more than that of FEM. In addition, the preparation of sufficiently sharp and stable tips is simple only in the case of a few metals, usually refractory metals, and difficult for semiconductors. Insulators are practically out of reach for FEM and FIM.

All these problems can be eliminated if the electric field is not used for producing the imaging particles and for image formation simultaneously but only for the imaging process. This is done in a third imaging method, immersion microscopy, which can also be performed both with electrons and ions. In this method, the charged particles originating from a point on a macroscopic specimen surface are accelerated in a (more or less) homogeneous electric field which produces a virtual image disc. The real image is obtained by a subsequent magnetic or electrostatic lens. This imaging method is as old as electron microscopy in general[7,8], but has been overwhelmed by transmission electron microscopy and nearly condemned to oblivion by SEM of surfaces. There are several reasons for the lack of interest in immersion microscopy in the past: inferior resolution to TEM and SEM, specimen limitations as compared to SEM (surface roughness, emission properties), lack of elemental analysis and less convenient specimen handling than in SEM being the most important ones. The tremendous growth of surface science in the past 25 years, however, has produced a large interest for laterally resolved surface analysis on flat macroscopic surfaces for which immersion microscopy appears particularly well suited. One of the first results obtained by Germer and Hartmann[9], after their revival of LEED, was the formation of streaked LEED patterns which were attributed to preferred adsorption on steps[10] and led to the development of immersion microscopy with slow reflected electrons, i.e. low energy electron microscopy (LEEM)[11].

This section discusses the possibilities and limitations of immersion microscopy with emitted and reflected electrons from the point of view of surface science, the rationale being that volume properties can be studied better with TEM of thinned specimens and rough surfaces such as fractured surfaces better with SEM. For the more general aspects of immersion emission microscopy, e.g. its application in biology and metallurgy, the reader is referred to recent reviews[12,13]. Where appropriate, comparisons with other imaging methods will be made.

THE IMMERSION OBJECTIVE

The heart of any microscope is the objective lens. Its aberrations determine the size of the aperture which has to be used in order to achieve the optimum resolution. The aperture in turn has a strong influence on the intensity available for imaging and the intensity determines the recognizable contrast due to signal to noise ratio considerations. Thus, primary attention has to be paid to the electron optical properties of the objective which in the present case is an immersion lens, also called a cathode lens because one electrode, the specimen, is the source of the electrons, whether they are emitted or reflected. The conceptually simplest configuration of such a lens consists of a homogeneous acceleration field and an electrostatic einzel-lens (fig.1a) or a magnetic lens (fig.1b). Acceleration and imaging can, however, also be combined into a three-electrode lens. For optimum access to the specimen it is useful to use a strongly asymmetric arrangement as shown in fig.1c, which can also be realized with a magnetic objective (see e.g. fig.10).

Inasmuch as objective lenses, at least of the magnetic type, with very small aberrations can be built, the limiting aberrations of a cathode lens are those due to the accelerating field. The calculation of its aberrations has to take into account two features foreign to conventional microscopy: an angular spread of the electrons up to 90° and an energy spread of the same order of magnitude as the mean energy. The problem was first solved by Recknagel[14] whose work was followed by many other investigations (for references see the reviews 12,13), with the major result that the resolution is given by $d \approx V/F$ where eV is the mean starting energy of the electron on the surface and F the field strength. This simplified expression has led to the widespread belief that good resolution is obtainable only at very low energies and high fields and much of the effort in emission microscopy was directed in this direction.

That this belief is incorrect can be seen by simple analytic calculations for an homogeneous electric field[15]. The trajectory of an electron starting with an energy $eV = e\rho U_O$ (U_O = acceleration potential) at an angle α with respect to the surface normal is given by

$$R(Z) = 2\rho\sin\alpha\cdot[(\frac{Z}{\rho} + \cos^2\alpha)^{1/2} - \cos\alpha] \qquad (1)$$

where $R = r/L$ and $Z = z/L$ (L length of field F). When the electron leaves the field it seems to come from points on the axis given by the intersection of the tangent R_t to $R(Z)$ at $Z = 1$ ($z = L$) with the Z axis.

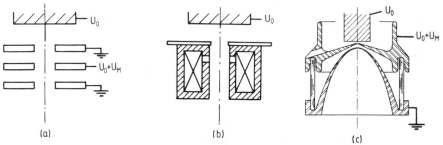

Fig. 1. Cathode lens configurations. (a) Homogeneous acceleration field plus electrostatic einzel-lens; (b) acceleration field plus magnetic lens; (c) combined acceleration and imaging field lens.

For $\alpha \to 0$ and the mean energy $eV_o = e\rho_o U_o$, R_t intersects the axis at

$$Z_G = -\{1 + 2\rho_o[1 - (\frac{1}{\rho_o} + 1)^{1/2}]\} \tag{2}$$

which determines the location of the "virtual Gaussian image plane". All other tangents ($\alpha \neq 0$, $\rho \neq \rho_o$) intersect this plane off axis in points $R(\alpha, \rho, \rho_o)$ thus producing a disc of confusion. Expanding $R(\alpha, \rho, \rho_o)$ in powers of $\sin\alpha$ yields the chromatic aberration

$$\delta_c = d_c/L = -(2/\eta)[\rho(\eta - 1) - \rho_o(\eta_o - 1)]\cdot\sin\alpha$$

$$\approx -2\varepsilon(1 - \eta^{-1})\cdot\sin\alpha \tag{3}$$

with $\varepsilon = (\rho - \rho_o)$ for $\rho \approx \rho_o \ll 1$, $\eta = (\frac{1}{\rho} + 1)^{1/2}$, $\eta_o = (\frac{1}{\rho_o} + 1)^{1/2}$ and the third order spherical aberration

$$\delta_S = d_S/L = [\rho(1 - \eta^{-1})^2]\cdot\sin^3\alpha \tag{4}$$

The maximum emission angle α_A is limited by the exit aperture with radius $r_A = R_A L$ (whose influence on the aberrations can be neglected for the ranges of R_A and ρ of interest here as numerical calculations show[16]) to

$$\sin\alpha_A = \frac{R_A}{2\sqrt{\rho}}\sqrt{\frac{1 + 2\rho + (4\rho^2 + 4\rho - R_A^2)^{1/2}}{1 + R_A^2}} \approx \frac{R_A}{2\sqrt{\rho}} \tag{5}$$

for ρ, $R_A \ll 1$.

The aperture cannot be chosen too small because of the diffraction disc of confusion

$$\delta_D = d_D/L = 1.2\Lambda/R_A \tag{6}$$

$$[\Lambda = \lambda/L = (\frac{1.5}{U_o + V})^{1/2}/L \quad ; \quad U_o, V \text{ in Volt}, \lambda \text{ in nm}]$$

which it causes. Therefore, at very low energies which are typical for thermionic and photoemission microscopy close to threshold (see below), it is difficult to limit α by an aperture. When $R_A \geq 2\sqrt{\rho}$ then $\sin\alpha_A = 1$ which leads to $\delta_c \approx -2\varepsilon$, $\delta_S \approx \rho$ [or $d_S \approx V/(U_o/L) = V/F$, the Recknagel formula; $\varepsilon = \Delta V/U_o$, where $e\Delta V$ is the width of the energy distribution]. With aperture limitation ($R_A < 2\sqrt{\rho}$), $\delta_c \approx -(\varepsilon/\sqrt{\rho})R_A$ and $\delta_S \approx (1/8\sqrt{\rho})R_A^3$.

Thus, the recipe for obtaining maximum resolution is not to make ρ or V/F as small as possible but just the opposite, keeping ε small. This is the concept used in LEEM. The improved resolution at larger ρ is due to the elimination from the imaging process of electrons leaving the surface at large angles which more or less reduces the intensity depending upon the angular intensity distribution. If sufficient intensity is available, the aperture may be optimized by minimizing $\delta = (\delta_c^2 + \delta_S^2 + d_D^2)^{1/2}$ ($d\delta/R_A = 0!$) which yields the energy dependence of the optimum aperture radius $r_A = R_A L$ and of the resolution limit d shown in fig.2 for $U_o = 25$ kV, $L = 3$ mm and $\Delta V = 0.25$ V ($\varepsilon = 1\times10^{-5}$). Such a small ΔV can be achieved by illuminating the surface with electrons from a field emission gun.

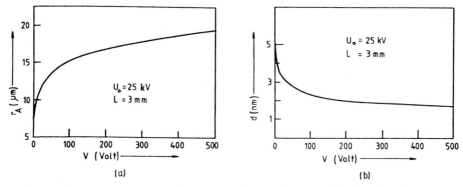

Fig. 2. Optimum aperture radius r_A and resolution limit of a homogeneous acceleration field with $U_O = 25$ kV, $\Delta V = 0.25$ V and L = 3 mm as a function of initial energy E = eV (from ref.17).

It is necessary to discuss briefly the difference between earlier resolution considerations and the one presented here. In the present treatment the electrons leaving the surface are classified according to their emission energy $E = eV = mv_0^2/2$ and their emission angle α. This is appropriate when the electrons have a narrow energy distribution about a mean energy as in LEEM or in emission microscopy with characteristic electrons. In the previous calculations the electrons were classified according to their axial and tangential starting energies. This is appropriate in the usual emission microscopy in which the width of the energy distribution is comparable to the mean or most probable energy. A second difference is in the image plane studied: here it is the Gaussian image plane; in the previous work it was the plane of least confusion. In this plane the resolution is always better than in the Gaussian image plane, depending upon angular and energy distribution by as much as a factor of 2, which is, of course, also true in LEEM.

The difference between the two situations ($\varepsilon \ll \rho$, $\varepsilon \approx \rho$) and their treatments can be seen best by comparing the approximate expressions for the optimum aperture and the resolution achieved with it in the two cases: a) $\varepsilon \ll \rho$, Gaussian image and b) $\varepsilon \approx \rho$, least confusion:

a) $R_{A,opt} \approx \sqrt{(1.2\lambda/\varepsilon)} \cdot \rho^{1/4}$ and $\delta_{opt} \approx \sqrt{(1.5\lambda\varepsilon)} \cdot \rho^{-1/4}$, (7)

b) $R_{A,opt} \approx \sqrt{(2.7\lambda)} \cdot \rho^{-1/4}$ and $\delta_{opt} \approx \sqrt{\lambda} \cdot \rho^{1/4}$. (8)

Here λ is the wavelength of the electron after exit from the acceleration field in units of the field length L, and ρ in case b) is V_m/U_O where eV_m is the most probable emission energy. ε does not appear in case b) because it is assumed to be related to $\rho = \rho_m = V_m/U_O$ via some kind of energy distribution. In the specific formulas shown, a Maxwellian distribution $I(E) \sim E \cdot \exp(-E/kT)$ has been assumed for which $eV_m = kT$ and $e\Delta V = 2.45$ kT (see next section), so that $\varepsilon = 2.45\rho$.

The important features of eqs a) and b) are: i). In case a) the resolution improves with increasing ρ, i.e. electron energy, while in case b) it worsens. ii). In case a) the optimum aperture increases with ρ while in case b) it decreases. This second aspect is important for intensity reasons which require larger apertures at higher energies because of the weaker deflection of more energetic electrons in the electric field and the resulting greater interception by the aperture. Even at low energies aperture limitations can be serious. Thus the optimum aperture diameter $2\sqrt{(2.7\lambda)}\cdot\rho^{-1/4}$ for thermionic imaging at T = 2300 K ($eV_m \approx 0.2$ eV) and for secondary electron imaging ($eV_m \approx 2$ eV) is 9.5 μm and 5.3 μm respectively for U_O = 25000 V and L = 3 mm. Therefore the optimum aperture cannot be used in most cases. For example, the emission images with the best point-to-point resolution reported to date were obtained with apertures from 30 μm to 50 μm diameter[18]. The theoretical resolution obtainable with non-optimum apertures is shown in fig.3 for conditions slightly different from those shown here. On the other hand, when 2 eV electrons with 0.25 eV energy width are used for imaging, the optimum aperture diameter $2\sqrt{(1.2\lambda/\epsilon)}\cdot\rho^{1/4}$ is 10 μm; for 10 eV electrons it is already 15 μm, so that optimum working conditions can be achieved more easily.

An even better resolution at low energies than that shown in fig.3 has been calculated for a special energy and angular distribution by selecting that image plane in which the intensity distribution is best confined into a small "image" region[13]. A resolution of about 3 nm including imaging lens aberrations and aperture diffraction was obtained[13]. It appears questionable, however, that the assumptions made in the calculations can be fulfilled in practice, so that the data shown in fig.3 should be considered more realistic for emission microscopy. Finally, it should be noted that an exact calculation of the resolution requires a quantum-mechanical treatment which also gives lower δ or d values than the particle trajectory calculations[19].

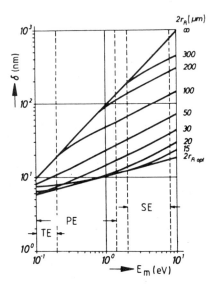

Fig. 3. Theoretical resolution limit δ of the homogeneous acceleration field as a function of most probable energy of a Maxwellian energy distribution for various aperture diameters $2r_A$. U_O = 50 kV, L = 4.2 mm. The upper and lower straight lines correspond to no aperture limitation and optimum aperture, respectively (adapted from ref. 18). TE = Thermionic emission; PE = Photoemission; SE = Secondary electron emission (see fig.4).

If the homogeneous field region were followed by a lens with low aberrations, the resolution discussed above and shown in figs 2 and 3 should be obtainable. Practical considerations such as UHV-compatible design or accessibility of the specimen, however, may require some compromise of resolution. This is the case for the immersion lens shown in fig.1c which is used in the instrument to be described later. It is a modified version of the lens described by Bartz for which a theoretical resolution of 8.5 nm at $U_o = 40$ kV, $\rho = 5 \times 10^{-5}$ and $r_A = 7$ µm was reported[20]. Detailed numerical calculations for several cathode-centre electrode distances and many ρ values showed that this lens (fig.1c) has a resolution which is by about a factor of 2 worse than that of the homogeneous field[15-17]. Nevertheless, a resolution of about 4 nm should be obtainable with 250 eV electrons.

An improvement of the resolution beyond that of the homogeneous field requires correction elements similar to those discussed for some time in transmission microscopy. These solutions are, however, impractical in the present state of instrument development, although plans in this direction have been discussed.

Summing up the resolution considerations, there are two basic imaging modes: one in which energy and energy width are comparable and very small, the other in which the energy width is much smaller than the energy. The first one is classical emission microscopy, the second one is LEEM and the hitherto undeveloped emission microscopy with characteristic electrons.

CLASSICAL EMISSION MICROSCOPY: INTENSITY AND RESOLUTION

This method makes use of thermionic, photoelectric or secondary electron emission. Exo-electron emission and field electron emission from MIM-structures have also been used but are limited to a small class of specimens and have impractically low intensity. The current density in thermionic emission is given by the Richardson-Dushman equation

$$j_{th} = AT^2 \exp(-\Phi/kT) \tag{9}$$

where A is a material constant and Φ the work function which, due to its appearance in the exponent, mainly determines j_{th}. A is generally in the range 10 - 500 $Acm^{-2}K^{-2}$ while Φ varies from about 2 eV to 6 eV with material, surface orientation and type and coverage by adsorbates. For a typical clean surface with $\Phi \approx 4.9$ eV and $A \approx 100$ $Acm^{-2}K^{-2}$ a temperature of 2300 K is needed to obtain a current density of 10^2 Am^{-2} which is sufficient to observe a properly magnified ($M = 10^3 - 10^4$) image. At this temperature the Maxwellian energy distribution $dj_{th}/dE \sim E \cdot \exp(-E/kT)$ (see fig.4a) has the full width at half maximum (FWHM) $\Delta E = 2.45$ kT ≈ 0.48 eV and the most probable energy $E_m = kT \approx 0.2$ eV mentioned earlier which results in a theoretical resolution of $\sqrt{\lambda \cdot \rho}^{1/4} \approx 8$ nm with optimum aperture, and of 1.2 $\rho \approx 30$ nm without aperture limitation (for $U_o = 25$ kV, L = 3 mm). Obviously most materials cannot be studied at such high temperatures and lower temperatures are needed also for improving the resolution. Sufficient emission at much lower T can be achieved by adsorption or deposition of low Φ materials, such as alkalis, alkaline earths or their oxides. With BaO a resolution of 40 nm was observed[18]. The need for high temperatures - and for Φ reduction by electropositive adsorbates in the case of most materials - makes thermionic emission microscopy rather useless for the study of well-defined surfaces in general but the technique has found some application in (bulk) materials science[12,24] and in special cases in which the thermionic emission distribution across a surface is of interest.

Photoemission microscopy is less restricted than thermionic emission because of the availability of intense light sources with photon energies above 4.5 eV which produce a photoemission current density given by the Fowler equation and its near-threshold approximation

$$j_{ph} = BT^2 f(h\nu - \Phi) \rightarrow B'T^2(h\nu - \Phi)^2 \qquad (10)$$

for metals near the threshold $h\nu_O = \Phi$. The photoelectric yield $Y(\nu) = N_{e\ell}/N_\nu$ is very small just above the threshold $h\nu_O = \Phi$, but rises rapidly with photon energy $h\nu$ as illustrated in fig.5. This suggests the use of photon energies above 12 eV but if this is done, the energy distribution – which reflects essentially the valence electron density of states in the material and contains also the secondary electrons generated by the excited electrons – becomes unacceptably wide from the point of view of resolution. An example is shown in fig.4d for excitation with HeI light ($h\nu = 21.2$ eV). Furthermore, low pressure light sources of this type have an intensity too low for microscopy. Therefore, the more intense though non-monochromatic

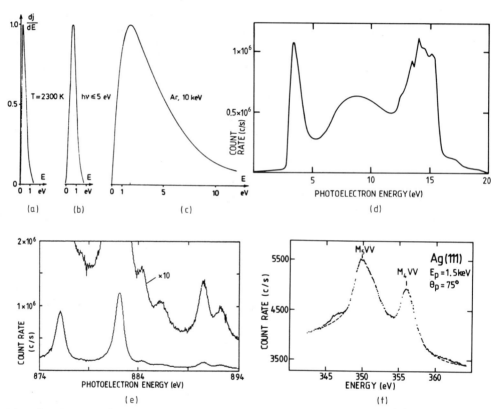

Fig. 4. Energy distributions in various emission modes. (a) Thermionic emission at 2300 K; (b) Photoemission excited by the light of a high pressure Hg lamp; (c) SE emission excited by 10 keV Ar atoms; (a) – (c) from contaminated polycrystalline Mo (adapted from ref.12); (d) Photoemission from Ag excited by HeI light ($h\nu = 21.2$ eV)[21]; (e) Photoemission from Ag excited by Mg K_α radiation [$h\nu = 1254$ eV, Ag $3d(M_{4,5})$] electrons (adapted from ref.22); (f) Auger electron emission from Ag excited by 1.5 keV electrons (Ag $M_{4,5}$) electrons)[23] with an incident angle $\theta_p = 75°$ to the surface normal.

high pressure sources (Hg, Xe) with lower range (hν < 5 eV) are used.   A
typical electron energy distribution obtained with such a lamp is seen in
fig.4b.   Its mean energy and energy width can be as small as in thermionic
emission (fig.4a) but is usually larger so that the resolution should be
worse.   Nevertheless a resolution of 12 nm has been obtained in the same
microscope in which thermionic emission from BaO gave only 40 nm resolution.   Most emission microscopes presently in use rely on photoemission,
mainly because of its nondestructive imaging characteristics, at least for
materials which are not modified by UV radiation.

    Secondary electron emission microscopy has, in principle, an even
wider application range than photoemission microscopy because there are no
work function limitations.   However a number of problems make this imaging

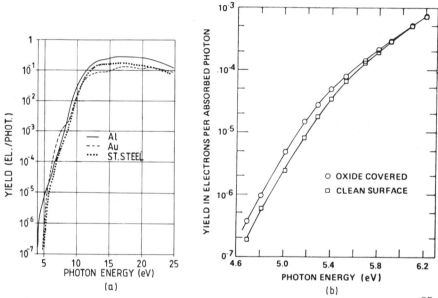

Fig. 5. Photoelectric yield for metals (Au, Al, stainless steel) (a)[25] and
    a Si(111) surface (b)[26] at near-normal incidence.   The metal
    surfaces are contaminated, the clean Si surface shows a sharp (7x7)
    pattern, the other Si surface is covered by a thin (< 10 nm) oxide
    layer.

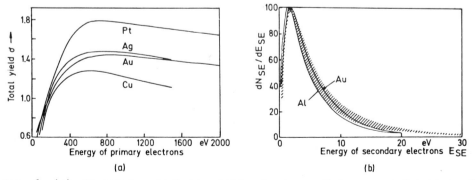

Fig. 6. (a) Secondary electron yield of some metals as a function of
    primary electron energy; (b) normalized energy distributions. Al,
    Au (solid lines) calculated, dashed region: variations between
    different metals, experimental (after ref.1).

mode less attractive. In the case of excitation by electrons, serious carbon contamination can occur in non-UHV systems due to cracking of adsorbed hydrocarbons from the residual gas by the incident electrons. In UHV systems adsorbed atoms or molecules may be desorbed (ESD = Electron Stimulated Desorption) (see chapter by Menzel) and ionically bonded materials may be dissociated. In the case of excitation by fast ion or atom impact ("kinetic" emission) the surface is continuously modified by sputtering so that no real surface imaging is possible but rather a "depth profile" imaging, which may be quite useful for metallurgical studies but not for surface science. A problem common to both means of excitation is surface charging in the case of insulating specimens which can be overcome by deflecting the ions in a mixed ion + fast atom beam - at the expense of intensity - or by suitable neutralization devices.

In spite of these problems secondary electron emission microscopy has been used extensively in the past and also should have some applications in the future, at least with electron excitation but possibly also with ion excitation, e.g. when the microstructure as a function of depth is of interest. The secondary electron emission yield is about 1 over a wide range of primary energies (see fig.6a) so that in principle imaging with high intensity should be possible. However, the energy distribution of the secondary electrons is very wide, both for ion/atom excitation (fig.4c) and electron excitation fig.6b), with peak energies from 1-5 eV and FWHMs of 3 - 10 eV. Therefore, rather small apertures have to be used in order to reduce chromatic aberrations and a significant fraction of the available intensity is lost. The best resolution obtained in this manner is about 25 nm[18].

The best resolution and highest intensity are useless unless the imaging process produces sufficient contrast for image recognition and focussing. There are two principal contrast mechanisms: i) topographic contrast due to the deviation of the surface from planarity, ii) material and crystal orientation contrast due to the local variation of the emission current density caused by variations in the probability of electron excitation, transport to and transmission through the surface. Mechanism ii) occurs both on smooth and rough surfaces but on rough surfaces it is usually overwhelmed by mechanism i). Topographic contrast has a purely geometrical and an electron optical component. The geometry has several consequences as far as excitation by photons, electrons, ions or atoms - which usually have a large polar angle of incidence β_0 - is concerned. Roughness produces shadows, causes variations in the incident current density ($j_0 \propto \cos\beta_0$) and variations in electron yield Y because of $Y = Y(\beta_0)$. In addition, the local inclination of the surface with respect to the optical axis (β) determines the emitted current density $j/\cos\beta$ and the angular distribution. The major effect of the topography is, however, frequently the local field distortion which changes the local electron optics and not only influences contrast but also reduces resolution. This is believed to be the major reason for the poor resolution of 40 nm - as compared to the expected 10 nm - in the thermionic image of a BaO-coated steel surface (fig.7a). The specimen shown in fig.7b is apparently smoother because the resolution is near the theoretically expected value so that the contrast is mainly material contrast. The specimen of figs 7c and 7d obviously is rough due to ion and fast atom bombardment and consequently the contrast is mainly topographic again.

In surface science studies one tries to work usually with surfaces as flat as possible so that topographic contrast occurs only at major surface imperfections such as etch pits, grain boundary grooves, etc. Material and orientation contrast is then decisive. In thermionic and near threshold photoemission, the work function ϕ plays a dominant role because it determines the escape probability, but it has also a significant influence on the transmission of secondary electrons through the surface. ϕ varies with the chemical composition, the structure and the surface orientation of the substrate and is strongly influenced by the kind i and coverage Θ_i of adsorbates; even for fixed Θ_i, ϕ can be strongly T-dependent. Besides ϕ - and to a lesser extent the factors A, B in j_{th}, j_{ph} - there are other factors which contribute to material contrast and which are connected with electron excitation and transport. This is immediately evident in fig.8: if the contrast were determined only by ϕ, the contrast between different grains should be the same in the different images. Of course, the comparison in fig.8 is hampered by different surface conditions (contamination, atomic roughness). A difference in surface conditions can be excluded in fig.9 in which only the dominant excitation energy was changed resulting in dramatic contrast changes. These changes are caused in part by the energy dependence of the optical properties of the material and of the propagation of the electron from the excitation region to the surface, but mainly by the energy dependence of the transition matrix elements which determine the primary excitation process. These matrix elements contain the vector potential of the exciting wave field and

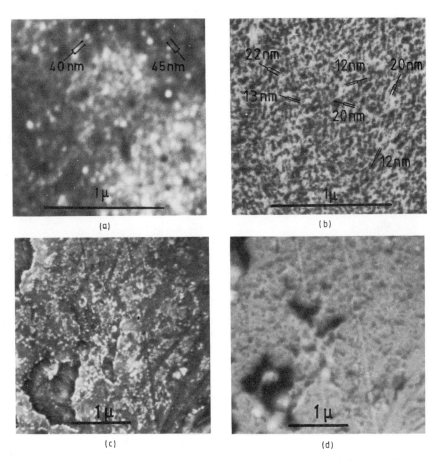

Fig. 7. Electron emission images of various surfaces. (a) Thermionic image of BaO-coated steel; (b) photoelectric image of beryllium bronze; (c) photoelectric and (d) (kinetic) secondary emission image of an aluminium alloy. In all cases the aperture was $\alpha = 1.45 \times 10^{-3}$ (from ref.18, courtesy of Dr. W. Engel).

205

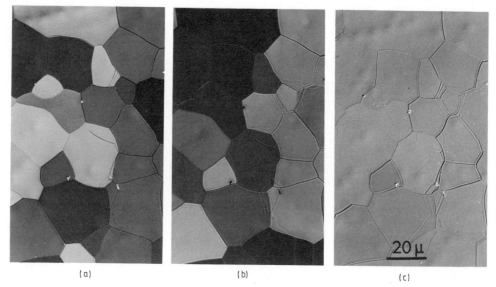

(a)   (b)   (c)

Fig. 8. Thermionic (a), photoelectric (b) and (kinetic) secondary electron
(c) emission images of polycrystalline Ta (from ref.18, courtesy of
Dr. W. Engel).

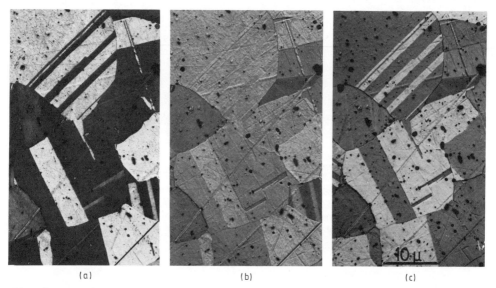

(a)   (b)   (c)

Fig. 9. Photoelectric electron emission images of polycrystalline beryllium
bronze as obtained with the low (a) medium (b) and high (c) energy
region of the UV spectrum of a Hg high pressure lamp (HBO 100)
(from ref.18, courtesy of Dr. W. Engel).

establish dipole and symmetry selection rules which can be strongly crystal
orientation and energy dependent[27]. Similar influences of the penetration
of the incident beam, of the excitation probability and, in particular, of
the transport of the excited electron to the surface occur in secondary
electron emission. However, the details of these influences vary from one
excitation mechanism to the other and with the excitation conditions (angle

of incidence, polarization, energy, etc.) so that many parameters would have to be varied in order to understand the contrast quantitatively. Unfortunately, most of the work has been done up to now with undefined surfaces and the few studies on well-defined surfaces[28,29] do not yet allow a formulation of the contrast mechanism in terms of basic theoretical models such as the band theory of photoemission.

The instrument with which the well-defined surfaces were studied is shown schematically in fig.10. It is a sputter-ion pumped bakeable ultra-high vacuum system with a base pressure of 2×10^{-10} Torr and contains i) an electron-optical system for photoemission microscopy consisting of magnetic cathode objective lens (1), projector lens (2), camera tube with fluorescent screen (to the right, not shown) and UV illumination system (3), ii) a cylindrical mirror analyzer (5) for chemical characterization of the surface, iii) a LEED optics (normal to the plane of drawing, not shown), a rotatable specimen manipulator (4) with up to 6 specimens, which can be heated by an electron gun (6), for positioning the specimen for PEEM, AES or LEED or in front of one of the other flanges in the yz-plane. On these flanges an ion gun for specimen cleaning, evaporators and other devices may be mounted. Thus, a thorough surface characterization is possible although only in the PEEM position with lateral resolution. With a cathode voltage of 50 kV and a distance of 4 mm between specimen and anode a field F = 125 kV/cm can be reached. From the focal length of 25.8 mm a spherical aberration constant of c_s = 80 mm can be derived and from c_s and F a resolution of 12 nm for PEEM. Up to now only low magnifications have been used, mainly for intensity reasons and no high resolution images have been reported.

For comparison with this modern, surface science-compatible instrument, an ill-fated commercial emission electron microscope is shown in fig.11. Although a masterpiece of instrument engineering and although equipped with many excitation sources [heating (2), multiple UV illumination sources (3) and neutral beam source (4)] as well as with a specimen preparation chamber (9), ion gun for cleaning (10), air lock for specimen exchange (11) and cold trap (8), this three-stage magnetic cathode lens system (5) never found acceptance in surface science because of the

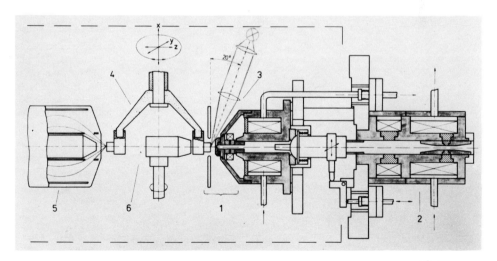

Fig.10. Schematic of ultra-high vacuum photoelectron microscope[28,29]. For explanation see text.

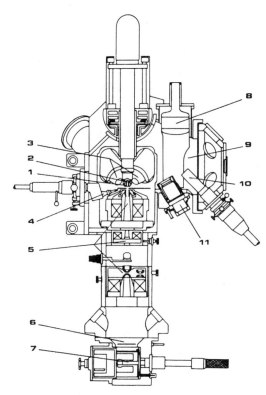

Fig.11. Cross-section of Balzers Metioscope KE 3 (from ref.12). For explanation see text.

hydrocarbon and H_2O-containing vacuum of about 10^{-7} Torr and the concomitant rapid contamination of the specimen (1). It also could not compete with SEM in the study of technological surfaces and, therefore, never gained significant practical importance.

Summarizing this section it can be stated that emission microscopy is a good tool to study the orientation and adsorbate dependence of emission processes if flat well-defined polycrystalline surfaces are studied whose grain orientation and surface structure is known. This suggests the combination of emission microscopy with micro-LEED which will be discussed in the next section. As will be seen, from the point of view of obtaining optimum resolution and surface characterization, LEEM is superior to emission microscopy so that classical emission microscopy may be relegated into a subsidiary technique to LEEM.

LOW ENERGY ELECTRON MICROSCOPY: INTENSITY AND CONTRAST

As shown before, a resolution in the 1 nm range can be expected with LEEM. In order to achieve such a resolution, sufficient intensity and contrast must be available for focussing and astigmatism correction. There are four contributions to the intensity: the reflection at the potential barrier between vacuum and condensed matter, the backward scattering from the ion cores, the three-dimensional diffraction effects due to the bulk periodicity and the two-dimensional diffraction effects involving waves localized in the surface.

The incident electrons experience in the specimen a mean inner potential V_O with typical values from 10 to 15 eV below vacuum level $V = 0$. If the transition from $V = 0$ to V_O were step-like then there would be a noticeable reflectivity for electron energies $E < V_O$ but little above V_O. However, the incident electron polarizes the material and thus experiences an image potential in front of the surface which leads smoothly from $V = 0$ to V_O. Such a barrier has a negligible reflectivity which for all practical purposes may be neglected even at $E < V_O$.

The backward scattering by the ion cores makes a much more important contribution to the intensity. At the low energies of interest in LEEM the Born approximation breaks down and partial wave analysis is the appropriate method[30]. In this theory the incoming plane wave and the scattered wave are expanded in spherical waves with different angular momentum ℓ and the phase shift η_ℓ between them, due to the interaction with the atom, is calculated. The scattering amplitude $f(\theta)$ is then given by

$$f(\theta) = \frac{1}{2ik} \sum_{\ell=0}^{\infty} (2\ell + 1) [\exp(2i\eta_\ell) - 1] P_\ell(\cos \theta) \tag{11}$$

Two θ-values are of interest in LEEM: $\theta = 180°$ and $\theta = 90°$, the second one because of the two-dimensional diffraction effects which can channel effectively intensity into $\theta = 180°$ by double scattering. The $P_\ell(\cos\theta)$-values at these angles are $(-1)^\ell$ for $180°$ and 0 for $\ell = 2n + 1$ for $90°$, compared with a value of 1 for $0°$. Thus all partial waves add up in the forward direction, only even partial waves contribute to $90°$ scattering and even or odd partial waves add in the backward direction. If only one partial wave is dominating, then

$$I(\theta) = |f(\theta)|^2 = \frac{(2\ell+1)^2}{k^2} \sin^2\eta_\ell \, P_\ell^2(\cos \theta) \tag{12}$$

which reaches its maximum value at $\eta_\ell = (2n + 1)\pi/2$. The phases η_ℓ for a given atom depend upon energy and can reach several π at low energies, e.g. $\eta_O \to 5\pi$ for $k \to 0$ in the case of Xe. Several phases reach at least $\pi/2$, e.g. η_4 for Xe. As a consequence, $180°$ and $90°$ scattering is not a monotic function of energy but can have maxima and minima, depending upon the η_ℓ combinations. Figure 12 shows $\eta_\ell(E)$ for $\ell \leq 5$ for W ion cores as represented by Mattheiss' V_2 potential[31,32]. It should be noted that the zero energy is 15 eV below the vacuum level and that no correlation and exchange has been taken into account. The backscattering calculated for such ion core potentials is comparable to the forward scattering at low energies as shown in fig.13 for Ag, Al and Cu for 50 eV electrons. In fig.14 the non-monotonic energy dependence of the backward scattering can be seen clearly as well as the non-monotonic dependence upon nuclear charge Z.

Qualitatively similar results are obtained for scattering from free atoms[34] and from measurements on clean polycrystalline surfaces as shown in fig.15. Thus, it may be concluded that there is sufficient intensity for LEEM at low energies (see fig.13) and that the intensity depends in a complicated manner upon energy and atomic number.

When the specimen is disordered (amorphous or liquid) the backscattered intensity is smeared out over a wide angular range. If the

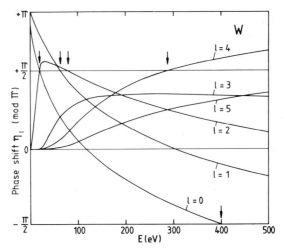

Fig.12. Scattering phase shifts (modulo π) of Mattheiss' V_2 ion cores of W atoms as a function of electron energy.

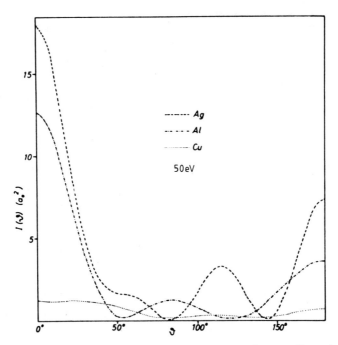

Fig.13. Differential scattering cross sections of Ag, Al and Cu ion cores for 50 eV electrons[33].

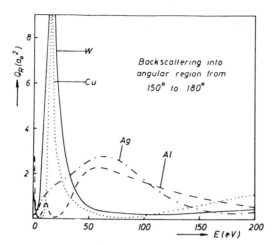

Fig.14. Backscattering cross-section Q_R of W, Cu, Al and Ag ion cores as a function of energy in atomic units[33].

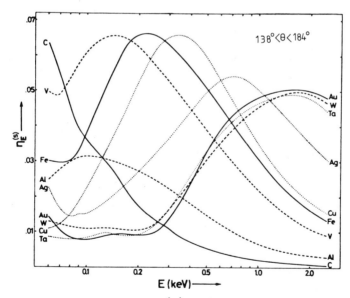

Fig.15. Backscattered fraction $\eta_E^{(s)}$ from polycrystalline surfaces as measured with LEED optics. Some inelastic scattering is also included due to the limited energy resolution of the detector[35].

specimen is crystalline, three-dimensional diffraction can concentrate much of this intensity into the back-reflected (0,0) beam and neighbouring beams. This subject, LEED, is discussed by Pendry[36], so that only an example of the magnitude of the effect of diffraction will be given here. Figure 16 shows the relative intensity of the specularly reflected beam at near-normal incidence as a function of energy. The stronger scattering of Ni compared to Cu is clearly seen, with reflectivities of > 0.5% up to 300 eV; at the lowest energies the reflectivities rise to about 30%. It is possible that on a cubic (100) surface the (0,0) intensity is enhanced at normal incidence because whenever the three-dimensional diffraction condition is fulfilled for the specular beam, it is also fulfilled for four beams parallel to the surface due to the symmetry of the surface. Double scattering can then enhance the (0,0) beam. This phenomenon is less pronounced on (110) surfaces - here only two beams in the surface are excited simultaneously - and absent on the (111) surface.

The beams diffracted parallel to the surface play a much more important role at very low energies (VLEED) where they can dominate the reflectivity. Figure 17 shows the normal incidence reflectivity of W(100) up to 20 eV. The comparison with the band structure in the [100] = Δ direction allows an assignment of the first peak to the band gap in which the electron finds no allowed states and, therefore, is reflected. The main peak at about 7.5 eV is due to a monolayer resonance[38,39]. At this energy, the diffraction conditions for the four {10} beams are fulfilled - taking into account the mean inner potential - and $\eta_2 \approx \pi/2$ (modulo π) with all other phases except η_0 close to 0 (modulo π) (see fig.12). As discussed above, this causes strong 90° scattering and, because of $\eta_0 \approx \pi/2$ (modulo π), at the same time strong overall scattering. The sharp peak in fig.14 is due to this phenomenon. The third peak in fig.17 at about 16.5 eV has a similar cause to the main peak, with the four {11} beams parallel to the surface being excited. Finally, the shoulder at about 10.5 eV is attributed to a surface barrier effect[40]. Thus, it is apparent that at very low energies the very surface has - via diffraction effects - a strong influence on the backscattered intensity although the inelastic mean free path λ is larger (> 1 nm) at such energies than at about 50 eV

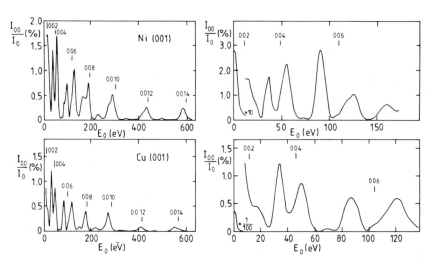

Fig.16. Intensity of the (0,0) beam from Ni(001) and Cu(001) at near-normal incidence as a function of energy. Top: Ni, bottom: Cu, left: 0 - 600 eV, right: low energy range (after ref.37).

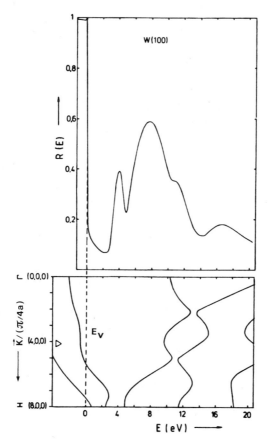

Fig.17. Normal incidence reflectivity of a clean W(100) surface at very low
        energies. For comparison the totally symmetric energy bands along
        the Δ = [100] direction to which the incident wave has to match are
        also shown[38].

where it is ≈ 0.5 nm[33]. In addition to the high surface sensitivity and
the high intensity - reaching more than 50% of I - there is an additional
reason for using very low energies in LEEM, albeit at the expense of
resolution (see fig.2). The weak inelastic scattering and the low
secondary electron yield make energy filtering unnecessary. For example up
to about 10 eV the total inelastic scattering for W is less than 10% of the
elastic scattering, most of which is in the (0,0) beam[38]. Inelastic
scattering is generally weak up to the threshold of plasmon creation which
is about $1.6E_p$, E_p being the plasmon energy[33] (see Howie, this book).

    In surface science the clean surface is usually not of interest -
phase transitions of clean reconstructed surfaces excepted - but a surface
modified by some adsorbate or condensate. It is, therefore, important to
know how sensitive the normal incidence reflectivity is to adsorbed atoms.
Figure 18 shows an extreme example. Hydrogen has such a small scattering
cross-section at standard LEED energies (> 20 eV) that it can be detected
only via the surface re-arrangement caused by it. Towards low energies,
however, the scattering cross-section of H increases strongly so that a H
monolayer has a significant reflectivity R_H reaching values above 50% (see
fig.18). Of course, in the real adsorbate, there is also diffraction and

substitute modification by the adsorbate which modulates R_H so that the reflectivity change ΔR caused by the adsorbate shows considerable structure as a function of energy. ΔR is in the 10% range and should be easily detectable by LEEM. It is important to note that ΔR is usually a strongly nonlinear function of coverage Θ so that Θ cannot be deduced in a simple manner from the image intensity. For example, 1/4 of a monolayer of oxygen on W(110) reduces the reflectivity by about 50% at 2 eV and 20% at 13.6 eV but further adsorption causes only changes of 10 - 20%[38]. Thus adsorption can cause either reflectivity increases as in the case of H on W(100) or decreases [0 on W(110)] which are large enough to be easily detected by LEEM.

From the theoretical and experimental data it can be concluded that about 50% to 0.5% of the incident intensity is reflected in the (0,0) beam in the energy range up to about 500 eV. With a well-designed gun it is no problem to achieve a current density $j_O = 10^2$ Am^{-2} at the specimen. Assuming a reflectivity $R = 10^{-2}$, a resolution $\delta_O = 10$ nm, and a detector resolution $\delta_D = 10^{-4}$ m so that a magnification $M = 10^4$ is appropriate, one obtains for the current density in the image $j_I = j_O R/M^2 = 10^{-8}$ Am^{-2}. j_I can be enhanced by a channel plate intensifier by a factor $V \approx 10^4$ so that a current density on the screen $j_S = 10^{-4}$ Am^{-2} can be achieved which is more than sufficient for observation. Of course, the image must also be bright enough to allow focussing. Here, the contrast is important.

Consider two neighbouring image elements with linear dimensions δ_D and signal difference $\Delta S = S - S'$. The usual condition for contrast recognition is $\Delta S/S > 3$ N/S. Assuming statistical noise $N = \sqrt{S}$, the signal needed is $S > 9(S/\Delta S)^2$. For $\Delta S/S = 1/10$ this yields $S > 900$ e/s. As the channel plate intensifier also multiplies the noise, the signal on its entrance has be considered. This is given by $S = j_I \delta_D^2 = 10^{-16}$ A = 1600 e/s so that the condition $S > 900$ e/s if fulfilled and the intensity is sufficient for focussing.

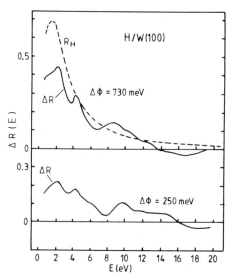

Fig.18. Reflectivity change ΔR of a W(100) surface as a function of energy for two hydrogen coverages Θ ($\approx 1/4$, $\approx 3/4$). The dashed curves show the calculated reflectivity of a monolayer of independently scattering hydrogen atoms[38,41].

The preceding considerations lead immediately to the question of contrast. In principle, there are four different contrast mechanisms: i) geometric phase contrast, ii) ion core scattering contrast, iii) diffraction contrast and iv) absorption contrast. Geometric phase contrast is due to surface irregularities such as steps and is the only contrast available on an otherwise homogeneous surface. It is caused by the geometrical path differences between incident and reflected waves which are connected with the irregularities. By proper defocus the phase contrast can be converted into an amplitude contrast so that the irregularity can be imaged. Thus it is possible, for example, to measure the height of steps with atomic resolution by determining the beam energies of maximum contrast[42,43].

Ion core scattering contrast is due to differences in the real and imaginary part of the scattering amplitude $f(\theta) = |f(\theta)| \exp[i\phi(\theta)]$ between different atoms and is the only contrast - in addition to contrast mechanism iv) - which is available on a disordered (amorphous or liquid) smooth surface. Little experience is available with such surfaces up to now. An amorphous NiZr alloy surface produced no contrast, probably due to the homogeneous distribution of its constituent atoms on the scale of the resolution of the instrument (> 10 nm). Also, the intensity was quite low because of the absence of intensity concentration into diffracted beams.

When diffraction can occur, i.e. in crystalline specimens, diffraction contrast is the dominant contrast mechanism. As discussed above, this is partially due to three-dimensional diffraction, which distinguishes not only different grain orientations but also different surface periodicities and compositions because of their influence on the wave field entering the crystal, and partially due to two-dimensional diffraction. As illustrated by the example H/W(100) and O/W(110), strong diffraction contrast can be expected even for weakly scattering adsorbates.

The last mechanism, absorption contrast, can be expected to occur only between chemically and/or structurally different regions which differ sufficiently in inelastic scattering cross-sections. These cross-sections depend mainly upon electron density[33] and the type of bonding which would have to differ significantly for absorption contrast to compete with diffraction contrast. In general, absorption contrast is probably unimportant, so that two major contrast mechanisms remain: diffraction contrast and geometrical phase contrast.

The inelastic scattering just mentioned can have significant influence on the specimen by electron stimulated desorption (ESD) of adsorbed atoms and molecules[44] or by dissociation of ionically bonded materials. The cross-sections for these processes have maxima at energies of several hundred eV so that this energy range should be avoided in the case of endangered specimens. This is another argument for doing LEEM at energies below 10 eV which usually are below the thresholds for these processes. When operated in the VLEEM mode, LEEM is in this respect far superior to SEM (and REM) where ESD is a well-known disadvantage.

LOW ENERGY ELECTRON MICROSCOPY: INSTRUMENT AND APPLICATIONS

After the conception of LEEM in the early sixties[11] there were several attempts, most notably that of the Brno group[44a], to build LEEM instruments but only the one described below has so far produced good quality images. In its basic construction[44b] it was completed in the late sixties by G. Turner but required considerable improvements before it became operational in the early eighties[17,42]. Since then a number of further modifications have been made. The present instrument is shown

schematically in fig.19 and its physical appearance in fig.20. The heart
of the instrument is a Bartz type cathode objective lens (shown also
schematically in fig.1c[20]) consisting of specimen 1, centre electrode 2,
anode 3, objective aperture 4 and magnetic stigmator 5. Incident and
diffracted beams are separated by a magnetic deflection field 6 with low
astigmatism. The incident beam is produced by a high brightness field
emission electron gun 16 and focussed by a magnetic lens 17 into the back
focal plane of the cathode lens. The diffracted beams enter the imaging
column whose main components are the intermediate lens 9, the projective
lens 10, the image intensifier 11 and the fluorescent screen 12.
Observation and/or recording is via a TV camera 13 which is connected to a
monitor or to a video recorder, or via direct photography. The
intermediate lens 9 is not only used to image the back focal plane of the
objective lens (with objective aperture 4 removed) onto the object plane of
the projective lens 10 for LEED, but also (with objective aperture 4
inserted) for more flexibility in imaging. The deflection coils 7 serve
for beam alignment, the quadrupoles 8 for aberration correction.

Emission microscopy is possible in three modes: i) in thermionic
emission by heating the specimen by electron bombardment from the back
side, ii) in photoemission by illumination with UV radiation from a high
pressure Hg lamp 15 and iii) in secondary emission by bombardment with
electrons from an auxiliary electron gun 14. The operation voltage U_0 is
in general 20 kV. In addition, the specimen can be bombarded with ions or
exposed to gas or condensable vapour beams (molecular beam epitaxy) in the
observation position or swung away from it. Specimen cleaning is done by
the usual surface science preparation procedures, i.e. heating without or
in reactive gases plus flashing, sputtering plus annealing, etc. The
instrument is pumped by a 270 ℓ/s sputter ion pump. After bake-out of the
complete instrument at 170°C, a base pressure below 1×10^{-10} Torr is
achieved. Because of the all-stainless steel construction and the absence
of magnetic shielding, DC and AC magnetic fields are compensated by large
Helmholtz coils surrounding the complete microscope. Air springs provide
vibration isolation.

As a first example of the application of LEEM in surface studies
fig.21 shows the image of a clean Mo (110) surface[42]. The lines are the
surface steps which were frozen in upon cooling from the high flashing
temperature (> 2000 K). They have mono-atomic height as determined from

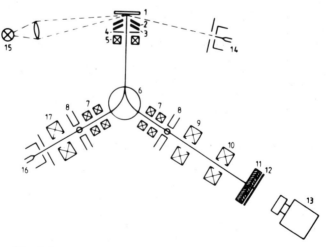

Fig.19. Schematic of LEEM instrument described in text[45].

Fig.20. The LEEM instrument with which the data described in this section were obtained[42],[43].

their appearance and disappearance with increasing energy. They are caused by geometric phase contrast - due to the different path lengths of electrons from the two terraces separated by the step - which is converted into an amplitude contrast by slight defocus. The step density is much larger in other regions of the surface although the crystal surface is oriented to within 0.05° from the (110) surface (see fig.4 in ref.42 and fig.5 in ref.45). Nevertheless, the Mo(110) surface is a rather perfect surface when compared to the Au(100) surface.

This surface appears so disordered in LEEM that one would not expect a good LEED pattern on first sight. Nevertheless an excellent LEED pattern of the reconstructed surface, which has an approximate (5x1) structure with two equivalent, 90° rotated domains, is obtained[45],[46]. Apparently, during the phase transition at T = 1050 K from the high temperature unreconstructed to the low temperature reconstructed surface a large number of terraces are formed whose dominant step edges determine the orientation of the (5x1) domains. This can be seen quite clearly in the image of an etch pit (fig.22) by comparing the bright field image a) which was taken with the aperture centred on the (00) beam (a) in fig.22b with the dark field images c) and d) which were taken with the respective superstructure reflections in the LEED pattern (fig.22b). The etch pit consists of flat terraces which are separated by facets or unresolved dense sequences of steps. The comparison between LEED and dark field LEEM pattern shows that the long dimension of the (5x1) unit mesh is preferentially aligned normal to the steps. It should be noted that dark field images taken with the {1/5,0} reflections have nearly the same resolution (≈ 15 nm) as the bright field images, but, with increasing n, resolution deteriorates in images taken with {n/5,0} reflections. The kinetics of the phase transitions at T_t = 1050 K can be studied quite well with LEEM. It is clearly a first order transition with instantaneous disappearance of many (5x1) domains at T_t. Other domains, however, decrease in size by sudden jumps with continuous shrinking in between. Whether this is an impurity or a strain effect can only be decided with additional studies.

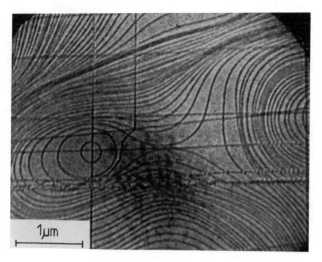

Fig.21. LEEM micrograph of a Mo(110) surface taken with 14 eV electrons[42].

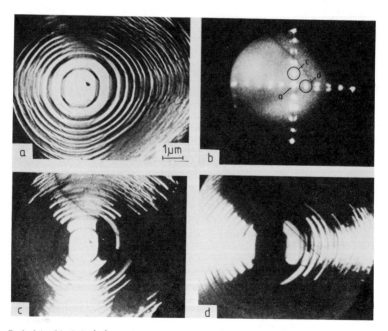

Fig.22. Bright field (a) and dark field LEEM images (c,d) of an etch pit on a Au(100) surface taken with the reflections a,c,d marked in the LEED pattern (b) with 16 eV electrons[46].

A third example is the Si(001) surface. It is also reconstructed but in an apparently much simpler manner: the dangling bonds of the atoms in the topmost layer lead to dimer formation which produces a (2x1) or (1x2) structure depending upon their orientation which changes by 90° from one to the next monolayer (see fig.23a). Simultaneously, subsurface atoms are also displaced but, for the understanding of the image contrast, only the (2x1) and (1x2) periodicity is important. At normal incidence of the illuminating beam, both domains are equivalent and only step contrast is seen. By tilting the beam in the [110] or [1̄10] directions the equivalence is destroyed and strong diffraction contrast occurs. Figure 23b shows the contrast between the two domains obtained by a tilt of 4° at E = 6 eV. Several features should be noted: i) there is no internal structure, i.e. no domain boundary within the terraces, ii) opposite ends of the terraces are smooth and rough, respectively, and iii) smooth edges turn rough and vice versa when the edge orientation changes by 90°. This can be understood by the existence of two types of steps on the surface (fig.23a), one in which the dimer bonds in the upper terrace are parallel to the steps (A), the other in which they are normal to the steps (B). From the contrast dependence on the tilt direction it can be concluded that type B steps are smooth, type A steps rough. Upon annealing, type B steps partially convert into type A steps by facetting[45]. At high temperatures (T > 1300 K) the steps move and the surface becomes rougher due to evaporation. The kinetics can be studied quite well at video rates. Formation of biatomic steps was not seen at any temperature, only strong preponderance of one terrace type for special local surface orientations. However, even in this case the smaller terraces had a width above the resolution limit.

The most interesting surface studied up to now with LEEM is the Si(111) surface. This surface has a "(1x1)" structure at high temperatures

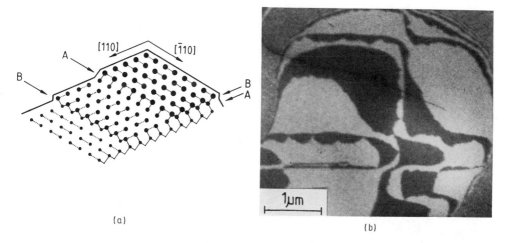

(a)   (b)

Fig.23. Terraces of mono-atomic height on a Si(100) surface. (a) Schematic illustration of type A and type B steps. The dimer bonds are indicated by dotted lines, the bonds connecting atoms in different layers by solid lines. Only the topmost atoms are shown. [Adapted from D.E. Aspnes and J. Ihm, Phys. Rev. Letters 57, 3054 (1986)]. (b) LEEM micrograph of terrace distribution taken with 6 eV electrons and tilted illumination[45].

and reconstructs into a (7x7) structure at 1100 K (see also the Chapters by Gibson and by Garcia). Although LEEM cannot resolve the (7x7) periodicity, it is ideally suited to study the details of the phase transition. Although contrast could be obtained by imaging with a superstructure reflection, this is not necessary because of the different $I_{oo}(E)$ dependence of the "(1x1)" and the (7x7) structure. By choosing the proper energy E, strong dark/light or light/dark contrast can be obtained between "(1x1)" and (7x7) regions. The microstructure of a Si(111) surface and the (7x7)↔(1x1) phase transition depend considerably on the surface treatment, although the LEED patterns are practically indistinguishable except for some difference in background. A surface heated to 1450 K shows well-developed terraces separated by mono-atomic steps which produce only weak step contrast in the "(1x1)" structure but are well visible below T_t = 1100 K because they are nucleation sites of the (7x7) structure. Thus in the early stages of the phase transition the (7x7) nuclei decorate the steps (fig.24a); in the later stages the steps are visible via domain boundary contrast (fig. 24b)[47,48]. On such a surface the phase transition proceeds until completion within $\Delta T = 6^{\circ}K$ below T_t. For $\Delta T > 12^{\circ}K$, homogeneous nucleation on the terraces is seen in addition to the heterogeneous nucleation on the steps. The two-dimensional (7x7) crystals have triangular shape with the corners always pointing in the $\langle 11\bar{2} \rangle$ directions. Again, the growth kinetics can be followed quite well with the video camera at low supersaturations ($\Delta T < 20^{\circ}K$).

If the Si(111) surface is heated for several hours at temperatures only slightly above T_t a completely different surface is obtained whose precise structure seems to depend upon impurity content. In the extreme case the surface appears structureless above T_t without any indications of surface steps and develops a very irregular and spotty distribution of (7x7) regions below T_t, which grow rather sluggishly with decreasing temperature and never cover the surface completely[43,45]. This type of surface and the apparent second order phase transition is probably the one studied in previous LEED phase transition work.

The examples given above and several other preliminary studies which have been made up to now clearly show the advantages of LEEM compared to

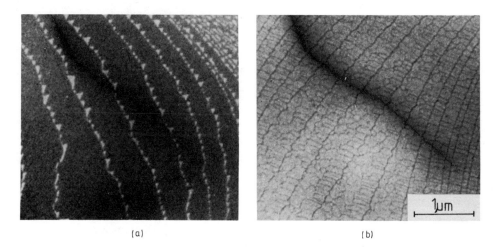

(a)   (b)

Fig.24. LEEM images of a Si(111) surface. (a) in the early stages of (7x7) formation at low supersaturation, (b) after rapid cooling. Electron energy 10 eV[47].

emission microscopy: i) step or diffraction contrast on chemically homogeneous surfaces which do not show any features (except larger imperfections such as precipitates, etch pits or very large, field-distorting steps) in emission microscopy, ii) superior resolution and intensity and, most of all, iii) the possibility of studying the kinetics of surface phenomena over a wide temperature range. Emission microscopy - usually photoemission microscopy - is used in the instrument, therefore, only occasionally for initial alignment after bakeout, for scanning the specimen and for monitoring the specimen cleaning. Furthermore, by combining LEEM and LEED, it is seen that, in spite of the lack of lateral atomic resolution, important information can be extracted on details on an atomic level, e.g. the stability of steps with different structure.

SPECTROSCOPIC EMISSION MICROSCOPY

Emission microscopy with characteristic electrons is a logical extension of LEEM for two reasons: i) emission processes involving electrons with energies characteristic of the material studied such as X-ray photoemission or Auger electron emission produce rather narrow peaks in the energy spectrum of the emitted electrons. Therefore, usually $\varepsilon \ll \rho$ and the resolution considerations made previously for LEEM are applicable which promise high resolution. ii) LEEM gives only structural information. Its combination with chemical imaging would be an invaluable addition to surface science. The idea of this type of imaging is not new[48,49] but no images with useful resolution have been obtained to date. Therefore, only the potential of this method and the achievable resolution can be discussed on the basis of what is at present known about the factors involved. Before turning to the emission microscopies it should be mentioned briefly that characteristic energy loss microscopy is also a possibility provided sufficiently sharp and intense losses exist. The best candidates for this type of microscopy are atoms with unfilled 4f and 5f electrons such as the rare earths and actinides in which intense sharp spin-flip excitations occur causing energy losses of the magnitude of the exchange splitting (several eV)[50]. Too little is known at present, however, about excitation functions and other parameters so that it is premature to speculate about the potential of energy loss microscopy. Furthermore, quantitative analysis is complicated by the coupling of the loss intensity to the no-loss intensity via loss-diffraction and diffraction-loss processes.

X-ray photoemission

The sharpest peaks of characteristic electrons and the lowest background B can be obtained by X-ray photoemission either with characteristic radiation from X-ray tubes or with monochromatized synchrotron radiation. Figure 4e illustrates this for the region of the Ag 3d electrons as excited by unmonochromatized Mg Kα radiation (hν = 1254 eV). The FWHM of the $3d_{5/2}$ peak at 880 eV is 1 eV, the S/B ratio (B at 6 eV above the $3d_{5/2}$ peak is about 20:1. The Mg Kα-excited electron energy distribution over a wider energy range in fig.25a shows a strong variation of the photoelectron intensity of the different core levels. The particularly high 3d photoelectron cross-section of Ag (Z = 47) can also be seen in fig.26 which shows the photoionization cross-sections σ as a function of atomic number Z for Aℓ Kα radiation (hν = 1487 eV). However, all atoms except H and He have nℓ shells with $\sigma_{n\ell} > 10^{-25}$ m² as compared to $\sigma_{3d} = 1.4 \times 10^{-23}$ m² for Ag, e.g. $\sigma_{3p} \approx 1 \times 10^{-24}$ m² for Si (fig.25b). The strong decrease of the cross-sections with decreasing Z is caused by the decrease of the excitation function with increasing E/E_i, E = hν being fixed and the ionization energy E_i decreasing with decreasing Z. The dependence of the $\sigma_{n\ell}$'s upon photon energy hν is illustrated in fig.27 for

Xe. For the optimum photon energy (hν ≈ 100 eV) the 4d cross-section (E_i = 68 eV), which is about 5×10^{-24} m^2 at hν ≈ 1500 eV, can be as high as 3×10^{-21} m^2. A less dramatic increase from 1.2×10^{-23} at hν = 1500 eV (see fig.26) to a maximum of about 5×10^{-22} m^2 (theory) or 2×10^{-22} m^2 (experiment) is seen for the Au 4f electrons in fig.28. The last two figures show that the nℓ photoelectron yield can be increased significantly

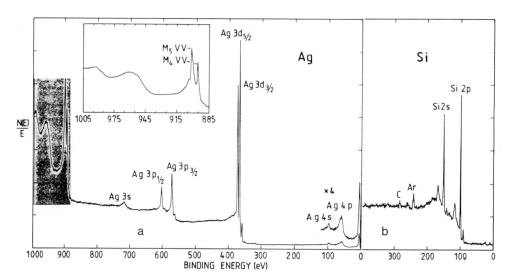

Fig.25. Electron spectra excited by Mg Kα radiation (hν = 1254 eV) in Ag (a) and Si (b). In (a) the peaks in the dark region and in the inset are due to Auger electrons, above 400 eV are due to photoelectrons; in (b) only the photoelectron region is shown. The two spectra have been recorded with different sensitivities. Note the different background above the $M_{4,5}$ Auger peak and the 3d photoelectron peak (adapted from ref.51).

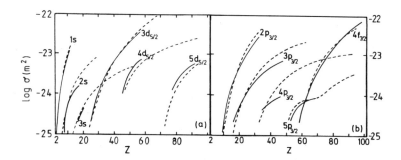

Fig.26. Photoionization cross-sections of the atomic levels as a function of atomic number at an excitation energy hν = 1.5 keV. These values apply not only to Aℓ Kα (hν = 1487 eV) but approximately also to Mg Kα (hν = 1254 eV) (adapted from ref.52). Dashed lines: theoretical values, solid lines: adjusted experimental values.

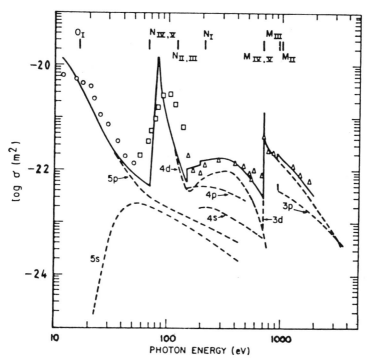

Fig.27. Photoionization cross-section of Xe as a function of photon energy.
The dashed lines are the calculated contributions of the $n\ell$
subshells, the solid line the calculated total cross-section. The
circles, squares and triangles are experimental total cross-
sections (adapted from ref.53).

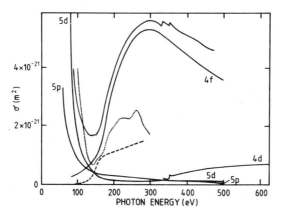

Fig.28. Photoionization cross-section of Au as a function of photon energy.
The solid lines are the calculated partial ($n\ell$) and total cross-
sections[53], the dotted line is the experimental total cross-section
as obtained from the absorption coefficient μ[54] and the dashed line
σ_{4f} in arbitrary units[55].

223

if the incident radiation can be tuned to near the maximum of the excitation function of the $n\ell$ electrons. This can be done with synchrotron radiation[56,57]. It has been stated that synchrotron radiation above 1000 eV from presently available sources cannot compete with characteristic X-ray lines in ordinary photoemission experiments[57]. However, this comparison assumes X-ray tube powers in the kW range, one order of magnitude above the power used in ordinary UHV photoemission experiments. Furthermore, in emission microscopy it is not the photon flux at the specimen which is of importance, but the flux density, which is larger with synchrotron radiation. Synchrotron radiation facilities will be even more superior once wigglers and undulators are available which are expected to increase the flux density/eV by 1 - 3 orders of magnitude[56,58].

Up to the present, Al Kα (1487 eV), Mg Kα (1254 eV) and, to a much lesser extent, Y Mξ (132.3 eV) characteristic radiation with line widths of 0.5 - 0.8 eV and photon fluxes of 3×10^{11} - 1×10^{12} photons/s[59], are the sources which have to be used as a basis for discussion. From the specifications of commercial X-ray photoelectron spectrometers (XPS) a maximum photoelectron brightness $B_O \approx 1\times10^{14}$ e/(s m^2 sterad eV) for 500 W power can be deduced for the Ag 3d$_{5/2}$ peak excited by Mg Kα or Al Kα radiation. If T_θ and T_E are the angular and the energy transmission, respectively, of the electron optical system then the current at the detector originating from a specimen area δ_O^2 (δ_O being the desired resolution) is given by

$$ I_D = 4\pi B_O \delta_O^2 T_\theta T_E \qquad\qquad (13) $$

The energy transmission is defined by $T_E = \Delta E/\Delta E_e$ where ΔE is the band width of the energy filter and ΔE_e the half width of the characteristic electron peak. For the Ag 3d$_{5/2}$ peak $\Delta E_e \approx 1$ eV with unmonochromatized X-ray excitation, so that $T_E = 1$ for a band width of 1 eV. The angular transmission T_θ depends upon the angular distribution $f(\theta)$ of the electrons and is defined by

$$ T_\theta = \int_0^{\alpha_A} f(\theta)\sin\theta d\theta \; / \int_0^\pi f(\theta)\sin\theta d\theta $$

For an isotropic distribution $T_\theta = (1 - \cos\alpha_A)/2 \approx \alpha_A^2/4$, for a $\cos\theta$ distribution, which appears more appropriate for a bulk clean specimen, $T_\theta = \sin^2\alpha_A/2 \approx \alpha_A^2$. With the optimum aperture for E = 880 eV, $\Delta E = 1$ eV (Ag 3d$_{5/2}$, Mg Kα) and $U_O = 25000$ V, which is about $R_A \approx 5\times10^{-3}$ (i.e. $r_A = 15\mu m$ for L = 3 mm), $\sin\alpha_A \approx 0.016$ and $\delta_O \approx 2.25$ nm for an electron optical system whose resolution is determined by the homogeneous acceleration field one obtains $T_\theta \approx 12.6\times10^{-5}$ and for the current per image element $I_D \approx 8\times10^{-21} B_O$ e/s. In the example given above, $I_D \approx 8\times10^{-7}$ e/s. Even for zero background signal and the best image intensifier, intolerable integration times are needed for image recording with acceptable contrast, e.g. between a surface region covered by a thick Ag layer and a Ag-free region, as can be seen by a S/N consideration similar to that made above for LEEM. The improvement in B_O by wigglers and undulators will be helpful but will not solve the intensity problem.

This intensity problem can be overcome only by increasing δ_O and T_θ, quantities which both increase with increasing R_A. For fixed R_A, δ_O and T_θ increase with decreasing energy (ρ) because of increasing $\sin\alpha_A$ (eq. 5). For example, for $R_A = 1\times10^{-2}$, but otherwise the same conditions as above, $\sin\alpha_A \approx 0.032$, $T_\theta \approx 5\times10^{-4}$ and $\delta_O \approx 3.5$ nm [from eqs 3, 4 and 6], resulting in $I_D \approx 7.8 \times 10^{-20}$ B_O e/s, i.e. one order of magnitude has been gained in intensity by accepting a deterioration of the resolution by a factor of 1.5 compared to the optimum theoretical resolution. At $R_A = 5 \times 10^{-3}$, $\delta_S \ll \delta_C \approx \delta_D$, while at $R_A = 1 \times 10^{-2}$ $\delta_S \approx \delta_C \approx 2.5 \ \delta_D$. A further increase of R_A causes a rapid increase of δ_S and therefore of δ_O: for $R_A = 2 \times 10^{-2}$, $\delta_O = 18.5$ nm and $I_D \approx 8.8 \times 10^{-18}$ B_O, for $R_A = 4 \times 10^{-2}$, $\delta_O = 143$ nm and $I_D \approx 2 \times 10^{-15}$ B_O. With the last aperture the intensity is sufficient for image recording in an acceptable time but not for focussing.

The best way to focus is via LEEM with reflected electrons of the same energy as the photoelectrons. In order to have sufficient intensity for focussing in LEEM, much lower energies are needed, e.g. 100 eV. Therefore, the X-ray source should be tuneable to photon energies from zero to several hundred eV above the photoionization threshold which is usually also the region with maximum photoelectron yield. Thus, synchrotron radiation is the ideal X-ray source for this kind of microscopy not only from the point of view of intensity (in the future) but also from the point of view of photon energy and its tuneability to optimum imaging energy and photoelectron yield. If the expected increases of the X-ray flux density/eV can be realized, resolutions in the 20 nm range appear feasible. For example, if the Ag $3d_{5/2}$ electrons are produced by 468 eV X-rays, so that their energy is 100 eV, then with $R_A = 2 \times 10^{-2}$, $\delta_O = 53$ nm is $I_D \approx 5 \times 10^{-16}$ B_O e/s. If the object area δ_O^2 is imaged onto the detector area, δ_D^2 ($\delta_D = 10^{-4}$ being the detector resolution, $M \approx 2 \times 10^3$) a current density of 8×10^{-27} B_O Am^{-2} can be obtained on the channel plate or 8×10^{-23} B_O Am^{-2} on the fluorescent screen. If B_O could be increased to 10^{17} e/(s m^2 sterad eV) the image could even be observed visually! Of course, all these considerations are based on a very favourable specimen such as a thick Ag specimen. Submonolayer coverages are far out of reach.

Auger electron emission

In view of these considerations, Auger electron microscopy will now be examined for its potential. Auger electron emission (for reviews see e.g. ref.60, 61) has three disadvantages compared to synchrotron radiation-excited photoemission: i) the signal sits on a strong, frequently sloping background, ii) only a few fixed energies are available for each element and iii) the characteristic signal is distributed in general over a much wider energy range. This last disadvantage is due to the fact that most intense Auger lines involve valence electrons and the intensity is spread, to a first approximation, over the self-convolution of the valence density of states in band-like spectra or over many multiplet states in atomic-like spectra. This, as well as disadvantage i), can be seen quite well by comparing the electron-excited Ag 3d Auger signal ($M_{4,5}$ VV electrons) in fig.4f with the photoemission spectrum in fig.4e or the corresponding X-ray excited features in fig.25. Therefore, although a much higher B_O value can be achieved by illuminating the surface with a high brightness electron gun, the influence of the disadvantages i) and iii) partially compensates this advantage. Again the favourable example of Ag will serve as an illustration and, in addition, imaging with Si Auger electrons will be considered, too. Beforehand, however, some general remarks have to be made. First of all, the inner shell ionization cross-sections are of the same order of magnitude for photons and electrons. The probability $a_{ik\ell}$, that the inner shell hole i is filled by an Auger transition is close to one for those levels i whose binding energy E_i is so low that their ionixation cross-section σ_i is acceptably large.

Figure 29 illustrates the Auger electron yields which can be expected from bulk materials for excitation with 3 keV electrons. With the same σ_i-functions as used in fig.29 one obtains for the Ag MNN (=MVV),

$E_p = 1.5$ keV, and for the Si LMM (-LVV), $E_p = 650$ eV bulk yields $Y^b_{Ag} \approx 1 \times 10^{-2}$ and $Y^b_{Si} \approx 3 \times 10^{-2}$, respectively. If isotropic emission is assumed, then the current at the detector originating from a specimen area δ^2_0 upon bombardment with an incident current density j_p is given by

$$I_D = Y j_p \delta^2_0 T_\theta T_E. \qquad (14)$$

This expression can be compared directly with eq.13 by replacing $4\pi B_0$ with $Y j_p$ and keeping in mind that now $T_E < 1$ because of the large width of the Auger peaks. With a properly designed electron optical system, j_p values of about 1×10^4 Am^{-2} = 1.6×10^{23} e/s m^2 appear reasonable, even in the unfavourable geometry of the immersion lens. In this case $Y j_p \approx 1.6 \times 10^{21}$ and $\approx 5 \times 10^{21}$ e/s m^2 for Ag and Si, respectively. Compared to the $4\pi B_0$ value for X-ray induced photoelectron emission quoted above, this is a gain of many orders of magnitude. This gain is only slightly reduced by the smaller energy transmission. For the same energy analyzer band width as discussed above, $\Delta E = 1$ eV, $T_E \approx 1$ for the Si LVV peak which has a FWHM of $\Delta E_e \approx 10$ eV (see fig.30a) and $T_E \approx 0.15$ if the 1 eV energy window is centred on the M$_4$VV peak (see fig.4f). This means that about one order of magnitude is lost in I_D due to the large width of the Auger peaks caused by the fact that Auger emission is a two-electron process and involves the valence band in the case of high intensity transitions. As an example, consider the Si LVV electrons with $E = 89$ eV for which approximately the 100 eV data from above may be used. In this case for $R_A \approx 2 \times 10^{-2}$ $I_D = 5 \times 10^{21}$ x $(53 \times 10^{-9})^2$ x 14×10^{-3} x 0.1 e/s $\approx 2 \times 10^4$ e/s and with $\delta_D = 10^{-4}$ m the current density on the channel plate is 3.2×10^{-7} Am^{-2}, orders of magnitude larger than the best expected X-ray sources! It should, therefore, be possible to reduce R_A, e.g. to 1×10^{-2} which for 100 eV electrons gives $\delta_0 = 11$ nm, $T_\theta = 3.6 \times 10^{-3}$ and $I_D = 2 \times 10^2$ e/s and thus still allows visual observation upon image intensification. Otherwise the

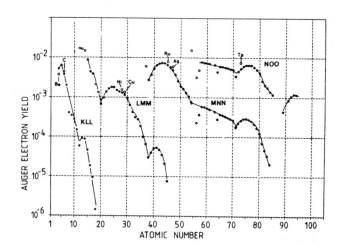

Fig.29. Calculated Auger electron yields of bulk materials, excited with 3 keV electrons (number of Auger electrons per incident electron). The calculations assume simplified ionization cross-section expressions, $a_{ik\ell} = 1$, a backscattering factor $r = 1$ and escape of all electrons from a surface slab with bulk density and with the thickness $d = \lambda_i$, λ_i being the inelastic mean free path. Experimental points for some indicated elements are also shown (adapted from ref.62).

considerations made above for photoelectrons may be immediately transferred to Auger electrons, with one important exception: the background may not be neglected (disadvantage ii).

This background has two consequences: i) the minimum Auger signal S_A needed for a given contrast must be based on the S/N considerations involving Auger signal <u>plus</u> background, $S = S_A + B$, and ii) two images must be taken for background subtraction, one at the Auger energy, the other with the background electrons above the Auger peak, e.g. in the case of Si, at 89 eV and about 100 eV (see fig.30a) and, in the case of Ag, at 350 eV and at 365 - 370 eV (see fig.4f). There is no quantitative information from theory concerning the background. Therefore the experimental data shown in fig.4f and fig.30 have to be used. They were obtained at normal emission with a primary beam incident at $\theta_p = 75°$ with $E_p = 650$ eV for the Si peak (fig.30a,b), $E_p = 1800$ eV for the Ag peak (figs 4f and 30c), using a 127° analyser. Normalized to the same incident current $I_p = 3 \times 10^{-6}$ A, the following average S_A and B count rates were measured for the clean Si(111) surface, the Si(111)-($\sqrt{3} \times \sqrt{3}$)Ag surface (Ag monolayer) and the thick Ag(111) layer, a) for the Si peak (89 eV) and background (100 eV): 10500 and 7000, 6000 and 7400, 9000 (no Si peak!) and 8500, respectively; b) for the Ag peak (350 eV) and background (365 eV): no data for clean surface, 1250 and 4600, 3500 and 5200, respectively[23]. Between 0 and 1 monolayer the Si and Ag signals decrease and increase linearly, respectively.

From these relative data and the Y_{jp}, T_E values given before, the aperture R_A necessary for achieving a desired sensitivity $D = \Delta C_a / C_{am}$ (ΔC_a = detectable adsorbate concentration difference, C_{am} = adsorbate monolayer concentration) may be estimated now. First of all, a comparison of the Si and Ag count rates for the Ag monolayer gives a much better signal/background ratio, 0.81 vs. 0.27 for the Si image compared with the Ag image and, consequently, also better contrast. However, the Si image gives no chemical information about the adsorbate and only indirect information on its coverage. Therefore, only imaging with Ag Auger electrons will be discussed. Consider two pixels (image "points" with area

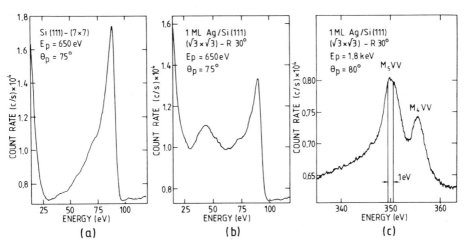

Fig.30. Angle resolved Auger electron spectra (a) of a clean Si(111) surface, (b) of a Si(111) surface covered with one Ag monolayer, both with $E_p = 650$ eV, $\theta_p = 75°$ in [101] azimuth, $\theta = 0°$; (c) Ag $M_{4,5}$ VV spectrum of one Ag monolayer with $E_p = 1800$ eV, $\theta_p = 80°$ in [10$\bar{1}$] azimuth, $\theta = 0°$[23].

$\delta_D{}^2$) i and k corresponding to specimen regions (with area $\delta_o{}^2$) with adsorbate concentration C_{ai} and C_{ak}. Because of the linear increase of the Ag Auger signal S_A up to one monolayer $D = (S_{ai} - S_{ak})/S_{am} = (n_{ai} - n_{ak})/n_{am}$ with n (= St) being the number of electrons accumulated during the image integration time t. The Auger signal S_A is superimposed on the background B which, to a first approximation, may be assumed to be independent of coverage. The total signal in pixel i is, therefore, $S_i = S_{ai} + B$, the corresponding number of electrons $n_i = n_{ai} + n_b$, with a standard deviation $\sigma_i = \sqrt{n_i}$. In order to distinguish two pixels i, k with a 99.7% confidence level, the condition

$$n_i - n_k \geq 3\sigma = 3(\sigma_i{}^2 + \sigma_k{}^2)^{1/2} = 3(n_i + n_k)^{1/2}$$

must be fulfilled. Because $n_i - n_k = n_{ai} - n_{ak}$, $D \geq 3(n_i + n_k)^{1/2}/n_{am}$. If the ratio of background to Ag monolayer signal $B/S_{am} = n_b/n_{am}$ is α and the average coverage in the specimen areas i, k is Θ, then $D > 3\sqrt{[2(\Theta + \alpha)/n_{am}]}$. The experimental data mentioned above give for the ratio of the monolayer signal to the signal from a thick Ag layer $n_{am}/n_{a\infty} = 0.295$, while theoretical estimates based on the inelastic mean free path $\lambda_i = 8 \times 10^{-10}$ m and the atomic density 7.9×10^{18} m^{-2} of the Ag monolayer on Si(111) gives 0.146. The discrepancy of a factor of 2 is probably due to an experimental problem: while every Ag atom is illuminated in the monolayer, a considerable fraction of the surface is shadowed in the thick film because of the surface roughness and the large angle of incidence ($\Theta_p = 75°$). Therefore, $n_{am}/n_{a\infty} = 1/4$ will be used in the estimate, together with $\alpha = 3.7$ from experiment; $n_{a\infty} = I_D t$, with $I_D = 3.2 \times 10^{21} \delta_o{}^2 T_\Theta \times 0.15$ (eq.14). Inserting all quantities into the detectability condition leads to the condition

$$\delta_o{}^2 T_\Theta \geq 18(\Theta + \alpha)(n_{a\infty}/n_{am})/Yj_p T_E D^2 t = 3 \times 10^{-19} (\Theta + 3.7)/D^2 t$$

In order to measure a coverage difference $\Delta\Theta = 1/10 = D$ at $\Theta = 0.3$, $\delta_o{}^2 T_\Theta$ must be $1.2 \times 10^{-16}/t = 1.2 \times 10^{-18}$ with an integration time t = 100 s. This large $\delta_o{}^2 T_\Theta$ value requires a rather large aperture. Calculations for 350 eV electrons similar to those made for photoelectrons show that for $R_A = 2 \times 10^{-2}$ (L = 3 mm), $T_\Theta = 4.5 \times 10^{-3}$ and $\delta_o = 30$ nm, which yields a sufficient $\delta_o{}^2 T_\Theta \approx 4 \times 10^{-18}$; for $R_A = 3 \times 10^{-2}$, $T_\Theta \approx 1 \times 10^{-2}$ and $\delta_o = 96$ nm, resulting in $\delta_o{}^2 T_\Theta \approx 9.2 \times 10^{-17}$ and a minimum integration time of 1.3 s. Inasmuch as with such large apertures the spherical aberration is dominating (here $\delta_S \approx 3\delta_C$ and $\approx 7 \delta_c$, respectively) the available intensity may be increased with a loss of resolution of less than a factor 1.5 by increasing the energy window and therefore the energy transmission T_E, but not more than a factor of 2 - 5 can be gained this way.

Another improvement may be achieved by increasing the energy of the incident beam which leads to an increase of the signal to background ratio. Finally, there is still the background subtraction by image subtraction (image taken at $E = E_A$ minus image taken at $E = E_A + \delta E$, δE several eV) but the best which can be achieved in this way is what would be observed in the absence of background ($\alpha = 0$), i.e. $D \geq 3\sqrt{(2\Theta/n_{am})}$, or in the present example a factor of about 4 in the necessary intensity or integration time.

This last expression is also valid for X-ray photoelectron microscopy where $\alpha \ll 1$ in general. Thus, it is seen that even when the intensity behind the image intensifier is high enough to observe the image at high resolution, signal to noise problems necessitate lower resolution and image recording times from 1 to 100 or - in photoelectron microscopy - 1000 s for sufficiently accurate coverage determination. Cruder chemical characterizations, e.g. distinction between a thick Ag film and clean Si, of course, requires much less stringent conditions.

SCANNING AUGER ELECTRON MICROSCOPY

A comparison with scanning Auger electron microscopy (SAM)[63] (see also Chapter by Cazaux) appears appropriate at this place. Typical good conditions for SAM are: $j_p = 10^7 - 10^8$ Am^{-2}, $\delta_0 = 100$ nm, t = 10^3 s. The two disadvantages compared to the non-scanning Auger electron microscopy discussed here are the high primary current density and the long integration time. They are an immediate consequence of sequential instead of simultaneous image acquisition. The much higher current density in SAM causes much greater damage to specimens which are sensitive to electron bombardment (desorption, dissociation, heating, charging); the longer integration times put more stringent requirements on specimen stability. Another disadvantage of SAM is the resolution limitation by the diameter of the diffusion cloud which can be significantly larger than the probe size[1]. Thus immersion Auger microscopy is far superior to SAM. However, if only a local analysis in a small selected area is desired, the fine high current density beam of a scanning instrument is superior to the broad illumination of the immersion system.

Up to now it was explicitly or implicitly assumed that the photoelectron or Auger electron angular distribution is isotropic or $\cos\theta$-like. This is, however, a serious oversimplification. Photoelectrons and Auger electrons are subject to diffraction in crystalline materials just as the incident electrons in LEEM are. In photoelectron diffraction (for references see ref.64) the relevant phenomenon is known as normal photoelectron diffraction which causes a non-monotonic variation of the photoelectron signal with surface orientation and photon energy and, in the case of adsorbates (at fixed surface orientation, coverage and photon energy), an adsorption site-dependent signal. Particularly strong diffraction and backscattering effects have been observed in Auger electron emission.

Some examples are shown in the polar plots of fig.31: i) The angular distribution of the $M_{2,3}VV$ Auger electrons of (a) a Cu(100) and (b) a Cu(111) surface in two selected azimuths. The characteristic shape, with a

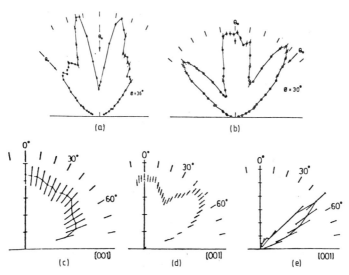

Fig.31. Polar plots of the angular distribution of Auger electrons from single crystal surfaces in selected azimuths. (a) Cu(100) and (b) Cu(111) ($M_{2,3}VV$ electrons)[65], (c) O (KVV), (d) S ($L_{2,3}VV$) and (e) Se ($M_{4,5}VV$) on Ni(100) at 1/2 monolayer coverage [c(2x2) structure], [001] azimuth[66].

deep minimum at θ = 0° on Cu(100) and with maxima at θ = 0° and θ ≈ 35° on Cu(111), is also seen in other azimuths. A specimen consisting of (100) and (111) orientated grains will, therefore, show strong contrast in the Auger image although the chemical composition is constant. ii) The second example concerns adsorbates. The chalcogenides O, S and Se form at 1/2 monolayer coverage a c(2x2) structure. The angular distribution of the O-KVV (515 eV), S-$L_{2,3}$VV (150 eV) and Se-$M_{4,5}$VV (42 eV) Auger electrons in the Ni[001] azimuth are depicted in (c),(d) and (e) respectively. While O and S emit also normal to the surface, Se practically emits only into off-normal directions. Thus, Se may not be detectable in immersion Auger electron microscopy, at least not with $M_{4,5}$VV Auger electrons.

Fortunately, in many cases the angular variations are not so serious. Frequently, they are of the order of a few 10%, e.g. in adsorbates (O, Fe, Cu, Au) on W[67,68], and in the examples considered here, Ag and Si, the angular distribution shows no large deviations from a monotonic change with polar angle (fig.32): on a clean (a) and on a Ag monolayer-covered (b) Si surface the $L_{2,3}$VV signal decreases nearly linearly with increasing θ; the Ag monolayer $M_{4,5}$VV signal (b) changes smoothly though not cosθ-like with θ; the $M_{4,5}$VV signal from the thick Ag layer varies approximately $\cos^2\theta$-like with θ, with somewhat stronger oscillations in the [1$\bar{1}$0] azimuth (c) than in the [11$\bar{2}$] azimuth (d). Nevertheless, quantitative analysis has to take these variations into account and this problem is not restricted to spectroscopic emission microscopy but occurs with all other spectroscopic techniques as well.

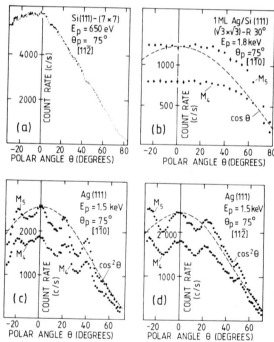

Fig.32. Angular distributions of (a) Si $L_{2,3}$VV Auger electrons from a clean Si(111)surface in the [11$\bar{2}$] azimuth, E_p = 650 eV, θ_p = 75° and of Ag $M_{4,5}$VV Auger electrons (b) from 1 monolayer of Ag on Si(111) with (√3x√3)-R30° structure in the [1$\bar{1}$0] azimuth, E_p = 1800 eV, θ_p = 75°; (c) and (d) from thick epitaxial Ag layer with (111) surface in [1$\bar{1}$0] and [11$\bar{2}$] azimuth, respectively, E_p = 1500 eV, θ_p = 75°[23].

Finally, a brief remark is needed about the possible imaging energy analyzers which can be used in this type of microscopy. Two types of instruments appear suitable: the Castaing analyzer[69] and Ω type filters[70]. While the first one is already used very successfully in a commercial transmission microscope[71], the second type is still under development[72]. It appears that the Ω filter, in spite of its greater complexity, will have a better electron optical performance. However, too little is known at present about it so that a discussion in connection with spectroscopic emission microscopy is premature.

SUMMARY

Immersion electron microscopy which has long been limited to thermionic, near UV photoelectric, secondary and mirror electron microscopy with its limited resolution, intensity and applications, has recently entered a new phase of evolution in the form of low energy electron microscopy (LEEM). The theoretical resolution considerations which lead to the development of LEEM pointed at the same time to new possibilities of emission microscopy, inner shell photoelectron and Auger electron microscopy. This spectroscopic emission microscopy, previously unattractive on the basis of Recknagel's resolution criterion, opens the way to high resolution chemical surface characterization in conjunction with LEEM. LEEM has already demonstrated its capabilities on the 10 nm resolution level as the examples given in this paper show. Improvements in instrument design should push the resolution into the 1 nm range in the future. In spectroscopic emission microscopy the attainable resolution is limited mainly by the much lower available intensity, but 10 - 100 nm appears feasible in Auger electron microscopy and 100 nm in X-ray photoemission microscopy with future wiggler and undulator synchrotron radiation.

ACKNOWLEDGEMENTS

The authors wish to thank Dr. W. Engel for making some of the unpublished photographs of his thesis available.

REFERENCES

1. L. Reimer, "Scanning Electron Microscopy", Springer, Berlin (1985).
2. J. Kirschner, T. Ichinokawa, Y. Ishikawa, M. Kemmochi, N. Ikeda, and Y. Hosokawa, "Scanning Electron Microscopy 1986/II", SEM Inc., Chicago, p.331.
3. T. Ichinokawa, Y. Ishikawa, M. Kemmochi, N. Ikeda, Y. Hosokawa, and J. Kirschner, Surface Sci., 176:397 (1986).
4. R. Gomer, "Field Emission and Field Ionization", Harvard University Press, Cambridge (1961).
5. D.W. Turner, I.R. Plummer and H.Q. Porter, J. Microscopy, 136:225 (1984); Phil. Trans. Roy. Soc. London, A318:219 (1986).
6. E.W. Müller and T.T. Tsong, "Field Ion Microscopy", American Elsevier, New York (1969).
7. M. Knoll and E. Ruska, Ann. Phys., 12:607 (1932).
8. E. Brüche and H. Johannson, Naturw., 20:49, 353 (1932).
9. L.H. Germer and C.D. Hartmann, J. Appl. Phys., 31:2085 (1960).
10. E. Bauer, Phys. Rev., 123:1206 (1961).
11. E. Bauer, "5th Intern. Congr. Electron Microscopy", Academic Press, New York (1962), p. D-11.
12. R.A. Schwarzer, Microscopica Acta, 84:51 (1981).

13. O.H. Griffith and G.F. Rempfer, in: "Adv. Opt. Electron Microscopy", R. Barer and V.E. Cosslett eds, Academic Press, London, Vol.10 (1987) p.269.
14. A. Recknagel, Z. Physik, 117:689 (1941); 120:331 (1943).
15. E. Bauer, Ultramicroscopy, 17:51 (1985).
16. D.R. Cruise and E. Bauer, J. Appl. Phys., 35:3080 (1984) and unpublished work.
17. W. Telieps, Ph.D. thesis, Tech. Univ. Clausthal (1983).
18. W. Engel, Ph.D. thesis, Berlin (1968).
19. W. Glaser, "Grundlagen der Elektronenoptik", Springer, Wien (1952).
20. G. Bartz, "Proc. Intern. Kongr. Elektronenmikroskopie", Berlin (1958), Vol. 1, p.201; Optik, 17:135 (1960).
21. P. Cyris and W. Koch, private communication.
22. Leybold-Heraeus GmbH, technical data information.
23. K.-D. Hermbecker, Ph.D. thesis, Tech. Univ. Clausthal (1979).
24. E. Eichen, in "Techniques of Metals Research", R.F. Bunshah, ed., Interscience Publ., New York (1968), Vol. II, Part 1, p.177.
25. B. Feuerbacher and B. Fitton, J. Appl Phys., 43:1563 (1972).
26. R.M. Broudy, Phys. Rev., B3:3641 (1971).
27. N.V. Richardson and A.M. Bradshaw, in: "Electron Spectroscopy", C.R. Brundle and A.D. Baker, eds, Academic Press, New York (1981).
28. H. Bethge and M. Klaua, Ultramicroscopy, 11:207 (1983).
29. H. Bethge, Th. Krajewski and O. Lichtenberger, Ultramicroscopy, 17:21 (1985).
30. N.F. Mott and H.S.W. Massey, "The Theory of Atomic Collisions", Clarendon, Oxford (1965).
31. L.F. Mattheiss, Phys. Rev., A139:1893 (1965).
32. H.N. Browne and E. Bauer, unpublished.
33. E. Bauer, J. Vacuum Sci. Technol., 7:3 (1970).
34. M. Fink, M.R. Martin and G.A. Somorjai, Surface Sci., 20:303 (1972).
35. R. Schmid, K.H. Gaukler and H. Seiler, "Scanning Electron Microscopy 1983/II", SEM Inc., Chicago, p.501.
36. J.B. Pendry, "Low Energy Electron Diffraction", Academic Press, London (1974).
37. S. Andersson, Surface Sci. 18:325 (1969); 25:273 (1971).
38. H.-J. Herlt, Ph.D. thesis, Tech. Univ. Clausthal (1982).
39. K. Kambe, Surface Sci., 20:213 (1970).
40. H.-J. Herlt, R. Feder, G. Meister and E. Bauer, Solid State Commun., 38:873 (1981).
41. H.-J. Herlt and E. Bauer, Surface Sci., 175:336 (1986).
42. W. Telieps and E. Bauer, Ultramicroscopy, 17:57 (1985).
43. E. Bauer and W. Telieps, "Proc. XIth Int. Congr. on Electron Microscopy", Kyoto (1986), p.67; "Proc. 5th Pfefferkorn Conf.", SEM Inc., Chicago (1987), p.99.
44. D. Menzel, this book.
44a. A. Delong and V. Kolařik, Ultramicroscopy, 17:67 (1985).
44b. W. Koch, B. Bischoff and E. Bauer, "Proc. 5th European Congr. Electron Microscopy", (1972), p.58.
45. W. Telieps, Appl. Phys., A44:55 (1987).
46. W. Telieps, M. Mundschau and E. Bauer, Optik, 77:93 (1987).
47. W. Telieps and E. Bauer, Surface Sci., 162:163 (1985); Ber. Bunsenges. Phys. Chem., 90:197 (1986).
48. E. Bauer, Leopoldina Symposium "Physik und Chemie der Kristall-oberfläche", Halle (1978) unpublished.
49. J. Cazaux, "Scanning Electron Microscopy 1984/III", SEM Inc., Chicago, p.1193.
50. E. Bauer and J. Kolaczkiewicz, Phys. Stat. Sol., (b)131:699 (1985).
51. C.D. Wagner, W.M. Riggs, L.E. Davis, J.F. Moulder, G.E. Muilenberg, "Handbook of X-ray Photoelectron Spectroscopy", Perkin-Elmer Co., Eden-Prairie (1979).

52. R.C.G. Leckey, Phys. Rev., A13:1043 (1976).
53. S.T. Manson and J.W. Cooper, Phys. Rev., 165:126 (1968).
54. From Physics Data 18-2 (1981), FAZ (Fachinformationszentrum) Karlsruhe, p.58.
55. I. Lindau, P. Pianetta, K.Y. Yu and W.E. Spicer, Phys. Rev., B13:492 (1976).
56. C. Kunz, "Synchrotron Radiation", (Topics in Current Physics, Vol. 10), Springer, Berlin (1979); in ref. 57, p.299.
57. L. Ley and M. Cardona, "Photoemission in Solids II", Springer, Berlin (1979).
58. A. Goldmann, W. Greulich and E. Dreisigacker, Physikalische Blätter, 43:80 (1987).
59. M. Cardona and L. Ley, "Photoemission in Solids I", Springer, Berlin (1978).
60. E. Bauer, Vacuum, 22:539 (1972).
61. J.C. Fuggle, in: "Electron Spectroscopy", C.R. Brundle and A.D. Baker, eds, Academic Press, New York (1981), Vol. IV, p.85.
62. P. Staib and G. Staudenmaier, "Proc. 7th Intern. Vac. Congr. and 3rd Intern. Conf. Solid Surfaces", Vienna (1977), p.2355.
63. A. Mogami, Surface and Interface Analysis, 7:241 (1985); JEOL News 24E:Nr.3, p.45 (1986) and references therein.
64. S.Y. Tong and C.H. Li, in: "Chemistry and Physics of Solid Surfaces", R. Vanselow and W. England, eds, CRC Press, Boca Raton, Florida (1982), Vol.III, p.287.
65. L. McDonnell, D.P. Woodruff and B.W. Holland, Surface Sci., 51:249 (1975).
66. T. Matsudaira, M. Nishijima and M. Onchi, "Proc. 7th Intern. Vac. Congr. and 3rd Intern. Conf. Solid Surfaces", Vienna (1977), p.2285; Surface Sci., 61:651 (1976).
67. T.M. Gardiner, H.M. Kramer and E. Bauer, Surface Sci., 121:231 (1982).
68. T. Koshikawa, T. von dem Hagen and E. Bauer, Surface Sci., 109:301 (1981).
69. R. Castaing and L. Henry, J. Microscopie, 3:133 (1964).
70. H. Rose and W. Pejas, Optik, 34:171 (1971).
71. W. Egle, D. Kurz and A. Rilk, MEM (Magazine for Electron Microscopists) 3:4 (1984); Opton Feintechnik, Oberkochen.
72. S. Lanio, H. Rose and D. Krahl, Optik, 73:56 (1986).

SCANNING TUNNELING MICROSCOPY AND SPECTROSCOPY

Nicolás García

Dpto. de Fisica de la Materia Condensada, C-III
Universidad Autonoma de Madrid
Cantoblanco, 28049-Madrid, Spain

INTRODUCTION

Scanning tunneling microscopy (STM) is a recent surface imaging technique discovered and put forward in 1982 by Binnig, Rohrer, Gerber and Weibel[1]. Their achievement was recognised in a short period of time by many prizes culminating in the Nobel Prize for Physics in 1986. Here we employ the acronym STM to denote either scanning tunneling microscopy or the scanning tunneling microscope.

The basic principle is quite simple. A sharp conducting tip, usually of tungsten is held at a distance s of a few Ångströms away from the surface of a conducting crystal and a bias voltage V applied so that a tunneling current flows either from tip to sample or in the reverse direction. As the tip is slowly scanned in the x,y plane of the surface, the separation s is controlled by a servo-mechanism which keeps the tunneling current I constant. The motion of the tip in all 3 directions is generated by piezo-electric drives. In favourable cases a display of tunneling current as a function of tip position yields a surface image with atomic resolution. Given sufficient stability, further information can be obtained by measuring at any given point (x,y), I vs V at fixed s or I vs s at fixed V.

In this chapter we review the STM technique from a historical as well as from an experimental and theoretical point of view. This is already a formidable task, since surface scientists all over the world have swiftly adopted STM and are making numerous modifications and improvements leading to many new applications. A number of excellent reviews[2,13] have already been written as well as more technical papers dealing with specific aspects of the subject and, with the kind permission of the authors and journals concerned, we have decided to reprint or quote extensively from some of these.

HISTORICAL BACKGROUND AND EARLY DEVELOPMENT

The Nobel Lecture[3] given by Binnig and Rohrer and entitled "Scanning Tunneling Microscopy - from Birth to Adolescence" is a fascinating and authoritative account of events leading up to and following the start of their collaboration in late 1978. In particular the links to other subjects like tunneling spectroscopy of thin films and field emission microscopy are illuminated.

"The original idea then was not to build a microscope but rather to perform spectroscopy locally on an area less than 100 Å in diameter."

"<<On a house-hunting expedition, three months before my (G.B.) actual start at IBM, Heini Rohrer discussed with me in more detail his thoughts on inhomogeneities on surfaces, especially those of thin oxide layers grown on metal surfaces. Our discussion revolved around the idea of how to study these films locally, but we realized that an appropriate tool was lacking. We were also puzzling over whether arranging tunneling contacts in a specific manner would give more insight on the subject. As a result of that discussion, and quite out of the blue at the LT15 Conference in Grenoble - still some weeks before I actually started at IBM - an old dream of mine stirred at the back of my mind, namely, that of vacuum tunneling. I did not learn until several years later that I had shared this dream with many other scientists, who, like myself, were working on tunneling spectroscopy. Strangely enough, none of us had ever talked about it, although the idea was old in principle.>> Actually, it was 20 years old, dating back to the very beginning of tunneling spectroscopy[4]. Apparently, it had mostly remained an idea and only shortly after we had started, did Seymour Keller, then a member of the IBM Research Division's Technical Review Board and an early advocate of tunneling as a new research area in our laboratory, draw our attention to W. A. Thompson's attempting vacuum tunneling with a positionable tip[5]."

"We became very excited about this experimental challenge and the opening up of new possibilities. Astonishingly, it took us a couple of weeks to realize that not only would we have a local spectroscopic probe, but that scanning would deliver spectroscopic and even topographic images, i.e. a new type of microscope. The operating mode mostly resembled that of stylus profilometry[6], but instead of scanning a tip in mechanical contact over a surface, a small gap of a few Ångströms between tip and sample is maintained and controlled by the tunnel current flowing between them. Roughly two years later and shortly before getting our first images, we learned about a paper by R. Young et al.[7] where they described a type of field-emission microscope they called a "topografiner". It had much in common with our basic principle of operating the scanning tunneling microscope (STM), except that the tip had to be rather far away from the surface, thus, on high voltage, producing a field-emission current rather than a tunneling current and resulting in a lateral resolution roughly that of an optical microscope. They suggested to improve the resolution by using sharper field-emission tips, even attempted vacuum tunneling, and discussed some of its exciting prospects in spectroscopy. Had they, even if only in their minds, combined vacuum tunneling with scanning, and estimated _that_ resolution, they would probably have ended up with the new concept, scanning tunneling microscopy. They came closer than anyone else."

To obtain a conclusive demonstration of vacuum tunneling, let alone tunneling microscopy, Binnig and Rohrer, with the expert skill of Gerber and Weibel, had to overcome formidable problems of vibration, sample positioning and tip fabrication, as the Nobel lecture goes on to make clear.

"During the first few months of our work on the STM, we concentrated on the main instrumental problems and their solutions[8]. How to avoid mechanical vibrations that move tip and sample against each other? Protection against vibrations and acoustical noise by soft suspension of the microscope within a vacuum chamber. How strong are the forces between tip and sample? This seemed to be no problem in most cases. How to move a tip on such a fine scale? With piezoelectric material, the link between electronics and mechanics, avoiding friction. The continuous deformation of piezomaterial in the Ångström and subångström range was established only later by the tunneling experiments themselves. How to move the sample on a fine scale over long distances from the position of surface treatment to within reach of the tip? The "louse." How to avoid strong thermally excited length fluctuations of the sample and especially the tip? Avoid whiskers with small spring constants. This led to a more general question, and the most important one: What should be the shape of the tip and how to achieve it? At the very beginning, we viewed the tip as a kind of continuous matter with some radius of curvature. However, we very soon realized that a tip is never smooth, because of the finite size of atoms and because tips are quite rough unless treated in a special way. This roughness implies the existence of minitips as we called them, and the extreme sensitivity of the tunnel current on tip-sample separation then selects the minitip reaching closest to the sample."

"Immediately after having obtained the first stable STM images showing remarkably sharp monoatomic steps, we focused our attention onto atomic resolution. Our hopes of achieving this goal were raised by the fact that vacuum tunneling itself provides a new tool for fabricating extremely sharp tips: The very local, high fields obtainable with vacuum tunneling at only a few volts can be used to shape the tip by field migration or by field evaporation. Gently touching the surface is another possibility. All this is not such a controlled procedure as tip sharpening in field-ion microscopy, but it appeared to us to be too complicated to combine STM with field-ion microscopy at this stage. We hardly knew what field-ion microscopy was, to say nothing of working with it. We had no means of controlling exactly the detailed shape of the tip. We repeated our trial and error procedures until the structures we observed became sharper and sharper. Sometimes it worked, other times it did not."

"But first we had to demonstrate vacuum tunneling. In this endeavour, apart from the occurrence of whiskers, the most severe problem was building vibrations. To protect the STM unit also against acoustical noise we installed the vibration-isolation system within the vacuum chamber. Our first set-up was designed to work at low temperatures and in ultrahigh vacuum (UHV). Low temperatures guaranteed low thermal drifts and low thermal length fluctuations, but we had opted for them mainly because our thoughts were fixed on spectroscopy. And tunneling spectroscopy was a low-temperature domain for both of us with a Ph.D. education in superconductivity. The UHV would allow preparation and retention of well-defined surfaces. The instrument was beautifully designed with sample and tip accessible for surface treatments and superconducting levitation of the tunneling unit for vibration isolation. Construction and first low-temperature and UHV tests took a year, but the instrument was so complicated, we never used it. We had been too ambitious, and it was only seven years later that the principal problems of a low-temperature and UHV instrument were solved[9]. Instead, we used an exsiccator as vacuum chamber, lots of Scotch tape, and a primitive version of superconducting levitation wasting about 20 ℓ of liquid helium per hour. Emil Haupt, our expert glassblower, helped with lots of glassware and, in his enthusiasm, even made the lead bowl for the levitation. Measuring at night and hardly daring to breathe from excitement, but mainly to avoid vibrations, we obtained our first clear-cut exponential dependence of the tunnel current I on tip-sample separation s characteristic for tunneling. It was the portentous night of March 16, 1981."

"So, 27 months after its conception, the scanning tunneling microscope was born. During this development period, we created and were granted the necessary elbowroom to dream, to explore, and to make and correct mistakes. We did not require extra manpower or funding, and our side activities produced acceptable and publishable results. The first document on STM was the March/April 1981 in-house Activity Report."

"A logarithmic dependence of the tunnel current I on tip-sample separation s alone was not yet proof of vacuum tunneling. The slope of ln I versus s should correspond to a tunnel-barrier height of $\phi \sim$ 5 eV, characteristic of the average work functions of tip and sample. We hardly arrived at 1 eV, indicating tunneling through some insulating material rather than through vacuum. Fortunately, the calibration of the piezo-sensitivity for small and fast voltage changes gave values only half of those quoted by the manufacturers. This yielded a tunnel-barrier height of more than 4 eV and thus established vacuum tunneling. This reduced piezo-sensitivity was later confirmed by careful calibration with H. R. Ott from the ETH Zurich, and by S. Vieira of the Universidad Autonoma, Madrid[10]."

"U. Poppe had reported vacuum tunneling some months earlier[11], but his interest was tunneling spectroscopy on exotic superconductors. He was quite successful at that but did not measure I(s) Eighteen months later, we were informed that E. C. Teague, in his Ph.D. thesis, had already observed similar I(s) curves which at that time were not commonly available in the open literature[12]."

The lecture goes on to describe the exciting progress towards producing the first STM images and the ensuing scramble by many laboratories to install the new technique. In a real sense, however, the demonstration of vacuum tunneling in the context of a scanning instrument marks the end of the beginning.

PHYSICAL PRINCIPLES OF OPERATION

As indicated by Binnig and Rohrer, both in the previously quoted Nobel lecture[3] and elsewhere[13], the quantum mechanical tunneling phenomenon which underlies STM has already been extensively exploited by solid state physicists in the context of thin oxide and other insulating films[14,15]. The situation for electron tunneling between two planar conducting electrodes in the presence of a bias voltage V is shown schematically in fig.1. A simple trapezoidal tunneling barrier, related to the work functions ϕ_1 and ϕ_2 is employed and for simplicity we assume that the bias voltage V is small compared to the effective barrier height $\phi \sim (\phi_1 + \phi_2)/2$. In these circumstances the tunneling current density j for free electrons is given by the expression[16]

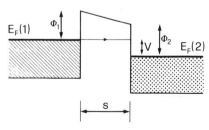

Fig. 1. Schematic energy diagram for a tunneling junction with a trapezoidal barrier (ϕ_1, ϕ_2) of thickness s.

$$j = (e^2/\hbar)(Q/4\pi^2 s) \cdot V \cdot \exp(-2Qs) \qquad (1)$$

where $Q = (2m\phi\hbar^2)^{1/2}$ is the inverse decay length of the wave function outside the surface. Compared with typical oxide film tunneling barrier heights of below 1 eV, the effective barrier height for vacuum tunneling is usually a few eV.

The simple expression (1) has to be modified at larger bias voltages, where the barrier height becomes a function of V but the dominating exponential dependence on barrier height and width remains. For non-free electron situations, the exponential pre-factor has to be altered to take account of density of state effects. It can be seen from fig.1 that the main contribution to the current depends on the density of filled states just below the Fermi level, E_{F1}, and the density of empty states just above the Fermi level, E_{F2}, since these electrons experience the minimum barrier height. This energy selectivity allows the density of states on either side of the Fermi level to be explored by observing the current voltage characteristic and is exploited in classical tunneling spectroscopy of insulating films[14,15].

The STM technique depends crucially on the extreme sensitivity of the tunneling current to the separation distance s (it typically decreases by about one order of magnitude for each 1 Å increase in s). Of course the planar geometry assumed in fig.1 no longer strictly applies. It is assumed that the tunneling current is confined to a filament between the apex of the tip and the specimen surface, whose own deviations from strict planarity are often the essential subject of interest. Further discussion of more sophisticated and realistic tunneling theories for the STM situation is given in a subsequent section.

The most frequently used mode of operation is illustrated schematically in fig.2, taken from Binnig and Rohrer[13]. In this mode the tip is scanned laterally across the surface, whilst its z position is controlled in order to keep the tunneling current I constant. Contours of constant tunneling current can then be obtained which, on a surface which is chemically and electronically uniform will yield an image of the surface topography limited by the instrumental resolution. Examples are shown in figs 3, 18 and 19 below.

Electronic or chemical inhomogeneities will also produce features in the scanned images. Such inhomogeneities can arise at foreign surface atoms as indicated schematically in fig.2 or at special surface features such as steps or other defects. The local electronic or chemical variations can be distinguished from simple height changes by I vs V or I vs s measurements (see later section on localised spectroscopy). These measurements provide for example information on local variations in tunneling barrier height or density of states which are vital in analysing the images.

Another procedure which can be useful in some situations is to check the reproducibility of the image scan or I vs s scan. Although the STM generally causes no appreciable specimen damage (see later section), its occasional occurrence may be signalled by apparent image changes or hysteresis effects.

BASIC INSTRUMENTATION

A schematic diagram of the instrumentation for STM is shown in fig.2 due to Binnig and Rohrer[13]. As indicated there the equipment must be well

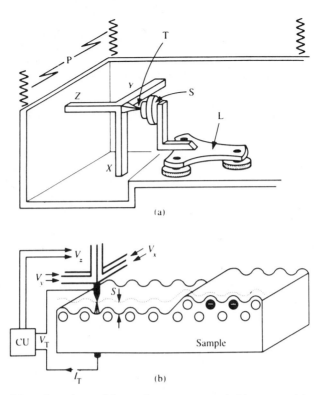

(a)

(b)

Fig. 2. Schematic of a tunneling microscope and its operation. The tip T
of the microscope depicted in (a) is scanned over the surface of a
sample S with a piezoelectric tripod (X,Y,Z). The rough positioner
L brings the sample within reach of the tripod. A vibration filter
system P protects the instrument from external vibrations. In the
constant tunneling current mode of operation, a voltage V_Z is
applied to the Z piezoelectric element by means of the control unit
CU depicted in (b) to keep the tunneling current constant while the
tip is scanned across the surface by altering V_x and V_y. The trace
of the tip, a y-scan, generally resembles the surface topography.
Electronic inhomogeneities also produce structure in the tip trace,
as illustrated on the right above two surface atoms having excess
negative charge. Figure reproduced from ref.13, courtesy of IBM
J. Res. Develop.

protected from external vibration by a vibration filter system. The most
serious vibrations are those which excite resonances of the system itself
and change the tip-sample separation s. With increasing design experience,
particularly the move towards microscopes of smaller physical size, the
vibration problem is now less severe. In one of the successful later
designs[17], the necessary mechanical isolation is achieved simply by
mounting the instrument on a stack of metal plates and separated from its
neighbour by a viton damper. It is also useful in the tip-position control
loop to employ a low-pass filter, limiting the tip motion to frequencies
below the lowest mechanical mode frequency. This can restrict the imaging
speed particularly on rough surfaces.

The sample positioning and scan drive system has obviously also to be
designed to minimise vibration effects. This is difficult to reconcile

240

with the desire to be able to select from a reasonably large surface of perhaps millimetre dimensions, a specifically chosen small area for high resolution imaging. Coarse positioning of the sample has been achieved in a number of ways including the piezo-electric walker or louse[8] shown in fig.2a, as well as mechanical pushers[18] and screw reduction drives. The fine, three-dimensional, highly controlled motion of the tip which is required for high resolution imaging is usually provided by some form of the piezo-electric tripod shown schematically in fig.2a. A more compact design, which may have advantages if the area to be scanned is not too large, replaces the tripod by a single piezo-electric tube fitted with several electrodes to produce the necessary deflections[19]. If distortion-free images are to be produced, careful attention to calibration of the piezo response at different scanning speeds is often required.

The tip is evidently a very crucial component. Its structure can influence the image, not only in determining the resolution attainable but also, in extreme cases, it may contribute to the final result just as significantly as the structure of the sample under investigation. Tips are generally made by grinding or etching wires of tungsten (and on occasion stainless-steel, iridium or gold). Further in-situ tip shaping experiments can often be useful, particularly after damage through accidental contact with the sample. These procedures usually depend on removing the tip to a few 100 Å of the surface and applying a pulse +100 V or more. More systematic studies of tip structure and its influence on STM images are possible by field ion microscopy[20,58].

Computer control of the various STM imaging and microscopy operations is essential for efficient operation. Ideally, however, this should also be integrated with a data acquisition, processing and display system. The basic requirements for such a system, together with examples of the significance of background subtraction, filtering and enhancement have been presented by Becker[21]. Anyone who studies the STM literature will quickly appreciate the enormous part played by sophisticated image processing and display in improving the quality of the published images. As always however, these procedures have their danger unless are used intelligently.

IMAGING APPLICATIONS

The number of significant and high quality STM images already published means that it would be impractical to give a comprehensive review of the results here. We will simply outline what has been achieved in the main fields explored so far, drawing attention to points which seem to be of wider significance either for surface science or for general STM imaging.

The technological importance of semiconductor surfaces has ensured that a considerable effort has gone into the study of their structures and reconstructions. In many of these cases, including the best known example of the 7x7 reconstruction of the (111) surface of Si (for details see the chapter by Gibson), the basic elements of the surface structure had already been worked out on the basis of LEED, photoemission, ion-channeling or transmission diffraction data. Nevertheless even for regions of perfect structure, the STM images have supplied important confirmatory information on points of detail, particularly over the outermost ad-atoms in the 7x7 structure. Furthermore, the images obtained from such known structures have been extremely useful both experimentally and theoretically in separating the topographical and electronic contributions to the STM image contrast (see following two sections). In imperfect regions such as where the surface reconstruction encounters a surface step[22,23] (see fig.3) or where different domains of surface reconstruction form in the same region[24], the high resolution information provided by the STM is unrivalled. However, other high resolution techniques such as REM imaging (see chapter by Cowley) and LERM imaging (see chapter by Bauer and Telieps) may retain an advantage where it is desired to follow in real time the dynamics of surface phase transitions at temperature.

Surface structures and reconstructions have now been observed on a considerable variety of other surfaces particularly of metals such as Au[25] and Pt[26,27] and including situations with monoatomic surface steps[59], close-packed atomic corrugations[60] or where ad-atoms are present, e.g., O on Ni[28], Au on Si[29]. With metal surfaces or when relatively large sample-tip distances are employed, the images are largely due to surface topography. However as indicated schematically in fig. 4 adapted from Binnig and Rohrer[13], the contours of equal tunneling current need not follow the surface topography exactly. In semiconductors the image contribution from electronic effects such as varying local density of states (LDOS) at the surface is generally more significant. For instance, there is an apparent but not real height difference between the two halves of the unit cell in the Si 7x7 structure. As shown in the next section, these effects can sometimes be unravelled by careful local spectroscopy. An even more striking case of spurious topography are the giant corrugations observed in graphite[30,31]. These may be due to the extremely low density of states at the Fermi level in this case. However they may also be a consequence of actual physical contact between tip and surface since the typical barrier height measured is only a few tenths of a volt or less.

The STM has already been applied to a number of specimens not previously accessible by any kind of high resolution imaging. These

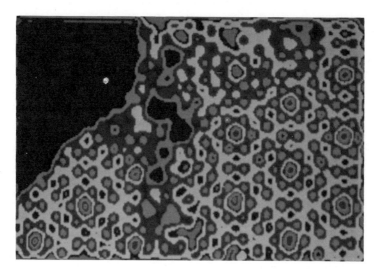

Fig. 3. STM image of the 7x7 reconstruction at the Si(111) surface (courtesy Drs J.E. Demuth, R.J. Hamers, R.M. Tromp and M.E. Welland[18]). The distance between the main 6-fold symmetry centres is 2.6 nm. Colour enhancement of the intensity scale reveals fine details of atomic heights and bonding states at the surface. A surface step is visible as a discontinuity running vertically near the centre of the figure. A colour print of this figure will be found in the frontispiece.

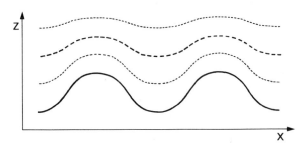

Fig. 4. Expected STM-equicurrent lines (dashed lines) above an atomically
corrugated surface (solid line). The different spacings of the
equicurrent lines at the protrusions and in the valleys can be
described by a varying effective tunneling barrier height.

include surfaces under air at atmsopheric pressure[32,33], under water and
saline solutions[19] and under various other electrolytes[34]. Another novel
and exciting new application is to the imaging of organic molecular
structures[32] which would be severely damaged by more energetic electrons.
More recent work has included imaging of stearic acid, purple membrane[35]
and of Langmuir-Blodgett films[36,37].

LOCALISED SPECTROSCOPY IN THE STM

As mentioned above, given sufficiently stable conditions, the
feed-back loop employed to maintain the tip-sample distance can be briefly
interrupted to permit measurement of tunneling current I vs separation
distance at constant bias V, or of I vs V at constant s. The first of
these measurements gives a value for the effective barrier height. The
second measurement yields density of states information near the Fermi
level and is a kind of spectroscopy based on elastic tunneling.

A most impressive demonstration of the power of elastic tunneling
spectroscopy was given by Hamers et al.[38] who were able to probe the
density of states with a resolution of 3 Å inside the unit cell of the
Si(111) 7x7 reconstruction. Their results are given in fig.5 and show the
very striking variations which occur within the unit cell. Hamers et al.
were able to relate the measurements to the presence of surface states
previously observed in UV and inverse photoemission spectroscopy. Detailed
analysis of these results indicated not only the presence of the electron
state responsible for the difference in STM contrast in the two halves of
the unit cell but also provided direct evidence for the presence of a
stacking fault in one half, as required by the Takayanagi model (see
chapter by Gibson).

In principle it may be possible also to carry out inelastic tunneling
spectroscopy[14,15] in the STM. In this case the tunneling electron loses
some characteristic amount of energy, e.g. by exciting a vibration in a
molecule adsorbed on the surface. It might then be feasible to identify
various adsorbed species from their spectra. Unfortunately the expected
signal strengths are very weak and although the technique is thought to be
possible[39], it will undoubtedly be extremely difficult.

MORE REALISTIC THEORY OF STM

Imaging theories. A quantitative theory of imaging in the STM is obviously
an essential priority for full interpretation of the results obtained. In

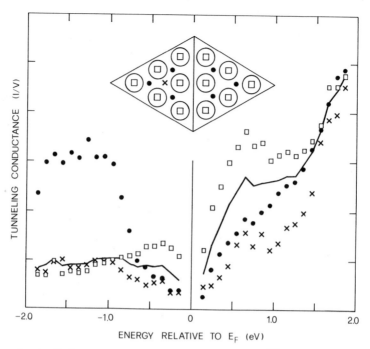

Fig. 5. Constant-distance conductance vs V spectra[38] for the Si(111)-(7x7) surface averaged over one unit cell (solid line) and at selected locations in the unit cell (other symbols). Crosses are from faulted half only; others are averaged over both halves of unit cell.

going beyond the rudimentary picture of tunneling employed so far, we must take account not only of the non-planar geometry of the tip and sample, but also of the form of the tunneling potential arising from more detailed consideration of electron interactions.

Progress towards the required theory has in fact proceeded along two parallel paths. In the first of these, the perturbation theory of Bardeen[40], based on a transfer Hamiltonian for tunneling between planar electrodes, has been extended and adapted by Tersoff and Hamman[41] to deal with a spherical electrode in s-wave approximation. In the second method, followed by Garcia et al.[42] as well as by Stoll et al.[43], a transmission scattering solution has been developed for three-dimensional tunnel barriers with free electron metals as the electrodes. This theory yields a more specific picture of the current distribution in the tunneling filament. Baratoff[44] has compared these two approaches, analysing their advantages and disadvantages. For tips of small radii < 2 Å, they yield very similar answers, but for larger tips, corrections to the s-wave approximation are required[56]. Ultimately, of course, one would like to do detailed tunneling calculations for various specific models of the specimen surface structure. As a first step in this direction, Lang[45] has calculated the tunnel current with an intercalated atom placed between two plane electrodes.

Potential and image force. We now outline more specifically some of the factors which arise in constructing a more sophisticated model of the tunneling barrier potential. In fig.6, taken from a survey of the problem by Garcia[46], we show the various contributions to the barrier potential in the case of a planar barrier. The simple electrostatic potential $V_{e\ell}$ has to be modified by the addition of an exchange and correlation potential V_{xc}, which within small distances of a free electron surface, can be approximated by the form of Lang and Kohn[57]. Since the tunneling time is not too short, V_{xc} must tend to the (static) classical image potential $q/2x$ at greater distances x from the surface. The significance of this image potential and its ability in the case of suitable surfaces to produce bound states and other surface resonances is well documented.

For the case of two surfaces, separated by a distance s, the image potential is more complicated and must be computed from an infinite series of multiple image charges q_i at positions $\underline{r}_i$ using the expression

$$V_{im}(\underline{r}) = \frac{1}{2} \sum_i \frac{q_i}{|\underline{r} - \underline{r}_i|} \tag{2}$$

The image planes for this calculation are located approximately where the $V_{e\ell} + V_{xc}$ curve cuts the Fermi energy as indicated in fig.6. They have a separation $d = \ell(s) = (s - 1.5)$ Å. Note that s here is the distance between the free electron jellium edges and not the distance between atom centres in the outermost layers. The relation between these distances can be affected by surface structure and reconstructions. Because of the image potential, the height of the barrier $\phi(s)$ will, as indicated in fig.6, depend somewhat on the barrier separation and approximately follows the relation[46]

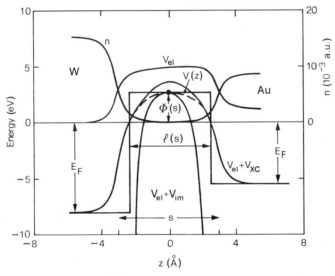

Fig. 6. Electron density n(z) and various potentials for a W-Au vacuum planar tunneling junction with s = 6 Å. $V_{el}(z)$, electrostatic potential; $V_{xc}(z)$, local density exchange and correlation potential; $V_{im}(z)$, image potential; V(z), tunneling potential. The dashed line is the approximate square barrier; $\phi(s)$ is the height and $\ell(s)$ is the width of the barrier. Reproduced from ref.46 by courtesy of IBM Journal for Research and Development.

$$\phi(s) = \phi(0) - \alpha/d \tag{3}$$

where $\alpha = 9.97$ eV/Å.

In fig.7, again taken from Garcia[46], we show the way in which these results for the tunneling potential barrier are extended for the case of a tip, modelled as a hemispherical protrusion on a flat surface. The apex of the tip is placed at a distance of 6Å from the specimen surface which is represented by a plane.

Plots of the potential on two different z at distances z = 6.5 Å and 7.5 Å from the tip centre, i.e. 0.5 Å and 1.5 Å from the top apex are given in the figure. We quote from Garcia's commentary[46]:

"This potential shows the reduction of the barrier height $\phi(s)$ due to the geometrical configuration and the image force contribution. The potential also shows a kind of channel where the electrons may tunnel easily around the axis of the tip because of the reduction of the effective tunneling barrier. A similar result was reported by Lucas et al.[47]. This "channel effect", as will be discussed, increases resolution and tends to reduce, $-d(\ln \sigma)/ds$, the slope of $\ln \sigma$ versus s."

"Figure 7b shows a plot of the potential in two lines perpendicular to the flat electrode, one between flat surfaces (dashed line) and the other along the axis of the tip (continuous line) from the tip apex to the sample surface. The geometrical effect which localizes the tunneling region is again clear. There is also a non-negligible image force concentration, as can be seen in fig.7c, where the variation of the maximum height of the tunneling barrier $\phi(s)$ along the tip axis (continuous line in fig.7b) is plotted against s; note its variation with s and the large decrease in $\phi(s)$ near the tip apex."

"The lateral resolution L_{eff} is defined by the diameter of a circle which has a constant current density j equal to that obtained in the direction of the tip axis, i.e., j at R = 0 (see fig.8a), and results in the same current, I(s), as that provided by the entire tip-sample junction. In other words, it is the effective surface area illuminated by the beam of the tunneling electrons[42]. This implies that two defects can be resolved if their separation is greater than L_{eff}. Figure 8b is an illustration of the effective area and resolution. The formula for L_{eff} is

$$\pi(L_{eff}/2)^2 \, j(R = 0) = I(s) \tag{4}$$

where L_{eff}, j, and the current I are functions of s. For small applied voltages V_a, I(s) is given by Ohm's law, namely

$$I(s) = V_a \cdot \sigma \tag{5}$$

but for $V_a \geq 1$ V, Ohm's law is not obeyed[48]."

"The current I(s) can be determined by solving the following equation:

$$I = (e^2/\pi\hbar) \int_0^{V_a} dV \sum_i T[\theta_i, (E_{Ft} - V), s] \tag{6}$$

where e is the electron charge, $\hbar$ is Planck's constant divided by 2π, and $T(\theta_i, s)$ is the transmission probability for the total potential of an electron with energy $(E_{Ft} - V)$ and at angle θ_i with the normal to the tip apex. The above formula is valid for large s or $T \ll 1$. If these conditions are not satisfied, the formula of Büttiker et al.[49] must be

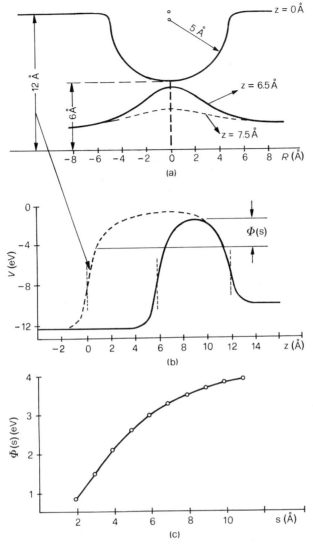

Fig. 7. (a). Jellium model for a hemispherical tip on a flat surface (thick line). The other lines are plots of the tunneling potential for two values of z (z = 6.5 Å, continuous line; z = 7.5 Å, dashed line) as a function of R. R is the distance measured from the intersection of the axis of the hemisphere with the plane at constant z; z is measured from the planar extension of the hemispherical tip. (b). Potential lines between the flat regions of the junction (dashed line) and along the tip axis at R = 0 for the geometry of fig.6 with varying z. ϕ is the barrier height for s = 6 Å. (c). Variation of $\phi(s)$ versus s, the distance from the apex of the hemisphere to the sample surface. Notice that the electron hole charge has been assumed to be spherical such that at the image plane [vertical dashed lines in (b)]. V_{ex} and V_{im} are the same, namely ≃ 8 eV. Reproduced from ref.46, courtesy of IBM Journal of Research and Development.

247

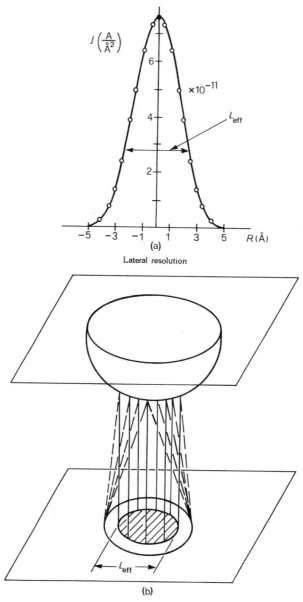

Fig. 8. (a). Lateral resolution L_{eff} corresponding to the diameter of the effective surface area illuminated by the beam of the tunneling electrons, as shown in (b), a schematic drawing of the effective illuminated area (shaded).

used. To calculate $T(\theta_i,s)$, Schroedinger's equation for the potential $V(R,z)$ must be solved."

"Solving Eq.6 and Schroedinger's equation for a square potential well with a constant barrier height ϕ_0, independent of z but dependent upon junction geometry, yields for the tunnel conductance[42,43,50,51]

$$\sigma(d) = K_0 \cdot r_t \cdot \exp(-\beta kd) \tag{7}$$

In Eq.7, $K_0 = 1.84 \cdot 10^{-5}$ A/(Å/eV), r_t = tip radius, and

$$k = 1/\hbar(2m\phi_0)^{1/2}$$

where m is the electron mass. By fitting values of s, E_F, and ϕ into Eq.7, one obtains $\beta = 2.15$, which accounts for the fact that the maximum contribution to the tunneling conductance comes from within a solid angle of $20°$. (Note: σ is proportional to r_t and not to r_t^2 as reported in ref. 41). Equation 7 fits well for $2 \leq s \leq 10$ Å [recall that d = (s - 1.5) (Å)], $2 \leq \phi_0 \leq 6$ eV, and $4 \leq E_F \leq 11$ eV."

"Values of L_{eff} calculated using Eqs 3 and 4 are plotted in fig.9, curve A, and L_{eff} can be approximated[41,44,52] by

$$L_{eff} = (\pi/2)[(d + r_t)/k]^{1/2}$$

In the above discussion, it was assumed that $\phi(s) = \phi_0$; however, this is not the case because of the image potential, and as shown in Eq.3, $\phi(s) = \phi_0 - \alpha/d$. Calculations which use Eq.3 for the dependence of ϕ upon s, with tip diameter $r_t = 5$ Å, $\phi_0 = 4.8$ eV, and $E_F = 8$ eV for the sample and 5.5 eV for the tip, show that σ is 100 times larger than the σ calculated for $\phi(s)$ = constant = $\phi_0 = 4.8$ eV. Moreover, the calculations also show that $-d(\ln \sigma)/ds$ (the negative slope of ln σ versus s) is practically constant and equal to $\beta\sqrt{\phi_0}$. This result is in accord with the experimental observation that $-d(\ln \sigma)/ds$ remains constant even though $\phi(s) = 0$ at $s = 3$ Å. For small values of s, $\phi(s)$ is considerably reduced and is reflected in the dependence of L_{eff} upon s, as illustrated in curve B of fig.9, where L_{eff} has a shallow minimum at $s = 5$ Å. Note that L_{eff} only

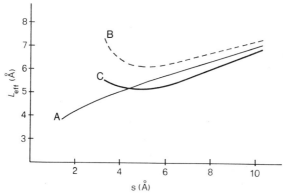

Fig. 9. Values of L_{eff} versus s for the different barrier approximations. Notice in curves B and C the minimum at $s = 5$ Å in L_{eff} and the increase in resolution due to the image force focusing effect. Curve A represents $\phi(s) = \phi_0 = 4.8$ eV, square barrier approximation. Curve B corresponds to $\phi(s) = (\phi_0 - \alpha/d)$ (Eq.3), square barrier approximation. Curve C shows the calculated results for the three-dimensional barrier. Reproduced from ref.46 with permission of IBM Journal of Research and Development.

varies from ~ 6 to 7 Å for 3.5 ≤ s ≤ 8 Å, which seems to be in accord with experimental observations[50]."

"For three-dimensional barriers, calculations of $\phi(s)$ and σ were accomplished using other techniques[53]. The L_{eff} thus obtained are illustrated by curve C in fig.9. Examining curve C with respect to curve B indicates that the three-dimensional calculation yielded an increased resolution compared to the square barrier calculation. This is because of the focusing effect of the image forces, creating a channel around the tip axis where the electrons can tunnel more easily[48]."

"Conductance Slope; $-d(\ln \sigma)/ds$. An interesting problem in STM theory concerns the measured $-d(\ln \sigma)/ds$ [$= \sigma'(s)$]. Although constant over a large range of s, 4 ≤ s ≤ 11 Å, in accord with theory, $\sigma'(s)$ is occasionally very small, corresponding to $\phi = 0.5$ eV. This latter value of ϕ is to be compared with $\phi = 2$-3.5 eV for metals and for semiconductors $\phi = 4$-5.5 eV. As yet there is no theoretical explanation for small $\sigma'(s)$ values. Coombs and Pethica[54] have presented evidence which proposes that small values of $\sigma'(s)$ are caused by the presence of tip irregularities in the junction."

"Calculations for the three-dimensional potential[48] for $V_a ≤ 0.1$ eV show that $\sigma'(s)$ can be reduced from the values corresponding to $\phi = 4.8$ eV at distances s > 12 Å to values of $\sigma'(s)$ corresponding to $\phi = 3.5$ eV for s ≈ 5 Å and then can be increased again if the equations of Büttiker et al. are used for calculating the conductance[49]. Small $\sigma'(s)$ values for work functions typical of metals are not obtained from this three-dimensional calculation."

The main conclusions from the tunneling theory, as presented by Garcia[46], are that the tunneling current is a delicate function of the tip and surface geometry as well as of the separation s, and that the influence of the classical image potential in the barrier region is very strong. Clearly more realistic calculations taking more specific account of different surface properties and structures would be very useful but are likely to be rather complex.

FORCES IN STM

Particularly in the case of poor conductors or, more strictly, materials with a low density of states at the Fermi surface, the tip-sample operation distance can be very small so that Van der Waals forces can come into play. This interaction can produce deformations of the tip or sample surface which affect the image, particularly in the case of soft layer materials like graphite, the typical example studied in many laboratories.

Now (courtesy of Physical Review Letters) we include an extract from the work of Soler, Baró, Garcia and Rohrer[55] which indicates how quantitative studies of these effects can be made and provides a basis for more detailed understanding.

"In the constant-current mode of STM[13], the tunnel tip traces contours of constant density of states[41-45,51]. On clean metal surfaces, in general, such an STM image is representative of the surface topography: on semiconductor surfaces, the varying local density of states (LDOS) mixes features of purely electronic origin into the STM images[13,38]. In graphite[30] and laminar compounds[31], the STM image was ascribed to LDOS features alone[61,62]. We show here that the giant corrugation amplitudes observed in these materials are due to a huge enhancement of the electronically based corrugations by local elastic deformations."

"Flatness and inertness make the cleavage (0001) plane of graphite an ideal substrate in surface-science and STM applications[63]. In a recent report, Selloni et al.[61] showed that the contours of constant LDOS at the Fermi level exhibit a corrugation of about 0.8 Å in contrast to the 0.2 Å of that of the total charge density. This should lead to a purely electronically derived corrugation in an STM image which would flatten out with increasing voltage, to the remaining total charge corrugation, as was observed experimentally."

"In the preceding Letter, Tersoff[62] attributes the huge corrugation observed in laminar compounds and occasionally in graphite to an anomalous corrugation of the contours of constant LDOS at the Fermi level due to the special electronic structure of these materials. He further shows that the contours are equally spaced and independent of distance from the surface, and concludes that the STM image is likewise independent of distance, read current, at constant voltage. However, new experiments do show that the corrugations of the STM image increase strongly with tunnel current at constant voltage, as shown in fig.10. If we consider the small density of states at the Fermi level, and the short decay length of the charge density outside the surface[61,62], the tunnel tip has to be extremely close to the surface at the imaging conditions of $I \sim 1$ nA and $V \sim 1$ mV. Tip displacements of a few Ångströms then imply "physical " contact of tip with surface, at least at the closest approach. It is then clear that the actual tip displacement cannot be solely a matter of the electronically induced corrugation."

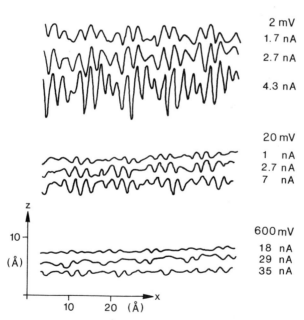

Fig.10. Graphite STM traces obtained at ambient-air pressure and room temperature with a "pocket-size" STM[17] with scanning speeds between 1 and 5 s per scan. The varying corrugation within a scan is due to the mismatch between crystallographic and scanning directions[30]; plateaus in the traces indicate the saddle points in the LDOS[30,61]. Reproduced from ref.55 with permission of Physical Review Letters.

"In fig.11a, we sketch one of the possible paths of the tip with respect to the surface while tracing a contour of constant LDOS. At point A, the force between tip and surface is repulsive, zero at B, and attractive at C (see fig.11b). Figures 11c and 11d illustrate the situations at points A and C of fig.11a. The tip-surface distance, d (tip jellium edge to top-layer carbon position), is determined by the tunneling parameters, and the local elastic deformation u by the tip-surface potential and the elastic constants of graphite."

"To calculate the deformation u, we used standard elastic theory[64]. We assume a rigid, paraboidal tip with a radius of curvature at the apex of R = 2 Å. For a perpendicular force f(r) per unit area, with cylindrical symmetry with respect to the tip axis, we find that

$$u = C \int_0^\infty f(r)dr \qquad (8)$$

where r is the distance parallel to the surface from the tip centre, and C is a constant which depends only on the elastic properties of the sample. For a very anisotropic medium like graphite, it may be calculated by performance of a Fourier analysis and summation of the elastic deformations over all possible wave vectors. We find that the only important elastic constants are the compressibility perpendicular to the surface and the shear between different basal planes. The constants given by Nicklow, Wakabayashi and Smith[65] give C = 8x10^{-11} m^2/N. We approximate f(z) by a Morse potential,

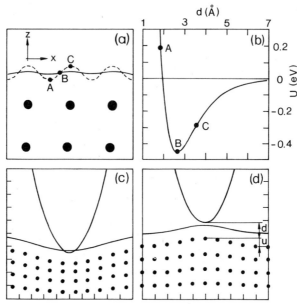

Fig.11. (a). Contour of constant local density of states (dashed line), and contour of total charge density (solid line). Filled circles indicate the positions of carbon atoms of the top two layers. (b). Potential used for the interaction of tip and surface. Schematic (c) compression and (d) expansion of graphite for the tip at points A and C, respectively, of (a). Reproduced from ref.55 with permission of Physical Review Letters.

$$f(z) = 2bV_0\{\exp[-2b(z - z_0)] - \exp[-b(z - z_0)]\}, \qquad\qquad (9)$$

with parameters appropriate for the force between graphite planes[66], e.i. $V_0 = 0.28$ J/m^2, $b = 0.97$ Å^{-1}, and $z_0 = 3.35$ Å. This is expected to be a good approximation for a carbon atom or carbon cluster at the apex of the tip[30]. Inserting (9) into (8) with $z(r) = d + r^2/2R$, and neglecting the small, total charge corrugation, we obtain the simple relation

$$u(d) = -1.2 \{\exp[-2b(d - d_0)] - \exp[-b(d - d_0)]\} \qquad\qquad (10)$$

where $d_0 = 3.0$ Å, and u and d are both in Ångströms. In fig.12, we have represented the enhanced vertical displacement of the tip, $(d + u)$, as a function of the distance d between tip and sample. It is clear that, in the repulsive region, the amplification of a corrugation Δd is enormous."

"The average tip-sample distance d as a function of applied voltage V_a and tunnel current I was calculated with use of the density of states of Selloni et al.[61] and the tunneling-current formalism of Tersoff and Hamann[41], as modified by Garcia and Flores[42,43,51]. Note that the "effective tunnel barrier" is $\phi_{eff} = \phi + ak^2 \sim 4.7 + 7.3 = 12$ eV, since only states at the K point of the surface Brillouin zone contribute to the tunnel current at low voltage[61,62,67]; ϕ is the average of the work function of graphite and tip. Electron transport then still occurs via tunneling, even if ϕ is completely quenched by correlation and image-potential effects[68] at the touching condition. Once d has been obtained, we calculate the total corrugation amplitude $\Delta d + \Delta u$, as sketched in fig.12, using a constant $\Delta d^* = \Delta d - \Delta C = 0.6$ Å, which is the difference between the corrugation amplitudes of the contours of constant LDOS, $\Delta d =$

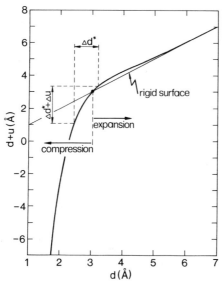

Fig.12. Deformation of graphite by the tunnel tip. The heavy solid line gives the total tip displacement $(d + u)$ vs tip-surface distance d. A tip motion Δd^* with respect to the surface results in an enhanced tip displacement $(\Delta d^* + \Delta u)$. A deformation by $u = 1$ Å corresponds to a force of 6×10^{-10} N for repulsion and 9×10^{-10} N for attraction, respectively. Reproduced from ref.55 with permission of Physical Review Letters.

0.8 Å, and those of constant force, ΔC = 0.2 Å[61]. The results for Δd + Δu are shown in fig.13, together with the experimental corrugations. The salient feature of the atomic-force model as applied to graphite is the dependence of the corrugation amplitudes on the tip-surface distance, d, only, with their dramatic increase at small d. Since in the voltage range considered I ∝ V as experimentally confirmed[69], d and thus the corrugation amplitude are functions of I/V. This behaviour, which is independent of the various assumptions made (e.g. choice of a particular tip-surface interaction), is convincingly verified by the experiment. Calculations and experimental data have not been collapsed to a single curve in fig.13 in order to better illustrate the ever stronger increase of corrugation amplitude with current at decreasing applied voltage. (The quantitative agreement of calculations with experiment in the medium voltage range is a matter of fortuitous choice of parameters, and should not be overrated). The strongly nonlinear amplification is also evident in the rapidly changing minima-maxima asymmetry in the traces of fig.10. At small currents, i.e. low amplification, the maxima – with respect to the flat saddle-point parts – are more pronounced; at large currents, the minima become the strong features of the traces."

"In the above analysis we have used a constant Δd. This is in the spirit of Tersoff's model, for varying current at fixed voltage. For varying voltages, on the other hand, the corrugation amplitude of the state density contributing to the tunnel current was predicted[61] to decrease with increasing voltage as apparently confirmed experimentally in the voltage range 50 mV to 1 V (see Fig. 3 of Ref. 30). In the light of the present model, however, we should view those experiments – in the voltage range 400 to 50 mV – rather as an increase of the corrugation from a bare electronic

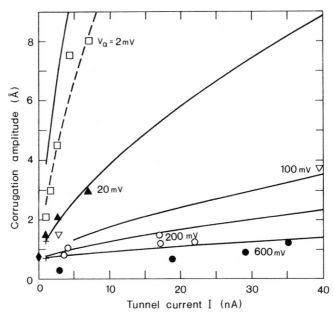

Fig.13. Measured (symbols) and calculated (solid lines) corrugations as a function of tunneling current and voltage. The dashed line was obtained with d* = 0.4 Å. The two crosses at 1 nA correspond to the measured corrugations[30] at 50 and 400 mV respectively; the diamond at zero current indicates the corrugation of the LDOS at the Fermi level[61]. Reproduced from ref.55, courtesy of Physical Review Letters.

value of 0.75 Å at 400 mV to a deformation-enhanced value of 1.3 Å at 50 mV, and only at higher voltages does the decrease of electronic corrugation become noticeable. The small experimental corrugations measured at 600 mV are attributed to such a reduction of Δd. The experimental values taken at 2 mV are likewise considerably smaller than predicted by our model. We believe that this is due to the close proximity of tip and graphite surface layer enhancing the density of states at the Fermi level. This enhancement is expected to decrease Δd, analogously to the effect of increased voltage. It further increases the tip-surface separation d and thus decreases the amplification factor."

"The measured corrugations can vary from experiment to experiment. We attribute this to changed tip conditions and thus to a change in d for a given current and voltage. However, the general dependence of the corrugation on current and voltage remains unchanged within an experiment."

"The corrugation amplification by elastic deformation rests on the conditions that (a) the tip traces a corrugation with respect to the surface which is different from that of the tip-surface interaction potential, (b) the tip-surface separation is very small, and (c) the material under investigation is sufficiently soft. With this in mind, we reinspect the pronounced difference in the corrugation observed in the charge-density-wave laminar compounds $1T\text{-}TaS_2$ and $2H\text{-}TaSe_2$[31]. Both compounds have elastic constants similar to that of graphite. But whereas $2H\text{-}TaSe_2$ possesses a metallic density of states, $1T\text{-}TaS_2$ is a semimetal with low density of states similar to that of graphite. This implies a very small tip-surface separation at the tunneling conditions of the experiment, and thus a large deformation enhancement of whatever LDOS corrugation is present."

"The fact that STM images can be dominated by elastic deformations can of course be exploited to study the very same deformations and thus interatomic potentials and local elastic constants. Separation of deformation - and tunnel-current - induced contributions to the total tip displacement can be done by performing experiments at different distances, i.e., tunnel currents. The analogy to the separation of electronic and structural features in STM via tunneling spectroscopy is evident. The role of voltage is played by the current-controlled distance; the quantities of interest are interatomic potentials and local elastic constants instead of electronic properties. This type of atomic-force spectroscopy and imaging is not restricted to compact materials of the kind described above; it will likewise be important in the study and imaging of any "soft" object ranging from simple adsorbates to biological material."

Soler et al.[55] directed their attention to <u>elastic</u> deformations of the sample. However, as Pethica[70] pointed out, irreversible <u>plastic</u> deformation may also occur, particularly in easily sheared layer structures such as graphite. Such deformation may be the basis of new forms of lithography, as described in a later section.

ATOMIC FORCE MICROSCOPY

We have shown that the STM can be used to measure not only tunneling currents but also Van der Waals and other close range atomic forces between tip and sample. In highly controlled experiments[71], these forces have been directly measured between such bodies at atomic-scale separation. As mentioned earlier, they are also utilised[7] for surface profiling with a lateral resolution of 1000 Å and a depth resolution of 10 Å although in this case the forces exerted by the stylus (radius ~1 μm) can be sufficient to cause plastic deformation of the surface[72]. The question therefore arises whether, by mounting the tip on a suitable spring, the atomic forces between tip and sample could provide another basis for atomic resolution microscopy of surfaces, thus extending the technique to include the surfaces of bulk insulators not accessible to the STM. We follow here the accepted practice of employing the acronym AFM for this new subject of atomic force microscopy as well as to denote the atomic force microscope itself.

The basic principles of AFM as well as the first working instrument, providing lateral resolution of 30 Å and depth resolution of 1 Å, were set out by Binnig, Quate and Gerber[73]. With kind permission of Physical Review Letters we now quote a part of this pioneering paper.

"The spring in the AFM is a critical component. We need the maximum deflection for a given force. This requires a spring that is as soft as possible. At the same time a stiff spring with high resonant freqency is necessary to minimize the sensitivity to vibrational noise from the building near 100 Hz. The resonant frequency, f_0, of the spring system is given by $f_0 = (1/2\pi)(k/m_0)^{1/2}$, where k is the spring constant and m_0 is the effective mass that loads the spring. This relation suggests a simple way out of our dilemma. As we decrease k to soften the spring we must also decrease m_0 to keep the ratio k/m_0 large. The limiting case, illustrated in fig.14, is but a single atom adsorbed at site A in the gap of an STM. It has its own mass and an effective k that comes from the coupling to neighboring atoms."

"The mass of the spring in manmade structures can be quite small but eventually microfabrication[73] will be employed to fabricate a spring with a mass less than 10^{-10} kg and a resonant frequency greater than 2 kHz. Displacements of 10^{-4} Å can be measured with the STM when the tunneling gap is modulated. The force required to produce these displacements is 2×10^{-16} N and this is reduced by 2 orders of magnitude when a cantilever with a Q of 100 is driven at its resonant frequency."

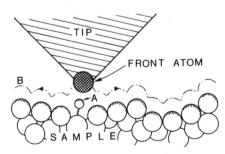

Fig.14. Description of the principle operation of an STM as well as that of an AFM. The tip follows contour B, in one case to keep the tunneling current constant (STM) and in the other to maintain constant force between tip and sample (AFM, sample, and tip either insulating or conducting). The STM itself may probe forces when a periodic force on the adatom A varies its position in the gap and modulates the tunneling current in the STM. The force can come from an ac voltage on the tip, or from an externally applied magnetic field for adatoms with a magnetic moment. Reproduced from ref.73 by permission of Physical Review Letters.

"AFM images are obtained by measurement of the force on a sharp tip (insulating or not) created by the proximity to the surface of the sample. This force is kept small and at a constant level with a feedback mechanism. When the tip is moved sideways it will follow the surface contours such as the trace B in fig.14."

"The experimental setup is shown in fig.15. The cantilever with the attached stylus is sandwiched between the AFM sample and the tunneling tip. It is fixed to a small piezoelectric element called the modulating piezo which is used to drive the cantilever beam at its resonant frequency."

"The STM tip is also mounted on a piezoelectric element and this serves to maintain the tunneling current at a constant level. The AFM sample is connected to a three-dimensional piezoeloectric drive, i.e. the x,y,z scanner. A feedback loop is used to keep the force acting on the stylus at a constant level. Viton spacers are used to damp the mechanical vibrations at high frequencies and to decouple the lever, the STM tip, and the AFM sample. The tip is brought in close proximity to the sample by mechanical squeezing of the Viton layers. High-frequency ($>$ 100 Hz) filtering of building vibrations is done as in the pocket-size STM with a stack of metal plates separated by Viton."

"We have operated the AFM in four different modes which relate to the connections of the two feedback circuits, one on the STM and the other on the tip. All four of these modes worked in principle. They each served to maintain a constant force, f_0, between the sample and the diamond stylus while the stylus followed the contours of the surface."

"In the first mode we modulated the sample in the z direction at its resonant frequency (5.8 kHz). The force between the sample and the diamond stylus – the small force that we want to measure – deflects the lever holding the stylus. In turn, this modulates the tunneling current which is used to control the AFM-feedback circuit and maintain the force f_0 at a constant level."

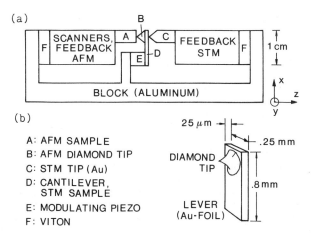

(a)

(b)

A: AFM SAMPLE
B: AFM DIAMOND TIP
C: STM TIP (Au)
D: CANTILEVER, STM SAMPLE
E: MODULATING PIEZO
F: VITON

Fig.15. Experimental setup. The lever is not to scale in (a). Its dimensions are given in (b). The STM and AFM piezoelectric drives are facing each other, sandwiching the diamond tip that is glued to the lever. Reproduced from ref.73 with permission of Physical Review Letters.

"In the second and third modes, the <u>lever</u> carrying the diamond stylus is driven at its resonant frequency in the z direction with an amplitude of 0.1 to 10 Å. The force, f_0, between sample and stylus changes the resonant frequency of the lever. This changes both the amplitude and phase of the ac modulation of the tunneling current. Either of these can be used as a signal to drive the feedback circuits."

"In the fourth mode we used one feedback circuit. It was connected to the AFM and it was controlled by the tunneling current in the STM. This system maintained the tunneling gap at a constant level by changing the force on the stylus."

"The fourth mode was further improved by reconnection of both feedback circuits in such a way that the AFM sample and the STM tip were driven in opposite directions with a factor α less in amplitude for the STM tip. The value of α ranged from 10 to 1000."

"In contrast to previous methods, the absolute value of f_0, the force on the stylus, was not well defined except at the beginning of the measurement. The deformation of the spring, Δz, is well calibrated at the starting point, but as the measurement proceeds each component of the system moves in an unknown way because of thermal drifts. These change the initial calibration. Additionally, we know that the three-dimensional motion of the AFM sample must produce modest amounts of change in Δz so as to compensate for the simultaneous motion of the stylus as it follows the topography of the surface. Therefore, even in the absence of thermal drifts the force f_0 will vary over a certain range that is dependent on both the roughness of the surface and the value of α."

"The fourth mode proved to be the most reproducible. We used it to record the results shown in fig.16, where we show the topography of an area of a ceramic (Al_2O_3) surface. The successive traces along the x axis are displaced from each other by the small, undefined thermal drift along the y axis. The vertical dashed lines of fig.16 indicate the smooth variation in the y direction of topographic features that can be followed from trace to trace. From these results and from the noise amplitude on the traces we can estimate that it should be possible to resolve a periodic corrugation on the sample with an amplitude below 1 Å when the period of the corrugation is between 1 and 100 Å."

Binnig, Quate and Gerber[73] then present a few further results on the Al_2O_3 surface before going on to analyse the future potential of the AFM as follows.

"The following improvements over the handmade version used here should increase the resolution to the point where we will be able to resolve the atomic features. Available microfabrication techniques[74] will allow us to reduce the mass of the stylus-cantilever unit by several orders of magnitude. When the instrument is mounted in an ultrahigh-vacuum chamber where clean surfaces can be well characterized we know from our STM experience that the stability will be improved by at least 2 orders of magnitude. With these optimum conditions the thermally induced vibrations of the cantilever at room temperature will limit the force sensitivity to 10^{-15} N. If the system is cooled below 300 mK we estimate that the lower limit will be 10^{-18} N."

"This level becomes interesting when we compare it to the interatomic forces[75]. In the strongest materials with ionic bonds the binding energy is 10 eV. It is 10 meV for those materials that are held together with the weak forces of van der Waals. If we arbitrarily equate the energy to a

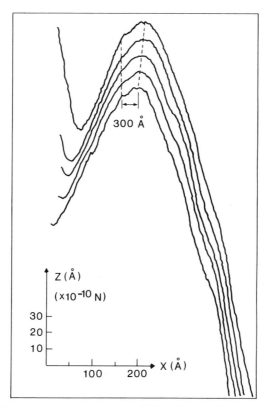

Fig.16. The AFM traces on a ceramic (Al_2O_3) sample. The vertical scale translates to a force between sample and tip of 10^{-10} N/Å. For the lower trace the force is near 3×10^{-8} N. The stability of the regulated force is better than 10^{-10} N. The successive traces are displaced by a small drift along the y axis. Reproduced from ref.73 with permission from Physical Review Letters.

force acting through a distance of 0.16 Å we find that a binding energy of 1 eV is equivalent to a force of 10^{-8} N. The interatomic forces therefore range from 10^{-7} N for ionic bonds to 10^{-11} N for van der Waals bonds[76] and down to perhaps 10^{-12} N for some of the weaker forces of surface reconstruction. The limiting sensitivity of our instrument is far less than these values. Therefore, we should be able to measure all of the important forces that exist between the sample and adatoms on the stylus."

It is interesting to note that some of the above prescient forecasts have already come to pass. Using the AFM, images have been achieved on graphite[77,78] and boron nitride at atomic resolution, as well as images of monatomic surface steps on an oil-covered surface of sodium chloride[78]. The improvements in AFM design, which made possible these spectacular results, are obviously still proceeding and many surface science laboratories are constructing such microscopes. The AFM shares the advantage of the STM in being readily compatible with standard surface science equipment and procedures, but is obviously complementary to the STM in the kind of information it can provide.

NANOMETER SCALE STRUCTURE FABRICATION

In most circumstances, the microscopist would prefer not to damage the sample he is observing and, because of the extremely low energy electrons employed, the STM offers considerable advantages in this regard over other forms of electron microscopy (see, for example, chapters on STEM by Howie and on HREM by Smith).   The possibility of controlled modification of surfaces on the nanometer scale has, however, many potential applications in lithography, device fabrication and information storage.

A few authors have now studied surface modifications produced by the STM in the tunneling mode[80,81], as well as in the field emission mode[82,83]. Becker et al.[84] have demonstrated that it is possible to write ~8 Å features on a Ge crystal surface by briefly applying a -4.0 V tip-to-surface bias which may transfer a tungsten atom or atoms to the sample. Another writing mechanism is thought to be local melting of the sample and has been reported by Staufer et al.[85].   With kind permission of Applied Physics Letters, we now reproduce some excerpts from their paper.

"This letter describes a new technique for creating nanometer scale structures with the STM. We use ion-etched metallic glasses as substrates because they provide atomically flat surfaces[86] and a small mean free path of the electrons."

"Figure 17 shows schematically the positioning of the tunneling tip to produce the nanometer structures described below. The sample was a glassy $Rh_{25}Zr_{75}$ alloy prepared by the splat cooling technique under high vacuum conditions. The splat was then transferred in air to our ultrahigh vacuum version of the STM. An atomically flat surface of the sample has been obtained after ion etching for more than 30 min with 5 keV Ar^+ ions[86]. We have observed a correlation between the energy of the ions and the formation of the nanometer structures.   The surface roughness is characterized by a rms value[87] which is less than 0.1 nm.   The detailed fabrication procedure is the following: We start with the tip at position A on the lower left corner of fig.17.   From there we move the tip in the constant tunneling current mode (1 nA, 100 mV) to position 1 where we produce the first nanometer structure. We increase the bias voltage up to 2 V (with the sample at positive polarity), as a result the tip to sample distance is increased.   Then we increase the tunneling current up to 315 nA. This current injected into a tiny region of the sample surface leads to a significant temperature rise and therefore to a thermal and electric field induced modification of the surface, perhaps even to local melting.   At a value of about 300 nA the current as a function of time starts oscillating.   These oscillations are attributed to the liquid

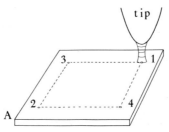

Fig.17. Schematic drawing of the positioning of the tip for producing the nanometer structures shown in figs 18 and 19.   Reproduced from ref.85 with permission of Applied Physics Letters.

surface. A Taylor cone is formed in the high electric field of 10^8 V/cm across the tunneling gap[88] while the tip is retracted because the feedback loop is still acting. The current is then lowered to 1 nA in approximately 1 s. A nanometer scale structure is formed upon cooling in the electric field which is still high. Afterwards the bias voltage is also decreased to 100 mV. Figure 18a depicts the STM image of the structure. In the next step we move the tip in the constant tunneling mode to position 2 where we produce the second hillock in the same way. The other structures are prepared at the positions 3 and 4 of fig.17. In Figs 18 a-d we show the XY recorder traces of the STM imaging the nanometer structures produced at the positions 1 to 4 in fig.17. The four structures are clearly visible as four hillocks with a diameter of about 35 nm and a height of about 10 nm.

In fig.19 the top view of the topographic image of fig.18d is shown on the left-hand side whereas on the right-hand side the local tunneling barrier height image[89] is presented. The small difference of 0.2 eV in the local tunneling barrier height between the substrate and the structures indicates that the chemical composition of the substrates and the hillocks is different. There are several possibilities to explain this fact. Contributions might arise from surface composition changes during ion etching as well as from segregation and electromigration during the fabrication."

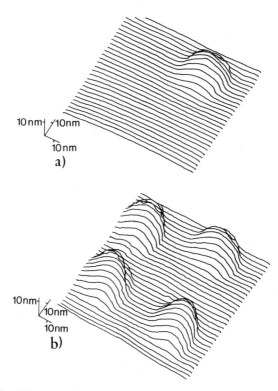

10 nm / 10 nm
10 nm
a)

10nm / 10nm
10nm
b)

Fig.18. (a), (b). XY recorder images of the surface of glassy $Rh_{25}Zr_{75}$ after the production of one and four nanometer structures respectively. The hillocks have a diameter of about 35 nm and a height of about 10 nm. A scale of 10 nm is indicated by the bars. Adapted from ref.85 with permission of Applied Physics Letters.

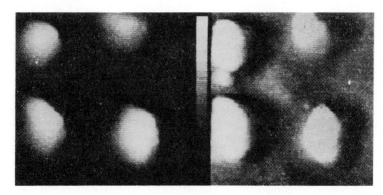

Fig. 19. Top view of the topographic image (left-hand side) and the local
barrier height (right-hand side) of the same region as that in
fig.18b. The total range in the topographic image is 9 nm; the
range of the total barrier heights is 0.22 eV. Reproduced from
ref.85 with permission of Applied Physics Letters.

Staufer et al.[85] also successfully demonstrated that with the same
process they could draw lines of width 20 nm and height about 2 nm at a
rate of about 3 nm per second. Similar structures could be produced on
other metallic glasses such as $Fe_{86}B_{14}$. They make some further remarks
about the temperature rise which may be occurring.

"The estimated increase of the temperature for tunneling currents in
the μA range and bias voltages of about 2 V is several hundreds of
degrees[90]. In our experiment the current is by a factor of 10 lower.
However, a similar temperature rise can also be expected by taking into
account the small mean free path of electrons in these metallic glasses
which reduces the area under consideration. In addition, there is a
considerable uncertainty to take the thermal conductivity from the
Wiedemann-Franz law in these d-electron transition metal glasses. We made
a control experiment with the same tunneling parameters but using an Ir
sample, which has a much higher melting point (bulk melting temperature
2683 K) than the RhZr splat (bulk melting temperature 1340 K). On this Ir
sample we could not produce any structures using the procedure mentioned
above. This result supports our interpretation that local melting rather
than, for example, a deposition process takes place on the RhZr surface."

CONCLUSIONS AND FUTURE APPLICATIONS

Six years after the discovery of STM it can be said that the technique
has been able to provide a new perspective on the topography of surfaces
for a wide range of materials. We do not believe that it is easy from STM
experiments to say precisely where the atoms are located in a
crystallographic sense, because the analysis of the data requires complex
theoretical analysis just as Low Energy Electron Diffraction data does.
However, the technique has been able to provide other additional
information including, for example, energy resolved surface spectroscopy
and is able to distinguish between surface topography effects and the
density of states contribution to the current. In our opinion the main
merit of STM is its versatility in yielding information from different
materials. To date, for example, useful results have been obtained from
metals, semimetals, semiconductors, laminar compounds, thin and growing

oxide layers, biological materials, electrochemical systems, plastics and oxides covered by very thin films of metals, etc. The instrument is also able to operate not only in ultra high vacuum but also in different surrounding atmospheres and liquids, giving information about liquid-solid interfaces.

In parallel with all this, the development of the atomic force microscope has provided a supplementary technique to investigate and develop force interactions, force microscopy and spectroscopy that can determine with atomic resolution the local elastic properties of materials.

The high resolution side of the application of STM has been developed and pushed by many laboratories studying all kinds of surfaces in ultra high vacuum. However, in our opinion, the major future of STM may be less in the high resolution side and more in its technological applications, i.e. as a complementary technique in an industrial laboratory that can provide high spatial resolution of novel properties to supplement other measurements. For example, we may introduce an STM inside a typical scanning electron microscope[17,23] or electrochemical cell.

Another possible and promising technological application is to use the tip of STM as surface modification technique to create nanostructures.

Finally, as Heini Rohrer is accustomed to say, STM is a technique for the imagination, easy to manage and to build within very reasonable means; in other words it is a nice toy to play with.

ACKNOWLEDGEMENTS

I am pleased to have been able to lecture in this school and have the opportunity to meet very enjoyable people in Erice. I am grateful to the Nobel Foundation, Physical Review Letters, Applied Physics Letters and I.B.M. Journal of Research and Development for permission to quote extensively from published papers and also to the authors of these papers for their agreement to this. Finally I wish to thank Professor A. Howie in editing the English text.

REFERENCES

1. G. Binnig, H. Rohrer, Ch. Gerber and E. Weibel, Appl. Phys. Lett. 40:178 (1982).
2. J.A. Golovchenko, Science 232:48 (1986); P.K. Hansma and J. Tersoff, J. Appl. Phys. 61:R1 (1987).
3. G. Binnig and H. Rohrer, Nobel Lecture 1987 (Nobel Foundation, Stockholm).
4. I. Giaever, Rev. Mod. Phys. 46:245 (1974).
5. W.A. Thompson, S.F. Hanrahan, Rev. Sci. Instrum. 47:1303 (1976).
6. B.P. Williamson, Proc. Inst. Mech. Eng. 182:21 (1967);
   K.H. Guenther, P.G. Wierer, J.M. Bennett, Appl. Opt. 23:3820 (1984).
7. R. Young, J. Ward, F. Scire, Rev. Sci. Instrum. 43:999 (1972).
8. Technical details: G. Binnig, H. Rohrer, Helv. Phys. Acta 55:726 (1982); Surf. Sci. 126:236 (1983); Sci. Am. 253:40, No 8 (1982); Spektrum Wiss. No 10, p 62 (1985).
9. O. Marti, Ph.D. Thesis, Eidgenossische Technische Hochschule Zurich (1986);
   O. Marti, G. Binnig, H. Rohrer, H. Salemink, Surf. Sci. 181:230 (1987).

10. H.R. Ott, H. Rohrer, unpublished; S. Vieira, IBM J. Res. Dev. 30:553 (1986).
11. U. Poppe, Verhandl. DPG(VI) 16:476 (1981).
12. E.C. Teague, Dissertation, North Texas State University, TA, USA (1978) (University Microfilms International, Ann Arbor, MI, USA, p 141); Bull. Am. Phys. Soc. 23:290 (1978); J. Res. Natl. Bur. Stand. (US) 91:171 (1986).
13. G. Binnig and H. Rohrer, IBM J. Res. and Dev. 30:355 (1986).
14. C.B. Duke, "Tunneling in Solids", Academic Press, New York (1969).
15. E.L. Wolf, Rep. Prog. Phys. 41:1439 (1978); also "Principles of Electron Tunneling Spectroscopy", Oxford University Press, Oxford 1985.
16. J. Simmons, J. Appl. Phys. 34:1793 (1963).
17. Ch. Gerber, G. Binnig, H. Fuchs, O. Marti and H. Rohrer. Rev. Sci. Instrum. 57:221 (1986).
18. J.E. Demuth, R.J. Hamers, R.M. Tromp and M.E. Welland, IBM J. Res. Dev. 30:396 (1986); J. Vac. Sci. and Techn. A4:1320 (1986).
19. R. Sonnenfeld and P.K. Hansma, Science 232:211 (1986).
20. H.-W. Fink, IBM J. Res. Dev. 30:460 (1986).
21. J. Becker, Surf. Sci. 181:200 (1987).
22. R.S. Becker, J.A. Golovchenko, E.G. MacRae and B.S. Swartzentruber, Phys. Rev. Lett. 55:2028 (1985).
23. J.E. Demuth, R.J. Hamers, R.M. Tromp and M.E. Welland, J. Vac. Sci. and Tech. A4:1320 (1986).
24. R.S. Becker, J.A. Golovchenko, G.S. Higashi and B.S. Swartzentruber, Phys. Rev. Letts. 57:1020 (1986).
25. G. Binnig, H. Rohrer, Ch. Gerber and E. Stoll, Surf. Sci. 144:321 (1984).
26. S.A. Elrod, A. Bryant, A.L. de Lozanne, S. Park, D. Smith and C.F. Quate, IBM J. Res. and Dev. 30:387 (1986).
27. R.J. Behm, W. Hoesler, E. Ritter and G. Binnig, Phys. Rev. Lett. 56:228 (1986).
28. G. Binnig, H. Fuchs and E. Stoll, Surf. Sci. 169:L295 (1986).
29. A. Baratoff, G. Binnig, H. Fuchs, F. Salvan and E. Stoll, Surf. Sci. 168:734 (1986).
30. G. Binnig, H. Fuchs, Ch. Gerber, H. Rohrer, E. Stoll and E. Tosatti, Europhys. Lett. 1:31 (1986).
31. R.V. Coleman, B. Drake, P.K. Hansma and G. Slough, Phys. Rev. Lett. 55:394 (1985).
32. A.M. Baró, R. Miranda, J. Alaman, N. Garcia, G. Binnig, H. Rohrer, Ch. Gerber and J.L. Carrascosa, Nature 315:253 (1985).
33. S. Park and C.F. Quate, Appl. Phys. Lett. 48:112 (1986).
34. R. Sonnenfeld and B.C. Schardt, Appl. Phys. Lett. 49:1172 (1986).
35. B. Travaglini, H. Rohrer, M. Amrein and H. Gross, Surf. Sci. 181:380 (1987).
36. H. Fuchs, W. Schrepp and H. Rohrer, Surf. Sci. 181:391 (1987).
37. J.H. Coombs, J.B. Pethica and M.E. Welland (in press).
38. R.J. Hamers, R.M. Tromp and J.E. Demuth, Phys. Rev. Lett. 56:1972 (1986).
39. G. Binnig, N. Garcia and H. Rohrer, Phys. Rev. B32:1336 (1985).
40. J. Bardeen, Phys. Rev. Lett. 6:L57 (1961).
41. J. Tersoff and D.R. Hamann, Phys. Rev. Lett. 50:1998 (1983), Phys. Rev. B31:805 (1985).
42. N. Garcia, C. Ocal and F. Flores, Phys. Rev. Lett. 50:2002 (1983).
43. E. Stoll, A. Baratoff, A. Selloni and P. Carnevali, J. Phys. C17:3073 (1984).
44. A. Baratoff, Physica B127:143 (1984).
45. N. Lang, Phys. Rev. Lett. 55:230 (1985) and 56:1164 (1986).
46. N. Garcia, IBM J. Res. Dev. 30:533 (1986).

47. A.A. Lucas, J.P. Vigneron, J. Bono, P.H. Cutler, T.E. Feuchtwang, R.H. Good and Z. Huang, J. de Physique C9:125 (1984).

48. R. Garcia, J.J. Saenz, J.M. Soler and N. Garcia, J. Phys. C19:L131 (1986).

49. M. Büttiker, Y. Imry, R. Landauer and S. Pinhas, Phys. Rev. B31:6207 (1985).

50. G. Binnig, N. Garcia, H. Rohrer, J.M. Soler and F. Flores, Phys. Rev. B30:4816 (1984)

51. N. Garcia and F. Flores, Physica 127B:137 (1984).

52. E. Stoll, Surf. Sci. 143:L4111 (1984).

53. R.G. Gordon, J. Chem. Phys. 51:14 (1969).

54. J.H. Coomb and J.B. Pethica, IBM J. Res. Dev. 30:455 (1986).

55. J.M. Soler, A.M. Baró, N. Garcia and H. Rohrer, Phys. Rev. Lett. 57:444 (1986).

56. M.S. Chung, T.E. Feuchtwang and P.H. Cutler, Surf. Sci. 187:559 (1987).

57. N.D. Lang and W. Kohn, Phys. Rev. B1:4555 (1970).

58. Y. Kuk and P.J. Silverman, Appl. Phys. Lett. 48:1598 (1987).

59. W.J. Kaiser and R.C. Jaklevic, Surf. Sci. 181:55 (1987).

60. V.M. Hallmark, S. Chiang, J.F. Rabolt, J.D. Swalen and R.J. Wilson, Phys. Rev. Lett. 59:2879 (1987).

61. A. Selloni, P. Carnevali, E. Tosatti and C.D. Chen, Phys. Rev. B 31:2602 (1985).

62. J. Tersoff, Phys. Rev. Lett. 57:440 (1986).

63. A.M. Baró, R. Miranda, J. Alaman, N. Garcia, G. Binnig, H. Rohrer, Ch. Gerber and J.L. Carrascosa, Nature 315:253 (1985).

64. L.D. Landau and E.M. Lifshitz, "Theory of Elasticity", Pergamon, London (1959).

65. R. Nicklow, N. Wakabayashi and H.G. Smith, Phys. Rev. B5:4951 (1972).

66. D.P. Di Vicenzo, J. Bernhole and M.H. Brodsky, Phys. Rev. B 28:3286 (1983).

67. J. Batra, private communication.

68. G. Binnig, N. Garcia, H. Rohrer, J.M. Soler and F. Flores, Phys. Rev. B30:4816 (1984).

69. O. Marti, private communication.

70. J.B. Pethica, Phys. Rev. Lett. 57:3235 (1986).

71. D. Tabor and R.H.S. Winterton, Proc. Roy. Soc. A312:435 (1979).

72. E.J. Davis amd K.J. Stout, Wear 83:49 (1982).

73. G. Binnig, C.F. Quate and Ch. Gerber, Phys. Rev. Lett. 56:930 (1986).

74. K.E. Peterson, Proc. IEEE 70:420 (1982).

75. B.H. Flowers and E. Mendoza, "Properties of Matter", Wiley, London, pp 22-55 (1970).

76. H. Krupp, W. Schnabel and G. Walter, J. Colloid Interface Sci. 39:421 (1972).

77. G. Binnig, Ch. Gerber, E. Stoll, T.R. Albrecht and C.F. Quate, Europhhys. Lett. 3:1281 (1987).

78. O. Marti, B. Drake and P.K. Hansma, Appl. Phys. Lett. 61:484 (1987).

79. T.R. Albrecht and C.F. Quate, Appl. Phys. Lett. 62:2599 (1987).

80. M. Ringger, H.R. Hidber, R. Schlögl, P. Oelhafen and H.-J. Güntherodt Appl. Phys. Lett. 46:833 (1985).

81. D.W. Abraham, H.J. Mamin, E. Ganz and J. Clarke, IBM J. Res. Dev. 30:492 (1986).

82. M.A. McCord and R.F.W. Pease, J. Vac. Sci. Technol. B3:198 (1985).

83. M.A. McCord and R.F.W. Pease, J. Colloq. Phys. (Paris) C2:485 (1986).

84. R.S. Becker, J.A. Golovchenko and B.S. Swartzentruber, Nature 325:419 (1987).

85. U. Staufer, R. Wiesendanger, L. Eng, L. Rosenthaler, H.R. Hidber, H.J. Güntherodt and N. Garcia, Appl. Phys. Lett. 51:244 (1987).

86. R. Wiesendanger, M. Ringger, L. Rosenthaler, H.R. Hidber, P. Oelhafen, H. Rudin and H.-J. Güntherodt, Surf. Sci. 181:46 (1987).

87. R.M. Feenstra and G.S. Oehrlein, Appl. Phys. Lett. 47:97 (1985).
88. J.A. Kubby and B.M. Siegel, J. Colloq. (Paris) C2:107 (1986).
89. G. Binnig and H. Rohrer, Surf. Sci. 126:236 (1983).
90. F. Flores, P.M. Echenique and R.H. Ritchie, Phys. Rev. B34:2899 (1986).

SPIN-POLARIZED SECONDARY ELECTRONS FROM FERROMAGNETS

Jürgen Kirschner

Institut für Atom- und Festkörperphysik
Freie Universität Berlin, Arnimallee 14, D-1000 Berlin 33
and
Institut für Grenzflächenforschung und Vakuumphysik
D-5170 Jülich, F.R. Germany

ABSTRACT

Motivated by the recent development of Scanning Electron Microscopes with secondary electron spin polarization analysis for magnetic surface domain structure imaging, we review the present understanding of the emission of spin-polarized secondary electrons. In particular, the following points are discussed:
(i) the relation between magnetization vector orientation of the sample and the spin polarization vector orientation of the secondaries;
(ii) the shape and origin of the polarization distributions as a function of the kinetic energy of primary electrons and of their spin orientation;
(iii) the depth from which the magnetic information stems.

INTRODUCTION

About a decade ago, Chrobok and Hofmann[1] discovered that secondary electrons from ferromagnetic EuS, excited by unpolarized primary electrons, are spin-polarized by up to 32%. This result initially went almost unnoticed, but time ripened, and with the growing demands of the magnetic data storage technology for analysing smaller and smaller magnetic bits, it was discussed by several authors[2-4] how spin polarization effects could possibly be used for magnetic structure analysis. There are several other techniques available for imaging magnetic structures, such as Lorentz microscopy[5], electron holography[6], magnetic contrast in the scanning electron microscope[7] or the magneto-optical Kerr effect[8]. However, those methods with good spatial resolution, like Lorentz microscopy and electron holography are restricted to thin films, observed in transmission. Those methods applicable to semi-infinite samples, like the Kerr effect, offer limited resolution only, e.g. about half the wavelength of visible light[8]. For magnetic structure analysis of such samples the superior resolution of the Scanning Electron Microscope (SEM) could be harnessed, if the spin polarization of secondary electrons were analysed and used as a source of contrast, much in the same way as the secondary electron yield in an ordinary SEM. The basic idea is sketched in fig.1. A fine primary electron beam hits the sample which is supposed to have a magnetic domain structure written in by a magnetic recording head. The orientation of the spin of the majority electrons is indicated by the arrows pointing upwards or downwards, assuming a vertical magnetic recording scheme. If the electrons detected outside the solid maintain their preferential spin orientation, the sign and magnitude of the measured spin polarization changes if the primary beam is scanned across the surface. This signal can be used for intensity modulation or colour coding on an SEM display tube. Moreover, if the spatial orientation of the spin polarization vector is the same inside and outside of the material, the orientation of the magnetization vector may be determined from that of the spin polarization vector. This is a particularly useful feature, since it allows a full 3-dimensional characterization of the magnetic domain structure. For reasons of space we refrain here from discussing how the spin polarization vector analysis may be done, but refer the reader to refs 9-13 instead. The first successful realization of this scheme was done by Koike and Hayakawa[9] in 1984. A very recent example from our group is shown in fig.2. The arrows indicate the magnetization along a particular direction. The constant grey level indicates zero polarization along the particular axis.

At present there are three such microscopes in operation (Hitachi Advanced Research Laboratory, National Bureau of Standards[14], and KFA Jülich[15]) and several others are being planned. The demonstrated lateral magnetic resolution at present is around 40 nm, that is already substantially better than in any other technique applicable to semi-infinite samples. There is hope that the resolution may be further improved by an order of magnitude in the not too distant future. For a more detailed discussion of the properties of this technique we refer to refs 16,17. Here we wish to discuss some fundamental aspects of the spin polarization of secondary electrons, which are vital for a quantitative interpretation of micrographs such as shown in fig.2.

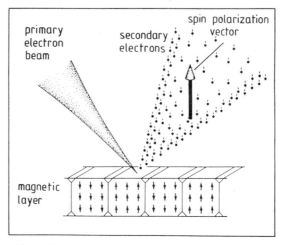

Fig. 1. Principle of magnetic structure analysis in a Scanning Electron Microscope (SEM). The finely focussed primary beam excites secondary electrons out of the surface of a magnetic material. The spin polarization of these electrons is analysed and used as a signal to the SEM display tube.

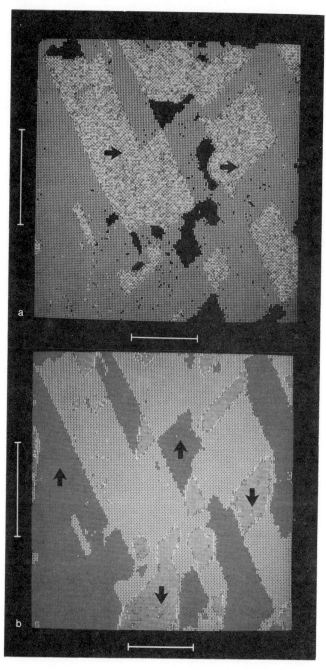

Fig. 2. Magnetic domain images from a Fe(100) surface, measured
simultaneously for the spin polarization vector component along the
horizontal axis [in a)] and the vertical axis [in b)]. Areas of
the same colour have the same magnetization (see arrows). Within
the grey areas in a) and b) the horizontal (a) and vertical (b)
polarization components, respectively, are zero, while the vertical
and horizontal components are maximum. The scale markers
correspond to 60 µm, each. A colour print of this figure will be
found in the frontispiece.

ORIENTATION OF THE SPIN POLARIZATION VECTOR RELATIVE TO THE SAMPLE MAGNETIZATION

Inside the sample the magnetization vector is oriented opposite to the spin polarization vector, because the magnetic moment of the electron and its spin are opposite to each other. It is by no means trivial that the polarization vector of the electrons in vacuum is directly related to the magnetization vector of the domain excited by the primary beam. This was implicitly assumed when drawing fig.1, and it is the basis for the assignement of the arrows in fig.2. The validity of this assumption will be tested in the following.

In very general terms, the very existence of a surface may induce a spin-dependence of electron transmission through that surface, since exchange interaction and spin-orbit interaction are present in the crystal, while they are negligible for secondary electrons in vacuum. In principle, the matching conditions of the electronic wavefunctions (two-component spinors) become spin-dependent because of this discontinuity. For pure exchange interaction the transmission of the one spin state may be different from that of the other, which may lead to a change of the magnitude of the polarization vector, but its orientation in space is conserved because in pure exchange interaction the individual spin states are conserved. By spin-orbit interaction, however, not only the magnitude but also the orientation of the polarization vector may be changed[18-20]. With complete integration over the total secondary electron flux into the empty half space such effects would largely cancel out, as well as by integration over a cone around the surface normal. To see, whether the spin-dependent surface transmission effects are detrimental to magnetic structure analysis in a general geometry, a "worst case" experiment was done[21], with emission into a narrow cone away from the surface normal (See the insert in fig.3). The acceptance angle was about ± 5° and the emission direction was 38° off the normal. Since the magnetization and the emission direction both were in a mirror plane of the crystal, the polarization

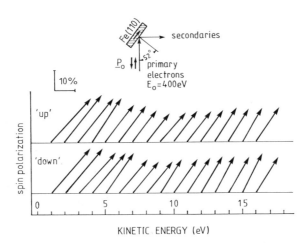

Fig. 3. Measurements of the orientation and magnitude of the spin polarization vector of secondary electrons from a clean Fe(110) surface for spin-polarized primary electrons [E_0 = 400 eV, polarization parallel ("up") or antiparallel ("down") to the vertical momentum]. The origins of the polarization vectors indicate the kinetic energy of the secondaries. The statistical uncertainty corresponds approximately to a circle around the arrow head with a diameter equal to the length of the head.

vector - also after transmission - lies in the emission plane for symmetry reasons, but because of spin-orbit interaction it might be rotated away from the magnetization direction. Since this would occur in an energy-dependent way, the rotation would supposedly be different for different kinetic energies. In fig.3 the polarization vector in the emission plane is plotted with the origin of the arrows indicating the kinetic energy of the secondaries. In addition, the experiment is done with spin-polarized primary electrons, with the polarization vector pointing either upwards ("up") or downwards ("down"). We comment on this aspect below. The experimental uncertainty in length and directions of the arrows, as given by statistics, approximately equals the length of the arrow head. Considering only the orientation in space, we see that the arrows all point into the same direction, within a root-mean-square angular deviation of 1.5°. Within the experimental uncertainty of the orientation of the crystal in space (± 1°), the polarization vector thus is exactly aligned antiparallel to the magnetization vector of the sample. Since we carried out a "worst case" experiment, its result provides a firm basis for the magnetization analysis in the SEM.

Strictly speaking, this holds only for 3-d ferromagnets. With the heavy rare-earth magnets disturbing effects of spin-orbit interaction might be stronger. In practice, however, the secondary electron flux is always integrated over as much of the empty half space as possible for intensity reasons. Any residual effects therefore are expected to be small. It should be pointed out, that the intrinsic effects described above are not the only ones that may lead to a distortion of the magnetic image. If there is a strong stray magnetic field above the sample, the magnetic moment of the electron may precess about the magnetic field lines (Larmor precession) and thus may change their spin orientation if the passage is not fast enough. So far no evidence for distortions due to this effect has been reported, but its possibility should be kept in mind. The only remedy would be to extract the electrons as fast as possible through the stray field region by applying strong electric fields.

A further remarkable result in fig.3 is that the secondary electron spin polarization is evidently completely independent of the spin of the primary electrons. This proves that in the higher generations of secondary and tertiary electrons any memory of the primary electron spin has been lost and that the secondary electron spin polarization is dominated by the majority spin electrons of the sample. For the applications this means that an unpolarized primary beam is as good as a polarized one, and that it is not worth the effort to try to build an SEM with a cathode producing polarized electrons.

A third finding evident in fig.3 is that the length of the vectors, though all pointing into the same direction, change as a function of kinetic energy. In particular, the degree of polarization seems to be largest at the smallest kinetic energies. While the magnetization direction can thus be read reliably, its magnitude appears not to be simply related to the magnitude of the spin polarization. This is the subject of the next section.

SPIN POLARIZATION IN SECONDARY ELECTRON SPECTRA

A collection of spin polarization and intensity spectra of secondary electrons from clean Fe(110) is shown in fig.4 and fig.5 with the primary energy as a parameter[22]. Only emission along the surface normal with an angular resolution of ~ ± 5° is accepted, while the angle of incidence is 48° with respect to the surface normal. The polarization spectra in fig.4

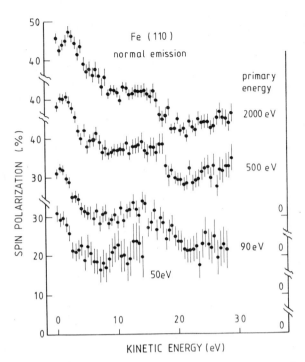

Fig. 4. Spin polarization distributions of secondary electrons emitted normally from clean Fe(110) as a function of their kinetic energy. The primary energy ranges from 50 to 2000 eV. The angle of incidence is 48° off the surface normal, and a bias potential of -10 V with respect to ground was applied to the sample.

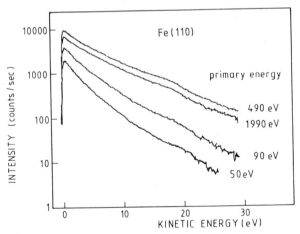

Fig. 5. Logarithmic intensity distributions of secondary electrons emitted normally from clean Fe(110) for various primary energies. The angle of incidence is 48° relative to the sample normal. The effects of a bias potential of -10 V at the sample and of the energy dependent transmission function of the analyser are not corrected. Note the intensity fine structure near 15-20 eV, and the similarity of the curves for 490 eV and 1990 eV primary energy.

show the highest polarization at very low kinetic energy, a gradual decrease towards higher kinetic energy, and also some polarization fine structure. The intensity spectra in fig.5 (on a logarithmic scale) show very similar shape for all primary energies, but different magnitude according to the dependence of the total electron yield on the primary energy. There is also some intensity fine structure, e.g. at 15-20 eV, which is correlated with structure in the polarization spectra. Further experimental results are given in fig.6, which shows, as a function of primary energy, the total secondary electron yield (left-hand-scale), and the ratio of majority ($N_\uparrow$) to minority ($N_\downarrow$) electrons at kinetic energies of 1 eV, 15 eV, and 20 eV (right-hand-scale). The ratio $N_\uparrow/N_\downarrow$ first increases with increasing primary energy, and then saturates at energies where the total yield starts to decrease again. These results shall now be discussed and explained.

When looking at the polarization spectra in fig.4, we see that fine structures seems to be rather independent of the primary energy. Therefore, we decompose, in a first approximation, the spectra into a smooth function $P_S(E)$ onto which the fine structure $P_f(E)$ is superimposed.

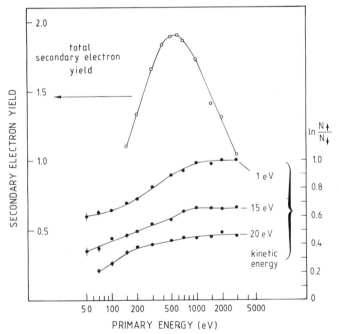

Fig. 6. Total secondary electron yield (left-hand-scale) and ratio of majority electrons $N_\downarrow$ to minority electrons $N_\uparrow$ as a function of the energy of the primary electrons. The angle of incidence is 48° with respect to the surface normal. The data on $N_\uparrow/N_\downarrow$ refer to electrons emitted along the surface normal within a cone of ~10° full angle with the kinetic energy indicated. The polarization P may be obtained from $\ell n\ N_\uparrow/N_\downarrow$ via the relation

$$\ell n\ N_\uparrow/N_\downarrow = \ell n\ [(1+P)/(1-P)] = 2(P + P^3/3 + P^5/5 + \ldots).$$

Note that the gain of low energy majority electrons over minority ones proceeds faster with increasing primary energy than for high energy secondaries.

We anticipate, that the smooth function P_S is related to the cascade of secondary electrons inside the sample, while the fine structure P_f is due to details of the unoccupied electronic structure of the sample. The main question to be answered on the continuous function is, why the spin polarization increases towards lower kinetic energy. The explanation is, that the secondary electrons produce Stoner excitations on their way out of the sample, whereby majority spin electrons are enriched over minority spin electrons.

Stoner excitations are elementary excitations specific to ferromagnets and characterised by an electron-hole pair with opposing spins. For example, a secondary electron with some 10 eV kinetic energy above the Fermi energy may scatter with a majority type electron below E_F. After scattering it may occupy an empty level of minority type above E_F, while the other electron, of majority spin type, may occupy an energy level far above E_F but somewhat below that previously occupied by the incident electron. Around the Fermi level we find the configuration of a Stoner pair, with a minority type electron above E_F and a hole in an otherwise occupied majority type band below E_F. The energy of the Stoner pair is the energy that is missing to the electron kicked up far above E_F, which is of majority type. Thus, for a particular minority type secondary electron moving towards the surface we may get out a majority type secondary at a somewhat smaller energy. Of course, the opposite process is possible also, but because there are predominantly empty minority states above E_F and filled majority states below E_F, majority electrons become enriched in the secondary electron flux oriented towards the surface. Because the generation of Stoner pairs involves a loss of kinetic energy, namely the energy of the Stoner pair, secondary electrons with lower energy are more likely to have majority spin than those of higher energy.

This is one reason, why the spin polarization increases towards the low energy end of the intensity distribution. A second reason is the following one: an electron starting at some depth and moving towards the surface is subject to a continuous loss of energy, not only because of Stoner excitations, but also because of "ordinary" electron-hole pair excitation within a spin system. Therefore, low energy secondaries on the average stem from larger depths than high energy ones. But if they come from large depths they have also had a high probability to excite Stoner pairs via exchange scattering, which results in an enhanced spin polarization. These two effects together explain the gradual increase of the spin polarization towards low kinetic energies.

We also observe a dependence of the spin polarization P_S on the primary energy (cf. fig.4), while the polarization fine structure P_f appears to be little affected. In fig.6 we have plotted as a function of primary energy, the ratio of majority spins to minority spins $N_\uparrow/N_\downarrow$ on a ℓn-scale for selected kinetic energies of 1 eV, 15 eV, and 20 eV (right-hand-scale). The logarithmic representation shows clearly that the shape of the curves is dissimilar, in particular that the ratio $N_\uparrow/N_\downarrow$ at 1 eV rises faster with increasing primary energy than at 15 and 20 eV. The drop in these curves at low energy is probably caused by a depression of the polarization by inelastically scattered primary electrons, which are unpolarized. These gain increasingly in weight with decreasing primary energy since the primary energy comes closer to the fixed secondary energy (i.e. a smaller energy loss is required) and because the yield of secondaries drops rapidly. This taken into account, the ratio $(N\uparrow/N\downarrow)$ at 20 eV is nearly independent of primary energy. This shows that the polarization enhancement effect discussed above plays a minor role, if any, at high kinetic energies of the secondaries.

It is interesting to compare the primary energy of the spin polarization at low kinetic energy with that of the total secondary electron yield (see fig.6). We note that the polarization at 1 eV reaches saturation at higher energy than that of the maximum yield. At the energy of the maximum yield the projected range of the primary electrons just overlaps the depth from where the secondary electrons stem. (For more details on the escape depth, see next section). At higher primary energy, the energy is deposited over larger depth and less electrons are able to reach the surface though more secondaries have been excited in total. Therefore the total yield decreases. Also, beyond the maximum yield, the shape of the intensity distributions becomes (almost) stationary, see fig.5. Though hardly visible on the logarithmic scale, a closer inspection reveals that at 1990 eV primary energy the relative intensity of the very slow secondaries is somewhat higher than at 490 eV, which is about the energy of the maximum yield. The peak position also shifts somewhat towards even smaller energy. This behaviour has also been observed on Ge (for example see ref.23). The reason is, that secondaries from large depths have lost almost all their energy when reaching the surface and are just able to escape into vacuum. Since these electrons travelled a long way, they almost inevitably suffered an exchange collision and consequently the electrons escaping into vacuum are highly majority-spin polarized. Since the relative contribution of these very slow, highly polarized electrons increases somewhat beyond the energy of the maximum yield, the spin polarization at 1 eV in fig.6 reaches saturation beyond the maximum of the yield curve.

Data similar to those of figs 4 and 6 but for the ferromagnetic glass $Fe_{83}B_{17}$ have been published by Mauri et al.[24]. The dependence on the angle of incidence and the energy of the primary electrons was studied by Koike and Hayakawa with polycrystalline permalloy[25]. The possibility of observing Stoner excitations in inelastic electron scattering was pointed out by Yin and Tosatti[26] and has been experimentally verified by several authors[27-29]. The role of Stoner excitations in secondary electron emission was discussed in refs 30-33. At present, it is not yet possible to predict a priori the magnitude of the spin polarization enhancement for a given material, but the qualitative aspects appear to be reasonably well understood.

The spin polarization fine structure P_f remains to be explained. A strong experimental hint for the origin of this structure is, that it is different for different single crystal faces of a particular element[34], and that for the same face it may depend on the direction of emission. The fine structure thus must be related to the crystallinity. It has been known for some time that the intensity fine structure in angle-resolved secondary electron emission is related to the structure of empty bulk and surface states above the vacuum level. The reason is, that kinetic electrons inside the sample are bound to populate the Bloch states of the solid, though these may be broadened and shifted by inelastic processes. Since the secondary electrons are detected in the vacuum, they may also populate surface states before being emitted. By arguments of current conservation and time reversal invariance it was shown[35], that for a given energy and parallel momentum, the secondary electron intensity is directly related to the intensity of elastically backdiffracted electrons, which may be obtained by a dynamical LEED calculation. Along this line, Tamura and Feder[36] showed, that a fine structure of the spin polarization from ferromagnets may arise because the electronic states above the vacuum level may also be spin-split, like the partly filled d-states, though to a lesser extent. Two examples are shown in fig.7 and fig.8 for normal emission from Fe(001) and Ni(100), respectively. The upper panels (a) show those bands of the spin-split bulk band structure along the direction normal to the

surface which may couple to free electron states. Panels (b) display the
spin-resolved theoretical intensity fine structure for majority and
minority-type electrons, from which the theoretical spin polarization fine
structures in (c) are derived (solid line). Experimental results of the
total spin polarization are given by dots. The dashed lines in panels (c)
denote the smooth polarization function P_S. We note reasonable agreement
of theoretical and experimental fine structures for the major features. It
was found in this study that surface effects, like an enhanced
magnetization or a surface layer contraction, had practically no effect,
and it was concluded that the polarization fine structure is dominated by
the bulk electronic states and their coupling to LEED states.

At this point we note that a reasonable qualitative and partly even
quantitative understanding of the spin polarization spectra of secondary
electrons has been obtained so far. What is missing, is a direct link
between the measured electron spin polarization at a certain kinetic energy
and material-related quantities like the magnetization or the magnetic
moment per atom. As a guide, the average polarization value at higher
kinetic energies, say around 15 or 20 eV (disregarding the fine structure)
may be taken as a rough measure of the spin polarization in the valence
band and of the magnetization. For example, Ni has a magnetic moment of
$0.6\,\mu_B$ (Bohr magneton) and a spin polarization of ~5% at 15 eV; Fe has a

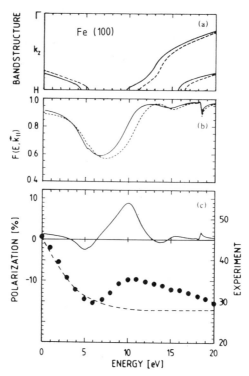

Fig. 7. Normal secondary electron emission from Fe(100): (a) majority spin
(solid line) and minority spin (dashed line) bulk band structure of
Δ_1 symmetry along ΓH; (b) theoretical intensity fine structure for
majority (solid line) and minority (dashed line) electrons; (c)
theoretical spin polarization fine structure P_f (solid line),
experimental total spin polarization (dots), and the smooth spin
polarization curve P_S (dashed). The experimental data are from
Taborelli et al., Phys. Rev. Lett. 55 (1985) 2599, shifted by 1 eV
towards lower energy.

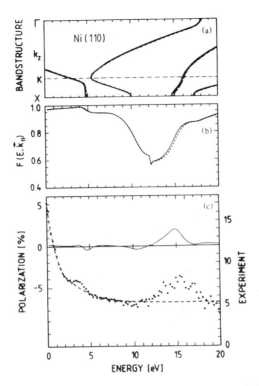

Fig. 8. Normal secondary emission from Ni(110). (a) - (c) as in fig.7 with
band structure of Σ_1 and S_1 symmetry along ΓKX in (a). The
experimental polarization is from ref.30.

magnetic moment of $2.2\mu_B$ and the spin polarization is 28%. Experience
shows (K. Koike, private communication) that the magnetic contrast in the
"spin-polarized SEM" scales roughly with the magnetization of the sample,
even when utilizing the very low energy secondaries only.

So far, we dealt with clean, homogeneous samples only. For applications, however, it is of major importance to know more about the depth of
magnetic information obtained by secondary electron spin analysis; also, we
need to know how an overlayer of foreign material (e.g. contamination)
affects the measured spin polarization. This is discussed in the next
section.

THE DEPTH OF INFORMATION

In this section we are concerned with the question of how deep we look
into the sample with the SEM with spin polarization analysis. For
simplicity, and because of their importance, we confine ourselves to
metals. First, we discuss secondary electron emission in general and
consider the electron spin later.

We assume the primary electron energy to be several keV or higher, so
that the penetration of the primary beam into the sample is much larger
than the escape depth of the slow secondaries. We further assume the
sample to be homogeneous, since it is well known that electrons
backscattered from high-Z inclusions deep inside the sample might generate
additional secondaries at the surface, thus increasing the effective depth

of information in the SEM[37]. These effects shall be disregarded here. The cascade of secondary electrons in the sample comprises all energies, from close to the Fermi energy (with very strong intensity) up to the primary energy in principle (though with very small intensity). We ask whether an electron with a certain energy, starting at a certain depth, may reach the surface and escape over the work function barrier into vacuum. There are two reasons, why the electron may not be detected outside: first, because it has lost too much energy on its way to overcome the surface barrier, and secondly, because it may have been scattered elastically out of its way towards the surface. Both effects presumably are important, but information on them is scarce at the low energies we are interested in.

In electron spectroscopy (cf. the papers by Cazaux and Dobson in this book) the concept of the inelastic mean free path (IMFP) has been introduced to evaluate the surface sensitivity. This quantity by definition gives the mean free path between two inelastic scattering events leading to a substantial discrete energy loss. The amount of energy loss is usually not specified, but assumed to be large enough to remove the electron from the "no-loss" line to which the spectrometer is tuned. The typical energy loss involved is several eV or more, corresponding to plasmon and/or interband transitions. Smaller quasi-continuous energy losses, such as electron-hole pair or phonon excitation are not taken into account. It is generally believed that elastic attenuation is of minor importance at the relatively high energies involved ($\gtrsim$ 100 eV) though this issue has not yet been settled[38]. The well-known "universal curve" for the IMFP has a minimum around 50-100 eV and increases towards lower energy, with a value of 3-5 nm at 5 eV above the Fermi-energy (see Dobson), i.e. roughly at the maximum of the secondary electron energy distribution. From this it was concluded by several authors[3,24,34,39-41] that the magnetic depth of information with very low energy electrons should be 3-5 nm.

In secondary electron emission, however, one has to deal with a continuous energy distribution of excited electrons, quasi-continuous energy losses and energy discrimination by a high-pass filter, given by the surface barrier. Therefore, the relevant quantity is not the IMFP but rather the electronic stopping power, expressed in energy loss per unit path length. The electronic stopping power is related to the IMFP, but in addition takes the amount of energy loss into account, in particular the frequent low-loss events at small kinetic energy. To the author's knowledge there is only the theoretical work by Ashley, Tung and Ritchie[42,43] dealing with this problem, though neglecting phonon, magnon, and Stoner excitations. The IMFP resulting from the statistical model employed by Tung et al.[43] was very recently shown by Penn[44] to be systematically overstimated at low energies by a factor of 2-3. With this in mind an electronic stopping power of 7...20 eV/nm for electrons at 10-20 eV above E_F may be estimated from ref.43, which shows immediately that the continuous slowing down in the cascade plays an important role. The elastic attenuation has been treated in a calculation by Ganachaud and Cailler[45] for Al, who showed that elastic collisions dominate over inelastic processes at all energies. Though these theoretical results depend on the particular choice of the scattering potential, they show that the attenuation of the electrons in the collision cascade is determined by elastic collisions and a more or less continuous slowing down, given by the electronic stopping power. This is a situation where a transport theoretical treatment is required to properly describe the secondary electron transport to and through the surface. Such a theory has been formulated by Schou[46] and Sigmund and Tougaard[47] based on the Boltzmann transport equation. It was shown that in the presence of continuous slowing down, the escape depth of secondaries is essentially determined by a transport decay length Λ, rather than by the inelastic mean free path

IMFP[47]. Thus the escape of secondary electrons may be described by an exponential law with a decay constant Λ, but this mean escape depth Λ is determined by the elastic scattering cross section and by the electronic stopping power simultaneously. Quantitative ab initio results have not yet become available, but from the above estimates of the electronic stopping power it becomes apparent that the escape depth of secondaries should not go far beyond 1 nm.

On the experimental side, a considerable body of information on secondary electrons has been accumulated, mainly in connection with scanning electron microscopy. According to a recent review by Seiler[48] in this context a distinction should be made between the escape depth Λ and the maximum escape depth. According to the exponential decay law, the latter may be 4-5 times larger than the former. Whether the one or the other quantity is the relevant one depends on the sensitivity of the observation. In many instances the contribution by about 30% of the electrons coming from a depth larger than Λ may be unimportant and the depth of information in SEM may be set equal to the escape depth Λ. According to Seiler[48] Λ lies between 0.5 and 1.5 nm for metals (10-20 nm for insulators because of smaller electronic stopping power in these materials). As has been analysed for example by Makharov and Petrov[49], the escape depth shows regularities with the periodic system of the elements. The larger values of Λ apply to simple metals, like alkalies, with a small density of states at the Fermi level. The smallest escape depths are found for d-band metals with half-filled shells. Because of the high density of filled and unfilled states around E_F the phase space for electron-hole pair excitations is largest in these materials and the electronic stopping power therefore is also large. Thus, minima in the escape depth are found around $Z = 25$, 43, 75 and maxima at $Z = 19$, 37, 55, in agreement with our expectation.

For samples with an overlayer, be it even as thin as a fraction of a monolayer, the effective depth of information may change because of the work function change associated with the deposition of this layer. For example, if the overlayer decreases the workfunction, a substantial increase of the total secondary yield may be noted (by up to a factor of 2-3[23]), which is to some extent due to secondaries excited in the overlayer itself, but mainly due to secondaries coming from larger depth. Because the surface barrier is lower, electrons that lost larger amounts of energy because of coming from deeper inside the sample are now able to escape into vacuum. The increase of the depth of information is difficult to determine because the energy distribution of the electrons inside the sample and the transmission function at the surface are not known, but from the shape of the energy distributions for various work function values[23] one might guess that the escape depth may easily increase by 30 to 50%.

Up to here we discussed the escape depth in general, without explicitly considering the electron spin. We recall from above that the largest polarization usually is found at near zero energy because of the polarization enhancement mechanism. Thus, if one tunes in for maximum polarization, the depth of information will be the same as when using the maximum of the intensity distribution to form an SEM picture. Experimentally, the magnetic probing depth in spin-polarized secondary electron spectroscopy has been studied very recently by Abraham and Hopster[39] on Ni(110). They found the magnetic depth of information to be three to four layers (0.4-0.5 nm), in excellent agreement with the figures quoted above for transition metals[48,50]. Thus, at least for homogeneous, clean samples the magnetic probing depth is the same as the depth of information of secondary electron emission in general.

However, with heterogeneous samples the situation may be more complex. Let us for example consider an overlayer of different material. The resulting spin polarization and with it the effective magnetic probing depth may depend strongly on the nature of this overlayer. The spin polarization may be unaffected, enhanced, weakened, or even reversed, depending on the existence or non-existence of magnetic moments in the overlayer and their orientation with respect to those of the sample. If we used an alkali metal, which has no magnetic moment, we would observe a slight decrease of the net spin polarization because of additional electrons from the overlayer which are unpolarized[51]. Of course, the total spin polarization will decrease with increasing thickness of the overlayer, but the essential point is that the spin-polarization of the secondaries from the substrate is not affected by the overlayer. This has been found for the alkalis but also for Au[52]. On the other hand, if the overlayer is ferromagnetic, and if its magnetic moments are aligned with that of the substrate, the polarization enhancement mechanism discussed above might increase the spin polarization of the secondaries from the substrate, just by being transmitted through the overlayer, if the latter has a higher intrinsic spin polarization than the substrate. The polarization will increase also because of the "trivial" effect of secondary emission from the overlayer with its higher polarization. In this case the magnetic probing depth is difficult to define. If the overlayer is magnetic, but not ordered magnetically, the total spin polarization will be less than that from the substrate. This seems to be similar to the situation with an alkali overlayer, but with a very important difference: because of the existence of magnetic moments, though not macroscopically aligned, the exchange scattering process leads to a strong depolarization of the secondaries from the substrate because its cross section is rather large. The mean free paths for this apparent "spin-flip" scattering have been found by Meier et al.[53] and Hüfner et al.[54] to be 1.25 nm for Ni, 0.38 nm for Gd and 0.32 nm for Ce. Thus, a monolayer of e.g. Ce on Fe may practically make the secondary electron spin polarization vanish. The magnetic probing depth would be virtually zero, though the secondaries predominantly come from the substrate.

An extreme but very instructive further example is given in fig.9, due to Taborelli et al.[41]. They evaporated a monolayer of Gd onto a Fe(100) surface and looked for the spin polarization of the Auger electrons from Fe and Gd. The sign of the polarization of the two Auger electron groups is opposite, which shows, first, that the Gd atoms are magnetically ordered, and secondly that their spin orientation is opposite to that of the Fe atoms (which is counted positive). The rather surprising result on the low energy secondaries below ~ 10 eV is that they also show negative spin polarization, while those from the clean Fe surface are positively spin polarized by up to 48%[55]. The sign reversal cannot be due to the secondaries from the Gd layer since its thickness is only 1 monolayer and since the escape depth is increased by the work function lowering when depositing Gd. The explanation is that the Gd layer acts as a very efficient "spin-filter" which passes only the minority electrons from Fe while the majority electrons are "trapped" in empty minority states of Gd, followed by ejection of negatively polarized majority electrons from Gd. This example shows that the magnetic depth of information may be completely different from the escape depth of secondary electrons: here the magnetic information is confined to the Gd layer.

We see that the magnetic probing depth in heterogeneous systems is not easy to determine without any knowledge of the composition. On the other hand, from the spin polarization of the secondaries a lot more information may be obtained than from their intensity distribution alone.

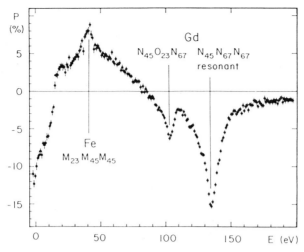

Fig. 9. Spin polarization vs. kinetic energy of secondary electrons from
Fe(100) covered with a monolayer of Gd (0.24 nm) at 150 K. Primary
energy is 2500 eV. The opposite sign of the spin polarization of
the Auger electrons from Fe and Gd show that the Gd monolayer coup-
les antiferromagnetically to the Fe substrate. Note that the polarization of the low energy secondaries, mainly coming from the Fe
substrate, has been reversed by a very efficient coupling to Stoner
excitations in Gd while passing through the overlayer (ref.41).

CONCLUSIONS

The analysis of secondary electron spin polarization in the Scanning
Electron Microscope, has lead to a very promising new technique for
studying magnetic domain structures at the surface of thick samples with
unprecedented resolution. In the absence of strong stray magnetic fields
the orientation of the spin polarization vector agrees well with the
(negative) magnetization vector, which provides a straightforward
measurement of three-dimensional magnetization structures. The magnitude
of the spin polarization is not as directly related to the magnitude of the
magnetization, though a rough scaling has been observed to exist. The
shape of the spin polarization distribution as a function of kinetic energy
of the secondary electrons is qualitatively understood to result from two
additive effects: (i) enhancement towards low energy by exchange scattering
processes involving Stoner excitations and, (ii) polarization fine
structure due to details of the bulk band structure above the work function
of the sample. The magnetic depth of information in homogeneous samples
equals the escape depth of the secondaries and is 0.5 to 1.0 nm for
metallic ferromagnets. The spin polarization from heterogeneous samples is
not always easily interpretable because of spin-dependent scattering
processes during transport to the surface and escape through the surface
barrier. A full exploitation of the information contained in the secondary
electron spin polarization then requires some knowledge of the elemental
and chemical composition of the sample. More work on such systems is
needed.

ACKNOWLEDGEMENTS

Sincere thanks are due to U. Valdrè and A. Howie for their invitation to
the Erice school. The support of this work by H. Ibach and helpful
discussions with H.P. Oepen and K. Koike are gratefully acknowledged.
Special thanks go to K. Koike and H.P. Oepen for providing results prior to
publication.

REFERENCES

1. G. Chrobok, M. Hofmann, Phys. Lett., A57:257 (1976).
2. R.J. Celotta and D.T. Pierce, in "Microbeam Analysis 1982",
    K.F.J. Heinrich ed., San Francisco Press, San Francisco p.469
    (1982).
3. J. Unguris, D.T. Pierce, A. Galejs and R.J. Celotta, Phys. Rev. Lett.,
    49:72 (1982).
4. J. Kirschner, "Scanning Electron Microscopy 1984/III", p.1179 (1982).
5. J.N. Chapman and G.R. Morrison, J. Magn. Magnet. Mat., 35:254 (1983).
6. A. Tonomura, T. Matsuda, H. Tanabe, N. Osakabe, J. Endo, A. Fukuhara,
    K. Shinagawa and H. Fujiwara, Phys. Rev., B25:6799 (1982).
7. D.E. Newbury, D.C. Joy, P. Echlin, C.E. Fiori and J.I. Goldstein,
    "Advanced Scanning Electron Microscopy and X-ray Microanalysis",
    Plenum Press, New York p.147 (1986).
8. W. Rave, R. Schäfer and A. Hubert, J. Magn. Magnet. Mat., 65:7 (1987).
9. K. Koike and K. Hayakawa, Jap. J. Appl. Phys., 23:L187 (1984).
10. J. Kirschner, Polarized Electrons at Surfaces, in: "Springer Tracts in
    Modern Physics", Vol.106, Springer-Verlag, Berlin (1985).
11. J. Kirschner, J. Appl. Phys., A36:121 (1985).
12. J. Unguris, D.T. Pierce and R.J. Celotta, Rev. Sci. Instrum., 57:1314
    (1986).
13. K. Koike, H. Matsuyama, H. Todokoro and K. Hayakawa, Jap. J. Appl.
    Phys., 24:1078 (1986).
14. J. Unguris, G.G. Hombree, R.J. Celotta and D.T. Pierce, J. Magn.
    Magnet. Mat., 54-57:1629 (1986).
15. H.P. Oepen and J. Kirschner, Jahresbericht 1986/87 der KFA Jülich p.11
    (1987).
16. G.G. Hembree, J. Unguris, R.J. Celotta and D.T. Pierce, Proceedings of
    the 5th Pfefferkorn Conference, Scanning Electron Microscopy,
    Suppl.1, p.229 (1987).
17. K. Koike, H. Matsuyama and K. Hayakawa, Proceedings of the 5th
    Pfefferkorn Conference, Scanning Electron Microscopy, Suppl.1,
    p.241 (1987).
18. R. Feder, in: "Polarized Electrons in Surface Physics", R. Feder ed.,
    World Scientific, Singapore chap.4 (1985).
19. B. Ackermann and R. Feder, Solid State Commun., 49:489 (1984).
20. E. Tamura, B. Ackermann and R. Feder, J. Phys., C17:5455 (1984).
21. J. Kirschner and S. Suga, Solid State Commun. 64:997 (1987).
22. K. Koike and J. Kirschner, in press.
23. P.W. Palmberg, J. Appl. Phys., 38:2137 (1967).
24. D. Mauri, R. Allenspach and M. Landolt, J. Appl. Phys., 58:906 (1985).
25. K. Koike and K. Hayakawa, Jap. J. Appl. Phys., 23:L85 (1984).
26. S. Yin and E. Tosatti, Report IC/81/129 (1981) Intern. Centre for
    Theoretical Physics, Miramare, Trieste (unpublished).
27. J. Kirschner, D. Rebenstorff and H. Ibach, Phys. Rev. Lett., 53:698
    (1984).
28. H. Hopster, R. Raue and R. Clauberg, Phys. Rev. Lett., 53:695 (1984).
29. J. Kirschner, Phys. Rev. Lett., 55:973 (1985).
30. E. Kisker, W. Gudat and K. Schröder, Solid State Commun., 44:591
    (1982).

31. H. Hopster, R. Raue, E. Kisker, G. Güntherodt and M. Campagna, Phys. Rev. Lett., 50:70 (1983).
32. D.R. Penn, S.P. Apell and S.M. Girvin, Phys. Rev. Lett., 55:518 (1985).
33. J. Glazer and E. Tosatti, Solid State Commun., 52:905 (1984).
34. M. Landolt, Appl. Phys., A41:83 (1986).
35. R. Feder and J.B. Pendry, Solid State Commun., 26:519 (1978)
36. E. Tamura and R. Feder, Phys. Rev. Lett., 57:759 (1986).
37. J. Kirschner, Scanning Electron Microscopy 1976/I, ITT Research Inst., Chicago, Ill., 215 (1976).
    H. Seiler, Scanning Electron Microscopy 1976/I, ITT Research Inst., Chicago, Ill., 9 (1986).
38. C.J. Powell, Surface Interface Anal., 7:256 (1985).
39. D.L. Abraham and H. Hopster, Phys. Rev. Lett., 58:1352 (1987).
40. R. Allenspach, M. Taborelli, M. Landolt and H.C. Siegmann, Phys. Rev. Lett., 56:953 (1986).
41. M. Taborelli, R. Allenspach, G. Boffa and M. Landolt, Phys. Rev. Lett., 56:2869 (1986).
42. J.C. Ashley, C.J. Tung and R.H. Ritchie, Surface Sci., 81:409 (1979).
43. C.J. Tung, J.C. Ashley and R.H. Ritchie, Surface Sci., 81:427 (1979).
44. D.R. Penn, Phys. Rev., B35:482 (1987).
45. J.P. Ganachaud and M. Cailler, Surf. Sci., 83:498 and 519 (1979).
46. J. Schou, Phys. Rev., B22:2141 (1980).
47. P. Sigmund and S. Tougaard, Inelastic Particle-Surface Collision, in: "Springer Series in Chemical Physics" vol.17, E. Taglauer and W. Heiland, eds, Springer-Verlag, Berlin, p.2 (1981).
48. H. Seiler, J. Appl. Phys., 54:R1 (1983).
49. V.V. Makharov and N.N. Petrov, Sov. Phys. Solid State, 23:1028 (1981).
50. H. Seiler, Z. Angew. Physik, 22:249 (1967).
51. H. Hopster, R. Raue, E. Kisker, M. Campagna and G. Güntherodt, J. Vac. Sci. Technol., A1:1134 (1983).
52. F. Meier, D. Pescia and M. Baumberger, Phys. Rev. Lett., 49:747 (1982).
53. F. Meier, G.L. Bona and S. Hüfner, Phys. Rev. Lett., 52:1152 (1984).
54. S. Hüfner, G.L. Bona, F. Meier and D. Pescia, Solid State Commun., 51:163 (1984).
55. R. Allenspach and M. Landolt, Surface Sci., 171:L479 (1986).

ELECTRONICALLY STIMULATED DESORPTION:

MECHANISMS, APPLICATIONS, AND IMPLICATIONS

Dietrich Menzel

Physik-Department E 20, Techn. Universität München
D-8046 Garching b. München, Fed. Rep. Germany

1. INTRODUCTION

To a varying degree, virtually all surface spectroscopies and microscopies using electrons, photons and/or ions suffer from beam damage effects. These can become so important that they effectively limit the obtainable resolution and/or sensitivity in a particular application. In a large fraction of applications, these beam-induced effects are the eventual consequence of primary electronic excitations, a term which will be used to include ionization. One of the possible final consequences is the release of surface particles into the vacuum, i.e. "Electronically Stimulated Desorption" (ESD; this term is used here to include electronic stimulation by photon absorption) or "Desorption Induced by Electronic Transitions" (DIET). This is a subject which has been studied in its own right for over 25 years now and has contributed considerably to the basic knowledge about surface and adsorbate excitations, their coupling to the bulk, their screening, delocalization, and deexcitation[1-4]. In the subject context of this volume, as well as in other practically important cases, such as total and partial pressure measurements, ultimate attainable vacuum or gas purity in UHV systems enclosing energetic particles - in particular plasma devices and accelerators/storage rings - these processes have a negative importance, as disturbing effects to be avoided: which is easier if their mechanisms are understood. It is hoped that the positive aspects will become appreciated a bit by the reader as well. It should also be emphasized that the concentration in this survey on desorption events as the ultimate effect of electronic surface excitations should not detract from the fact that many other beam-induced effects (such as creation of bulk disorder and other disturbances, or radiation-induced reactions as used in beam-induced etching or deposition and lithography; possibly even astrophysical effects such as creation of space molecules and evaporation of comet material, etc.) work on very similar principles and can probably be treated rather analogously.

2. SOME GENERAL CONSIDERATIONS; TWO LIMITING CASES

Electronic excitations can be induced by photon absorption, electron impact, or particle (ions, fast neutrals) impact. While photons, or electrons of low to medium energy (up to, say 50 keV) produce "clean" excitation/ ionization events, fast electrons as well as most ions also lead to momentum transfer which complicates things. We shall therefore concern ourselves only with photon or slow electron impact. Because of the energy dependence of these two types of excitation (sharp onset at threshold energy E_T with maxima close to it and subsequent fast decrease at higher energies for photons; linear increase above E_T, maximum cross section between (3-6) x E_T, slow decrease above this for electrons), electrons are better for practical applications while photons make it easy to distinguish the primary excitations and are thus preferable for studies of mechanisms.

Desorption from a surface necessitates the breaking of a surface bond; it is the surface analogue of dissociation of a molecule.

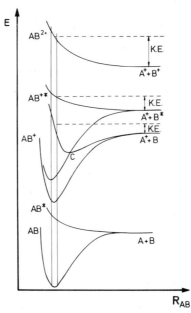

Fig. 1 Schematic illustration of the conversion of electronic excitation energy into nuclear motion in a Born-Oppenheimer picture. Potential energy curves of a diatomic molecule AB (ground state and various excited or ionized states) are shown. Electronic excitation by electron impact or photon absorption brings the system from the ground state into an excited state in the Franck-Condon range. If the latter is repulsive, the system can separate along the corresponding potential energy curve. Curve crossing into a bonding state (relative to the bond A-B) such as at C can abort this dissociation. After ref. 4.

Electronically stimulated dissociation can be viewed in a Born-Oppenheimer picture as the conversion of electronic excitation into motion of
the nuclei concerned via a Franck-Condon transition from the ground
state to an electronically excited state whose potential energy curve
has a sufficient gradient in the region of the ground state equilibrium
distance (fig. 1). That this can lead to breaking of the bond is most
easily seen in the case of a <u>repulsive</u> (with respect to the bond concerned) final state; however, even if the initial acceleration of nuclei
is <u>in</u>ward (attractive curve), a curve crossing and reflection of motion
can still lead to dissociation. The main point is that energy is
transferred from electronic to nuclear coordinates. If we transfer this
picture to a surface, many different situations can occur for which we
can formulate two important limiting cases. The first is closely
realized by the case of an adsorbate (submono- to monolayer) on a metal
surface. In this case the important feature is the high density of
possible excitations in the bulk which leads to extremely fast
degradation of excitations. In order to break a bond, the primary
excitation has to be localized on this bond, and to stay localized long

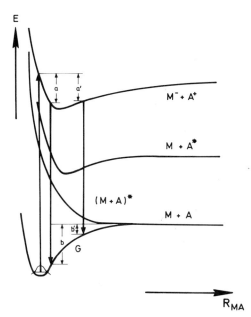

Fig. 2: Potential energy curves depicting an adsorbate A bound to a
metal surface M. G is the ground state, $(M+A)^*$ an excited state
separating into the ground state atom A and the metal, $M+A^*$ one
leading to an electronically excited atom, and $M^- + A^+$ an ionic
state. The system can be excited or ionized similarly to fig. 1.
However, at any position R_{MA} the system can fall back to G by
transfer of the excitation and/or the charge to the metal. As
shown for the ionic curve, this leads to quenching of the
excitation if it happens at distances smaller than a critical
distance (a < b); for larger distances a metal particle escapes,
as then a'>b'. This shows that delocalization of energy and
charge are not necessarily connected. After ref. 2.

enough to get the nuclei moving (fig. 2). Because of the fast degradation in the bulk, the excitation will never re-concentrate on the surface bond once it has hopped away. Thus in this case the localization of the electronic excitation on the one hand, its coupling into the bulk on the other are the decisive factors determining desorption. Compared to a similar molecule in the gas phase, the probability of desorption per excitation event will be decreased by this coupling, possibly by a very large factor.

The other limiting case is that of bulk insulators. Here a rather deep layer of the bulk can act as an accumulator of excitation which can migrate without degradation to the surface (for instance as an exciton) and dissociate there. In this case one primary electron can in fact lead to many desorption events as has indeed been observed.

To be sure, this subdivision is not perfect, even apart from the continuum extending between the extremes. On the one hand, photons or electrons impinging on a metal will create secondaries part of which may be energetic enough to create surface excitations in turn. This multiplication factor, however, is usually small compared to the attenuation factor by delocalization. On the other hand, localization may also be important in widegap materials. For instance, from a molecular solid, emission of ions will only occur if a surface molecule disintegrates; if the primary excitation migrates into the bulk intact, it may lead to radiation damage there, but this would not be seen as desorption. As always, classification has to be taken cum grano salis.

3. ESD OF ADSORBATES FROM METAL AND SEMICONDUCTOR SURFACES

This first limiting case has been investigated most extensively; the qualitative picture of localization vs. coupling has evolved over many years and from contributions of many authors. In the following we list the main experimental facts which were important in forming these concepts. We concentrate on chemisorbed systems and discuss physisorption only briefly at the end.

1) From atomic or molecular chemisorbates on metal surfaces, desorption of ions (mostly positive) and neutrals, fragments as well as whole adsorbates, occur under electron or photon irradiation.

2) The desorption signal is proportional to the electron or photon flux (provided correction is made for coverage changes under irradiation), i.e. a single-hit mechanism prevails.

3) Compared to ionization cross sections of comparable molecules, extreme variations of cross sections occur. Even for different binding states of the same molecule on the same surface, the desorption efficiency P can vary by orders of magnitude (where $P = q/q_0$ with q the desorption cross section and q_0 the excitation/ionization cross section). For adsorbed atoms, P is usually smaller than for molecules. Desorption of metal atoms from metal surfaces is immeasurably small.

4) There are very strong isotope effects, with the heavier isotope always desorbing less (record: for a certain state of hydrogen on W(100), $q_H/q_D > 150$)[5].

5) Primary thresholds are usually found in the valence excitation/ionization region (fig. 3), with strong increases at the

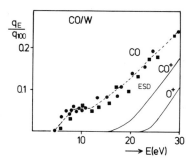

Fig. 3: Desorption of neutral CO, CO^+, and O^+ ions from CO/W(100) as a function of the energy of electrons used to excite the system. There are connections between structures in these threshold curves. After ref. 2.

energies of multiple excitations and further increases at core thresholds (even though their ionization cross sections are quite small compared to valence cross sections, fig. 4). With primary photons, resonant processes are often observed which can be analyzed in detail using polarization dependences. For different primary excitations, the contributions of the various desorbing species are often very different.

6) In the valence region neutral desorption predominates, in the core region that of positive ions.

7) Kinetic energy distributions of desorbing particles are broad (0.5 - 10 eV at maximum for ions, 0.1 - 0.5 eV for neutrals). Internal energies (rotational, vibrational) are nonthermal.

8) Angular distributions are narrow (neutrals 30 - 60° FWHM) to extremely narrow (ions 10 - 20° FWHM) (fig. 5).

9) In many cases the cross sections <u>drop</u> strongly with increasing coverage. In others, "minority states" are found, i.e. most of the desorption comes from a very small part of the coverage, which can often be linked to steps or other surface disturbances.

All these features can be semiquantitatively understood in terms of the Born-Oppenheimer picture with coupling, as discussed in Sect. 2 (fig. 2). If there were no coupling to the bulk, the excitation energy could stay localized on the bond during the 10^{-13} seconds or so which are necessary to make the nuclei move apart. On a metal surface, a considerable number of excitations can be transferred into the bulk before even any motion occurs; then the system directly falls back into the ground state. Those surviving these processes will evolve along the corresponding potential curve. During this motion there is still the possibility of <u>charge</u> transfer from the surface, so that more neutrals than ions finally make it away from the surface, even for an ionic primary excitation. An early semiclassical model (fig. 2)[8] of this qualitative picture and its quantum mechanical reformulation[9] were able

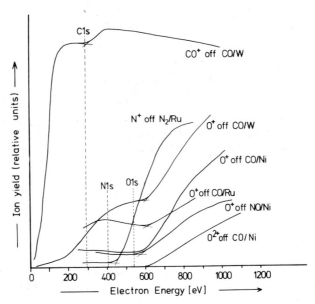

Fig. 4: Secondary thresholds observed at adsorbate core excitation thresholds in ion desorption by electron impact, for a number of adsorbate systems on metal surfaces. The ionization cross sections of these core excitations are two to three orders of magnitude smaller than the valence excitation cross sections. After ref. 6.

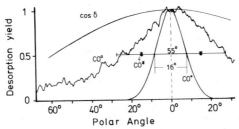

Fig. 5: An example for angular distributions of desorbing neutrals and ions, in the system CO/Ru(001) excited by 250 eV electrons. After ref. 7.

to explain many aspects semiquantitatively including the isotope effect ("recapture" by delocalization of the excitation, or reneutralization of the outgoing particle, are the more probable the slower the particle, i.e. the higher its mass). Generally we expect that desorption is the more probable the less probable is the transfer of excitations, or the stronger the <u>localization</u> of excitation. This explains the higher efficiency of:

- multiple excitations (for which the localization is enhanced by Coulomb interaction[10,11] although it can be decreased again by screening from the metal[12].
- core excitations (which lead, via Auger decay, to at least two-hole states or, if the primary excitation was a higher core excitation, to more complex states with correspondingly stronger correlation[12];
- low coverages (at high adsorbate densities an additional delocalization, apart from that into the bulk, exists <u>laterally</u> in the dense layer, as the adsorbate orbitals overlap and form a two-dimensional band); and
- adsorbates at surface imperfections (which lead to localized electronic states so that no additional localization mechanism is necessary)[4].
- That surface metal atoms are not desorbed is understandable from their strong coupling to the bulk metal.

The energy and angular distributions of desorbing particles are then connected to the three-dimensional shape of the ground and excited state potential surfaces. The fact that this connection contains not only the Franck-Condon overlap produced by the excitation, but also the coupling terms (the imaginary or absorptive part of the potential energy) has so far prevented a direct evaluation of such data to obtain potential parameters.

Physisorbed, i.e. Van der Waals-bound submono- or monolayers are an interesting special case. Here an ionized or excited particle is more <u>strongly</u> bound to the surface than the ground state, so that after

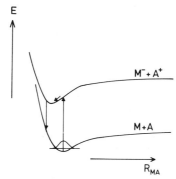

Fig. 6 Mechanism leading to desorption of <u>neutral</u> atoms from a physisorbed monolayer on a metal surface after ionization. The ion is more strongly bound and therefore accelerated towards the surface. Neutralization brings the system back to the ground state curve at too short distance and with kinetic energy, so that a neutral can desorb. After ref. 4.

excitation the particle is accelerated <u>towards</u> the surface (fig. 6). However, eventually it is reneutralized by an electron tunneling from the metal so that it finds itself back in the ground state potential at too short distance and with kinetic energy which after reflection can lead to desorption as a neutral[13]. This is a modification of a mechanism proposed originally in a different context[14]; it appears to work for ionic excitations and, with some suitable modification, for neutral excitations as well. Again, localization is important. If the coupling of the adsorbate to the metal is strong enough to make the primary excitation nonlocal, or the ion approaching the surface is reneutralized too early (i.e. before it has accumulated enough kinetic energy and/or been displaced sufficiently with respect to the ground state equilibrium distance), then the desorption efficiency is strongly degraded.

The qualitative picture given above emphasizes localization vs. coupling. Instead, another picture could have been used emphasizing the time domain, in which the characteristic time constants for the various processes (screening and electronic rearrangements; decay of higher excitations and transfer of energy and charge; nuclear motion) figure prominently. This shows that the study of such processes gives some access to time scales of concurring and competing events. The other side of the picture is that desorption events observed in such systems strongly select those excitations which remain localized (are long-lived locally), even if the majority of excitations is strongly delocalized. The observation of desorption thus allows the selective observation of localized excitation and bond-breaking[15].

The time scales we are talking about are in the range of 10^{-16} seconds for primary excitations and screening, 10^{-15} to 10^{-14} seconds for electronic rearrangement (Auger decay, etc.), the same range for energy and charge transfer to neighbours and the bulk, and 10^{-14} to 10^{-13} seconds for sufficient motion of nuclei. This again shows that in this limit the desorption events will contain the surviving tail of a broad distribution of lifetimes of excitations.

4. ESD FROM SURFACES OF BULK INSULATORS

In a wide-gap material, the degradation of excitation is very difficult. Thus, lifetimes of electronic excitations can become so long that not only is it easy for them to compete with nuclear motion, but the excitations can move through the crystal even under rather difficult situations. This situation constitutes the other of the two extremes mentioned above. The clearest case is probably that of Van der Waals crystals, of which the simplest are the rare gas solids. From these, desorption of neutral (ground state and electronically excited) atoms is observed. Their yield usually increases with increasing thickness of the layer over a certain thickness range, and can reach rather high efficiencies (e.g. several hundred Ne atoms per 2 keV electron from a thick Ne layer[16]). Selective measurements with photons show that surface and bulk excitons of various orders, ionic excitations, and two-step processes (creation of excitons by photoelectrons formed by ionization) are efficient primary processes[17,18] (fig. 7). This can be understood by creation of the excitation, for instance an exciton, in the bulk (though not too deep inside, the contributing depth being determined by the migration distance during the mobile life-time), its migration towards the surface in competition with selftrapping, and its conversion into a selftrapped excimer[19]. The latter radiatively decays into the ground state converting part of its binding energy into kinetic energy which expels a ground state atom. Similarities to the Antoniewicz mechanism

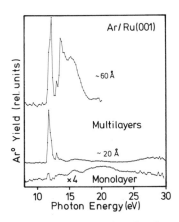

Fig. 7: Desorption yields of mono- and multilayers of Ar on Ru(001) showing contributions of surface and bulk excitons, ionic excitations, and two-step processes (after ref. 18, in which details can be found).

discussed above are obvious. Another possibility is the release at the surface of the mechanical stress connected to a selftrapped exciton which expels an electronically excited atom ("trampolin mechanism"[20]).

Not too different, although more complicated mechanisms appear to prevail in ionic insulators. In alkali halides, from which mainly neutral halogen and alkali atom desorption (the latter also in electronically excited states) is observed[21], the most important primary excitation appears to be the excitation of the halogen sublattice which can form a halogen molecular ion with a trapped electron (V_K+e state) which further decays into an interstitial halogen atom (H-centre) with kinetic energy and an electron in a halogen vacancy (F-centre). The former may start a replacement cascade or be mobile in an excited state, both finally leading to halogen desorption from the surface. If the temperature is high enough, the accumulating alkali (neutralized by the F-center electrons) may simply desorb thermally. Besides these thermal ground state atoms, electronically excited alkali atoms desorb which show up by their light emission. There exists controversy whether these come about by complicated electronic expulsion mechanisms involving Auger processes[22], or whether atom-electron collisions in the gas phase above the surface produce the observed electronic excitations; there may well be contributions from both[23].

A primary excitation mechanism which has been proposed in connection with ionic insulators and which may play a role in alkali halides and ionic oxides is the so-called Knotek-Feibelman mechanism[24]. Its main content is the assumption that for a maximally valent ionic compound, in which no valence electrons exist on the cation, a core hole created on a cation can only decay by an <u>interatomic</u> Auger process which transfers charge from an anion to the cation. As this reverses the charges on anion and cation, the formerly attractive Madelung potential may become repulsive and expel a neutral or positive "anion". This mechanism may have some importance in ionic oxides and halogenides, but the early implication that it will govern most core-initiated desorption events appears not to be warranted. It appears not even to be strictly applicable to its model compound, TiO_2[25].

These discussions have shown some of the important concepts, but
they are obviously far from comprehensive. The actual leading processes
may well vary strongly depending on the individual system concerned.
Very little is known at present about semiconductor surfaces, in
particular of compound semiconductors. In view of their importance, and
of the connection between ESD and beam-assisted etching and deposition
and similar processes, such surfaces should become a strong focus of
research in the future.

5. ELECTRONICALLY STIMULATED DESORPTION AS STRUCTURAL PROBE

Stimulated desorption is, because of its strongly selective nature,
not usable as a general surface analysis technique. Turning this into a
positive aspect, one can try to use ESD to selectively investigate
certain surface species. Also, in cases of very homogeneous surface
coverages, the ESD signals may be characteristic of the entire surface
layer. If we accept the necessity carefully to clarify for each system
and condition in which of these possible regimes we operate, we can try
to use the ESD signals for further characterization of surface layers.
Of the various possibilities for doing this, we here discuss two.

X-ray absorption fine structure measurements, both in the extended
range above a core excitation threshold (EXAFS) and in the threshold
region (NEXAFS/XANES), are well-established techniques to investigate
interatomic distances and directions, and symmetry properties and
orientations, respectively[26]. In order to apply them to a surface, a
signal has to be detected which strongly represents the surface
absorption only (SEXAFS). Total and partial electron yields are being
used for this purpose, with considerable success[27]. If one could be
reasonably sure that the ion yield under photon irradiation with
energies above a core ionization threshold would follow the absorption
cross section, a very surface-selective variety of N/SEXAFS were
possible. Indeed some systems have been investigated with this method[28].
Apart from the necessity to make sure that one knows from which state of
the surface layer the measured signal derives, two difficulties arise.
First, as discussed in Sect. 3, there are many cases in which multiple
excitations above core thresholds contribute strongly to ion desorption.
As these are not related to the fine structure in simple core ionization
(which is the basis of EXAFS), all such signals are not usable.
Furthermore, there may exist strong contributions from desorption events
not induced by primary photoabsorption in the surface layer, but by
electrons produced by photoabsorption in the bulk (Auger, secondary, or
photoelectrons) and causing desorption in a secondary process
("XESD"[29]). Then the ion-EXAFS will simply monitor the photoabsorption
in the bulk, and an EXAFS evaluation will yield bond lengths in the
bulk. The danger for this is obviously greatest at substrate core
levels. While cases may be found in which neither of these difficulties
prevails, this shows that ion-EXAFS is not likely to become a general
surface method.

The other, more established method is the so-called ESDIAD method[30]
which goes back to Madey and coworkers[31]. It is based on the finding
mentioned in Sect. 3 that ESD ions very often are found to desorb in
very sharply focussed beams of only 10-20° halfwidth. In the simplest
approach, these beam directions directly image the bond directions in
the ground state. This can then be used very directly to derive
structural models for the layer investigated (see fig. 8 for an example
and refs. 30 for further systems and a compilation). Obviously, this

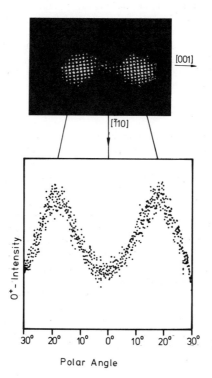

Fig. 8: An example of an ESDIAD measurement. For O$^+$ ions liberated by electron impact from a saturated layer of CO on Ni(110), the angular emission pattern is shown in a projection (with channel plates onto a fluorescent screen; top) and by a mass spectrometric scan along the [001] direction (bottom). The two focussed beams in this azimuth R18° from the surface normal indicate a tilting of the CO molecular axis perpendicular to the grooves on the (110) surface. After ref. 34.

interpretation cannot be strictly correct, as the angular distribution will be mainly determined by the three-dimensional shape of the repulsive excited state potential onto which the ground state distribution is projected; this will then be modified by trajectory-dependent neutralization/recapture, and deflection by the image potential. However, arguments have been given to show that the mixing of the excited and ground state potentials may in many cases preserve the ground state directions[32]; and the two other complications lead to small corrections into opposing direction directions as long as the polar angle of the trajectories is not too far (say, less than 30°) from the surface normal[33]. In a recent application, the inclination angle of tilted CO/Ni(110) under "crowding" conditions has been determined to within R1°-2° (fig. 8)[34], in very good agreement with independent methods[35]. Again, of course, the species "seen" in the ion signal has to be known well which may be a problem in case of strong coverage dependences of the ion signal and/or of minority species. Again, the latter effect can be turned into an asset, as shown by the investigations of adsorption at stepped surfaces[30].

6. SOME REMARKS ABOUT THE BEARING OF ESD ON BEAM DAMAGE

The obvious motivation for the inclusion of a discussion of stimulated desorption in this volume is the connection to beam damage. Therefore a few words about the relevance of the discussion given above for the surface methods discussed in this volume appears in place.

Many of the surface physical methods discussed here (LEED, XPS, UPS, Auger spectroscopy, electronic energy losses, MEED, RHEED, LEEM) are extensively used for the investigation of clean surfaces and well-defined adsorption layers. The discussions in Sect. 3 are therefore directly applicable. From what has been said there it is clear that the danger of damage by the probes used (primary electrons or photons; secondary electrons of various kinds) is negligible or small for clean metal and semiconductor surfaces, and becomes appreciable for insulators and for adsorbate layers. For the latter, damage is the more severe the weaker they are coupled to the (metal or semiconductor) surface. Generally, primary photons are less damaging than primary electrons under the conditions commonly used. This is in part due to the much higher electron fluxes normally used; but it would mostly be true also for comparable fluxes of electrons and photons because of the smaller number of secondaries produced with photons, and the cross section behavior (wide-band for electrons, resonant for photons which means that at an arbitrary energy electrons always "fit" while photons are unlikely to be in a resonance). There are many cases in the literature, not all recognized, where this has lead to erroneous interpretations. This is particularly true for Auger investigations of adsorbates where often high beam currents are used. Generally, electrons are the more dangerous the higher the energy so that as low energies as possible should be used, in particular if high current densities are necessary (LEEM!). Damage cross sections for adsorbates are between 10^{-21} and 10^{-16} cm^2, with the majority lying in the mid-range. It should be noted that desorption is but one possible damage effect, even in monolayer adsorbates; molecular adsorbates can be dissociated, and any species may convert to another binding state or structure after excitation. One example: 1/3 monolayer CO on a Ru(001) surface shows a certain ordered structure as evidenced by LEED ($\sqrt{3} \times \sqrt{3}$ R 30°). A 100 eV electron beam used to observe this structure desorbs CO molecules with a cross section of about 10^{-18} cm^2, dissociates them (knocking off O/O^+, leaving carbon on the surface) with about 10^{-19} cm^2, and disorders the $\sqrt{3}$ structure with about 10^{-17} cm^2 (ref. 36).

The situation is much more dramatic in high resolution electron microscopy of solid materials and surfaces. Fluxes and energies are so high there that all but the most stable adsorbates will be destroyed before even looking. The question then is shifted to the relative stability under these severe conditions of the bulk materials. While exact data are difficult to obtain in this range with the methods discussed above, the general criteria developed should still be usable for guidance. Thus, metals and elemental semiconductors should be safest because of the ease of energy degradation. In other systems, damage should be the less the easier the delocalization of excitation is. This would for instance mean that in organic matter, o-systems should be relatively stable compared to q-bound systems. Again, an interesting case are compound semiconductors which should have intermediate stability so that no problems would arise in low flux (10^{-2} to 1 el//^{2}s) surface physics investigations but damage might be seen with high fluxes (10^2 to 10^4 el//^{2} s).

An interesting special case is cryomicroscopy[37] of biological specimens embedded in amorphous ice. There, considerable structural damage appears to be connected with the stimulated evaporation of H_2O. The relevant mechanisms may be similar to those discussed for condensed layers but have not yet been investigated in detail.

7. CONCLUSIONS

Electronically stimulated desorption is a field which contains many challenges for physical understanding because of the complex nature of the important interactions which turn out to be of largely many-body, competing nature. Its obvious practical importance lends an additional appeal to its investigation. While it is difficult to give practical advice to the electron microscopist from the knowledge gained in fundamental investigations, some clarification may be derived nevertheless.

8. ACKNOWLEDGEMENTS

The excellent collaboration of many former and present coworkers, in particular P. Feulner, R. Jaeger, W. Riedl, R. Treichler, E. Umbach and W. Wurth, and valuable discussions with them and with W. Brenig, Z.W. Gortel, D.W. Jennison and H.J. Kreuzer have been decisive for our part of the work mentioned and for the shaping of my concepts of these processes. For this I am very grateful.

Our work has been supported financially by the Deutsche Forschungs-gemeinschaft and by the German Ministry of Research and Technology.

REFERENCES

1. A good survey of the state of the field with many references can be found in the volumes "Desorption induced by electronic transitions, DIET", of which DIET-I, M.M. Traum, N.H. Tolk, J.C. Tully, T.E. Madey, eds. (Springer, Berlin 1983), and DIET-II, W. Brenig and D. Menzel, eds. (Springer, Berlin 1985) have appeared; DIET-III (Springer, Berlin, in press) will appear in 1987.
2. For older surveys, see for instance T.E. Madey and J.T. Yates, Jr., J. Vac. Sci. Technol. 8:525 (1971); D. Menzel, Surface Sci. 47:370 (1975); dto., J. Vac. Sci. Technol. 20:538 (1982).

3. M.L. Knotek, Rep. Progr. Phys. 47:1499 (1984); T.E. Madey and R. Stockbauer, in "Methods of Experimental Physics", vol. 22:465 (1985); R.L. Park and M.G. Lagally eds.; Academic Press, NY.
4. D. Menzel, Nucl. Instrum. Methods B13:507 (1986).
5. W. Jelend and D. Menzel, Chem. Phys. Lett. 21:178 (1973)
6. P. Feulner, R. Treichler and D. Menzel, Phys. Rev. B24:7427 (1981).
7. P. Feulner, W. Riedl and D. Menzel, Phys. Rev. Lett. 50:986 (1983).
8. D. Menzel and R. Gomer, J. Chem. Phys. 41:3311 (1964); P.A. Redhead, Canad. J. Phys. 42:886 (1964).
9. W. Brenig, Z. Physik B23:361 (1976).
10. M. Cini, Solid State Commun. 24:681 (1977); G.A. Sawatzki, Phys. Rev. Lett. 39:504 (1977).
11. P.J. Feibelman, in DIET-I, p. 61 (ref. 1); D.R. Jennison, E.B. Stechel, and J.A. Kelber, in DIET-II (ref. 1), p. 24.
12. R. Treichler, W. Riedl, W. Wurth, P. Feulner and D. Menzel, Phys. Rev. Lett. 54:462 (1985); and in DIET-II (ref. 1), p. 68.
13. P. Feulner, Ph. D. thesis, p.62, Techn. Univ. München 1980; dto., DIET II (ref. 1), p. 142; Q.J. Zhang and R. Gomer, Surface Sci. 109:567 (1981); E.R. Moog, J. Unguris and M.B. Webb, Surface Sci. 134:849 (1983); Z.W. Gortel, H.J. Kreuzer, P. Feulner and D. Menzel, Phys. Rev. B 35:8971 (1987).
14. P.R. Antoniewicz, Phys. Rev. B 21:3811 (1980).
15. D. Menzel, P. Feulner, R. Treichler, E. Umbach and W. Wurth, Physica scripta T 17:166 (1987).
16. J. Schou and P. Borgesen, Nucl. Instrum. Methods B5:44 (1984).
17. G. Zimmerer, in "Excited State Spectroscopy in Solids", M. Manfredini, ed.; North Holland 1987, p. 37.
18. P. Feulner, T. Müller, A. Puschmann and D. Menzel, Phys. Rev. Lett. 59:791 (1987).
19. See articles by H. Haberland (p. 177), R. Pedrys, D.J. Dostra and A.E. de Vries (p. 180), and M.L. Brown, C.T. Reimann and R.E. Johnson, (p. 199) in DIET-II (ref. 1); F. Coletti, J.M. Debever and G. Zimmerer, J. Physique Lett. 45:L-467 (1984).
20. G. Zimmerer et al., to be published.
21. See the articles by N. Itoh in DIET-I, p. 229 (ref. 1) and by N.H. Tolk et al. (p. 152), and M. Szymonski et al. (p. 160) in DIET-II (ref. 1), and references therein.
22. N.H. Tolk et al., Phys. Rev. Lett. 46:134 (1981); dto., DIET-III (ref. 1).
23. R.E. Walkup, Ph. Avouris, and A.P. Ghosh, Phys. Rev. Lett. 57:2227 (1986).
24. M.L. Knotek and P. Feibelman, Surface Sci. 90:78 (1979).
25. R.L. Kurtz, R. Stockbauer and T.E. Madey, DIET-II, p. 89 (ref. 1), and DIET-III.
26. See e.g. "Principles, Applications, Techniques of EXAFS, SEXAFS and XANES", R. Prins, H. Konigsberger, eds. (Wiley, New York 1985).
27. J. Stöhr, R. Jaeger, S. Brennan, Surface Sci. 117:503 (1982); J. Stöhr, in ref. 26,
28. R. Jaeger, J. Feldhaus, J. Haase, J. Stöhr, Z. Hussain, D. Menzel and D. Norman, Phys. Rev. Lett. 45:1870 (1980); R. McGrath, I.T. McGovern, D.R. Warburton, G. Thornton, D. Norman, Surface Sci. 178:101 (1986).
29. R. Jaeger, J. Stöhr, and T. Kendelewicz, Phys. Rev. B28:1145 (1983) and Surface Sci. 134:547 (1983).
30. See e.g. T.E. Madey and J.T. Yates, Jr., Surface Sci. 63:203 (1977); T.E. Madey et al., DIET-I, p. 120; and DIET II, p. 104 (ref. 1); R. Stockbauer, Nucl. Instrum. Methods 222:284 (1984).
31. J.J. Czyzewski, T.E. Madey, and J.T. Yates, Jr,, Phys. Rev. Lett. 32:777 (1974).
32. W.L. Clinton, Phys. Rev. Lett. 38:965 (1977).

33. Z. Miskovic, J. Vukanic and T.E. Madey, Surface Sci. 141:285 (1984).
34. W.Riedl and D. Menzel, Surface Sci.163:39 (1985).
35. H. Kuhlenbeck, M. Neumann and H.J. Freund, Surface Sci. 173:194 (1986).
36. J.C. Fuggle, E. Umbach, P. Feulner and D. Menzel, Surface Sci. 64:69 (1977); and references therein.
37. "Cryomicroscopy and Radiation Damage", E. Zeitler, ed. (North Holland, Amsterdam 1984).

STRUCTURE AND CATALYTIC ACTIVITY OF SURFACES

Vladimir Ponec

Leiden University, Gorlaeus Laboratories
P.O.Box 9502, 2300 RA Leiden, The Nederlands

INTRODUCTION

In the last century and actually up to World War II, the power and wealth of industrialized countries could be measured by their steel and coal mining industries. For the period which followed, this function can be ascribed to the chemical and electronic industry and the wealth of the country is very well reflected by the level of the country's agriculture. The latter is again related to the agricultural-chemical industry which as well as the manufacturing of new or currently used materials of the electronic industry is dependent on many chemical processes.

It is only a very rough estimate, but probably ~70% of all chemical processes used has something to do with a phenomenon called "CATALYSIS". Typical catalytic processes are: production of sulphuric and nitric acid, ammonia synthesis, oxidation of olefines into ethylene-oxide, ammoxidation of propylene into acrolein, numerous polymerization processes and the whole production of gasoline from the crude oil is also "catalytic".

Catalysis is a kinetic phenomenon. Let us consider a hypothetic reaction of ν_A molecules of A and ν_B molecules of B giving a product $\nu_P P$

$$\nu_A A + \nu_B B \rightleftarrows \nu_P P \ . \tag{1}$$

This reaction achieves finally an equilibrium which is described by an equilibrium constant K. For a reaction at not too high pressure, K can be expressed by partial pressures

$$\ln K = \nu_P \cdot \ln(P_P) - [\nu_A \cdot \ln(P_A) + \nu_B \cdot \ln(P_B)] = -\Delta G^0 / RT \tag{2}$$

where G^0 is a standard free enthalpy (a change from P_A, P_B, P_P each equal 1 bar, to the equilibrium pressures). By definition: $\Delta G^0 = \Delta H^0 - T\Delta S^0$, being related to the enthalpy and entropy changes due to the reaction. A catalyst changes the rates $\vec{r}$ and $\overleftarrow{r}$. Since a catalyst (by definition) does not participate in eq.1 it does not change the thermodynamic parameters ΔG^0 and K. If the catalyst forms one phase with the reactants we call it a homogeneous catalyst, if it forms a separated phase, it is a heterogeneous catalyst. From now on we consider only heterogeneous catalysts.

In, probably, more than 90% practical cases, a heterogeneous catalyst works as follows. The gaseous or liquid components of a catalytic reaction are first adsorbed on the solid surface of the catalyst and in many cases transformed into different species by dissociation and rearrangements of certain bonds. Various adsorbed complexes then interact with each other, giving desorbable products which are observable by analytical means (gas-chromatography, mass spectrometry, etc.) in the gas or liquid phase. This is most frequently an easier way (catalytic reactions run under lower temperatures) leading to a smaller group of products (i.e. the selectivity is higher) and the overall reaction can be better controlled, than when the reactions run without any catalyst. Some reactions are even virtually impossible without any catalyst.

ADSORPTION COMPLEXES

The very essence of catalysis is thus the formation of proper adsorption complexes and, of course, in the past, all kinds of spectroscopies have already been mobilized to characterize the adsorption complexes: electronic spectra, vibrational spectra, magnetic resonance spectra, Mössbauer spectra, etc. It should not be forgotten that there are also important chemicals methods to characterize the adsorbed complexes and one of them (exchange reactions) is mentioned below.

With regard to the physical methods - a short remark. When an adsorption complex is visible, by one or another spectroscopy, it does not necessarily means that this species is an active intermediate of the catalytic reaction. It can be also a species at the end of a dead end street. This does not mean again that information on it is of no value at all, it only means that one has to think about which species are actually monitored by the techniques used.

Let us confine ourselves in what follows to metallic catalysts. Adsorption complexes then comprises several atoms originating from the reactant molecules and one or more atoms of the metal surface. We note immediately a very interesting problem: "how many metal atoms are involved directly, by forming chemical bonds, or indirectly by staying as ligands round the central coordinated site (consisting of one or more atoms), in the formation of catalytic intermediates or other adsorption complexes ?" Examples of information about this problem are presented below.

When more than one metal atom is involved in the formation of an adsorption complex, the structure, the geometrical arrangements of the surface, obviously plays a role. However, this is also true, when even only one single metal atom is involved, since properties of the surface atoms are also dependent on the coordinating (ligand) atoms in the nearest neighbourhood.

There are several ways to get hold of this aspect of the formation of complexes and to study the interaction of reaction components on varying surface structures. We shall review here three of the techniques.

FIELD EMISSION MICROSCOPY

For some reactions valuable information can be obtained by Field Emission Microscopy, introduced by the great German physicist E.W. Müller[1]. Since this technique is not a subject of other chapters in this book, a short description of it might be helpful[2].

302

The sharp tip (T in fig.1) 50 to 100 nm diameter is in most cases formed from a single crystal and it can be easily modelled by a hemisphere intersecting the single crystal. Under a high voltage, electrons tunnel into vacuum and produce on the screen a point projection of various crystallographic planes. A picture of this projection with the several planes identified (by comparison with a model) is shown in fig.2. Emission current is higher on places with lower work function ϕ. Work function differences $\Delta\phi$ mean that the most stables planes (with highest density of atoms) are dark and the changes caused by adsorption or reaction are thus most easily detected on the rough planes (due to the Smoluchowski effect they have the lowest ϕ)[3]. Techniques have been developed which allows one to follow $\Delta\phi$ caused by adsorption or reaction, on small spots on the surface, including the dark regions[5].

Let us mention at least briefly, some results obtained by this technique[5-8]. Oxygen adsorption follows an order in the reactivity of various planes which reflects the strength of the M-O chemisorption bond (M metal). The strength is the higher, the higher the coordination number of metal atoms which can potentially surround an - O - atom[6,7]. The reactivity of the M-O layer towards hydrogen follows just the opposite order, indicating that the relation between catalytic activity and chemisorption bond strengths is an antipathic one. The picture becomes

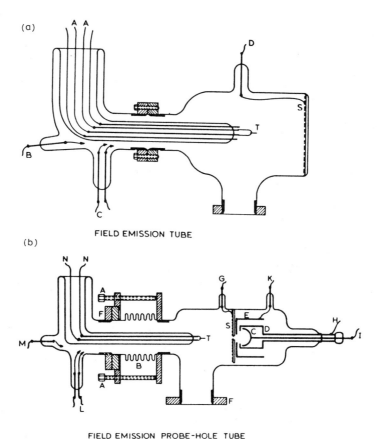

Fig. 1. (a) Field emission tube. (b) Field emission tube with a "probe-hole" facility. T, tip; S, screen; C, collector of electrons (from ref.2, with permission).

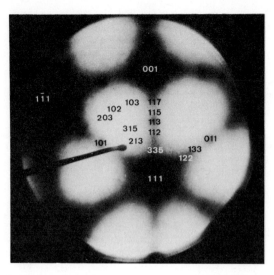

Fig. 2. F.E.M. picture of an Rh tip; various regions indexed (from ref.4, with permission).

more complex with a sligthly more complicated molecule such as $NO[8]$; the order in activities in which NO is dissociatively adsorbed on Rh is, for instance:

Rh(321) > Rh(331) > Rh(210) > Rh(100) > Rh(111)   .

Obviously, other factors, such as electronic structure of surfaces, play a role here in addition to the coordination of the adsorbed species.

ADSORPTION AND CATALYSIS ON SINGLE CRYSTAL PLANES

The development of the vacuum technique in the fifties made another approach possible in attacking the problems of crystallographic plane specificity (crystallographically different adsorption sites are assumed to vary in parallel): the use of well defined crystallographic planes cut out of single crystals. Not only the most stable lowest index planes can be prepared but also various high-index planes, with a varying population of steps, kinks, corners, etc.. In the beginning, these studies where limited to low pressures, but soon, Somorjai and his colleagues (a School which pioneered all this research) developed a technique which combines the preparation and characterization of well defined planes in UHV chambers, with the following of reactions at rather high pressures, up to several bars[9-10].

The first papers reported extreme differences in activity among various planes claiming that the flat planes are totally inactive[11]. Later research did not confirm those conclusions in all details, but important differences among different crystallographic planes have indeed been found.

Figure 3 can serve as an illustration of the results of such studies; but first a few words on the subject.

Hydrocarbon reactions on Pt form a complicated network with a typical pattern of products (selectivity) for each metal. With hexane as an example:

$$H_2 + C-C-C-C-C-C$$

C-C-C-C-C,   C-C-C-C-C (with methyl branches)
(isomerization)

C-C-C (with branches)
(dehydrocyclization, aromatization)

C, C-C, C-C-C, C-C-C-C, C-C-C-C-C
(hydrogenolytic cracking)

$$(3)$$

(hydrocarbons represented by the carbon-skeletons, hydrogen atoms omitted). A close inspection of product patterns and the use of labelled ($^{13}C-,$) molecules revealed[13,14], that the same products of isomerization can be formed from various adsorption complexes (involving 3 or 5 carbon atoms) or, through a consecutive reaction, via the gas phase: desorption and readsorption of methyl-cyclopenthane. It has been further established that there are several clearly distinguishable mechanisms of hydrogenolytic cracking, etc.. As can be deduced from fig.3, formation and reactivity of intermediates is related to the geometry of the adsorption sites.

Important catalysts of the reactions (reforming of naphta in the production of high octane number gasoline) modelled by scheme (3) are "bimetallics" (sometimes alloys). The most important bimetallics are the Pt/Re, Pt/Ir and Pt/Sn[15,16] combinations and a lot of fundamental work has been also done with model alloys such as Ni/Cu, Pt/Au, Pd/Ag, Pd/Au, Ru/Cu, etc.. (For review, see refs 15-19).

The importance of the surface geometric structure has been recently demonstrated again, but now with alloys. Somorjai et al.[20] studied the n-hexane skeletal reaction on the Au-Pt(111) and Au-Pt(100) planes. With

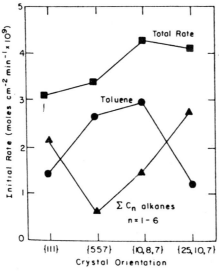

Fig. 3. Heptane conversion on various single crystal planes. Activity expressed as initial rate at 573 K (from ref.12, with permission).

the first mentioned plane, the isomerization rate (per Pt atom available at the start of the reaction) increases appreciably and monotonicaly with increasing Au content, although Au itself is inactive for this reaction, while with (100) this increase is only moderate. In contrast with this, hydrogenolysis decreases with increasing Au, for both planes. The authors conclude that alloying of Pt with Au in the (111) fcc plane symmetry, creates sites particularly suitable for isomerization.

Hydrocarbon reactions mentioned above are examples of structure sensitive reactions. However, not all hydrocarbon reactions belong to the groups of structure sensitive reactions. Hydrogenation of C=C bonds is a much less sensitive reaction[21] and there are also some other reactions, which when macroscopically monitored (i.e. by measuring the net overall rates), manifest themselves as structure insensitive reactions. Oxidation of CO seems to be one of them[21].

Let us mention already at this point that a surprisingly low structure or particle size sensitivity (see below) of certain reactions, has raised a question whether in all cases the sensitivity (or insensitivity) established experimentally, is an intrinsic property of a reaction, or is, perhaps, induced by some side effects. For example, with a molecule of di-tertiarybutyl-acetylene (in which the triple bond C≡C is sterically screened by bulky tert-butyl groups and yet its hydrogenation is particle size insensitive) a possibility has been discussed that the reacting molecule extracts a metal atom with which it reacts, out of the surface, making the original surface structure irrelevant for the running of the reaction[22].

HOW MANY METAL ATOMS CONSTITUTE AN ADSORPTION SITE ?

This is by no means a purely academic question. It is a question of great importance for a rational way of catalyst preparation. However, the fundamental value of an answer is of course, not negligible. The modern theory of chemical bonding in various species, etc. of molecules like benzene started in the last century by establishing the stechiometric ratio of elements: C_6H_6 and by writing down the simplest schematic molecular structure, the Kekulé form. A similar ratio $C_xH_yM_z$ comprising the surface metal atoms M, is not known for most of the adsorption complexes, which dramatically documents the real level of our knowledge on adsorption and catalysis. Needless to say, our current schematic description of "metal-adsorption complexes" bonds is probably even less correct than the Kekulé's form used to be a century ago.

A few molecules form an exception[23] and more is known about them. The CO molecule is one of them. With regard to this molecule quite detailed information on its adsorption site is available. Bradshaw and Hoffmann[24] suggested several conclusions on this subject which are now generally accepted. With Pd(111) as an example:
1) at lowest coverages $\theta(CO)$, CO occupies three-fold sites and absorbs IR at about 1920 cm^{-1} ($\theta < 0.35$).
2) At higher coverages, CO can be shifted out of its optimal position into the two-fold (bridge) sites; such sites are also populated on Pd(100); CO in these sites absorbs IR at 1880-2000 cm^{-1}.
3) At coverages $\theta(CO)$ nearing saturation (calculating θ in surface M-atoms this happens at $\theta > 0.5$) single coordinated CO appears with $\nu = 2050$-2100 cm^{-1}. Structures corresponding to various LEED patterns are shown in fig.4.

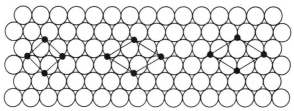

Fig. 4. CO adsorbed on Pd(111). Various coordinations of CO indicated, occurring at various surface coverages (see the text) (from ref.24, with permission).

PARTICLE SIZE EFFECTS

With only a very few exceptions (Fischer Tropsch synthesis of hydrocarbons, ammonia synthesis), the industry does not use "bulk" metals. Since catalysis is a surface phenomenon, one tries to enlarge the surface of the metal (i.e. tries to disperse it) as much as possible, until almost all metal atoms in the system are in the surfaces. Such systems must be prepared and then protected against the loss of surface area by aggregation and by sintering. To this end the metals are dispersed over the surface of "supports", such as SiO_2, Al_2O_3 and other oxides or other compounds. One can prepare metals with D, dispersion (= number of metal atoms in the surface/number of all metal atoms) nearing unity. The question immediately raises: by increasing dispersion do we only change the size of particles, keeping the metal chemically the same or, in other words, do the metal atoms of small particles react in the same way as those of the bulk metals? The answer is negative (for review, see 16,26).

Many reactions are particle size sensitive, the hydrocarbon skeletal reactions among them[16]. Although the question of the particle size sensitivity is at a first glance rather straitforward, it took quite a long time to learn to appreciate the particle size effects and to acknowledge in proper proportions all side effects of it.

It is a simple exercise to imagine how small particles grow, going from 1 to 2, 3, 4 and more and more metal atoms[27]. By involving rules minimizing the energy (maximizing the bonds and "bond-energy") one can derive the probable shapes of the smallest possible particles. This simple procedure has lead to several very interesting results[27].

Formation of a metal in a fcc structure probably follows this sequence. Three atoms form obviously a triangle, four atoms a tetrahedron. Then, the straitforward growth of fcc structure is interrupted and five atoms form a trigonal bipyramid. This grows further into a tripyramid and a seven atoms pentagonal bipyramid. Further growth follows the line to the 13 atom icosahedron, when coming atoms are regularly placed above the existing faces of the bipyramid. In the same way the "complete" bodies are formed as: the 33 atom dodecahedron (made from pentagonal planes) and the 55 atom icosahedron [all planes (111) like].

Some of these smallest particles with unusual structure have been already seen by electron microscopy or by EXFAS[28,29]. The larger particles are probably all shaped into bodies reminding of a sphere-cubo-octahedrons.

Needless to say, further analysis by EM and HREM of problems like the

Table I. Size and surface properties of stable platinum crystals

Edge length		Total No of atoms in crystal	Fraction of atoms on surface	Relative No of diatoms in different states on the surface			Average coordination No of atoms on the surface
No of atoms	$\overset{o}{A}$			in vertices	on edges	on facets	
2	5.5	6	1	1	0	0	4
3	8.25	19	0.95	0.33	0.67	0	6.0
4	11	44	0.87	0.16	0.63	0.21	6.94
5	13.75	85	0.78	0.09	0.55	0.36	7.46
6	16.50	146	0.7	0.06	0.47	0.47	7.76
7	19.25	231	0.63	0.04	0.41	0.55	7.97
10	27.5	670	0.49	0.02	0.29	0.69	8.31
15	41.25	2 255	0.35	0.01	0.20	0.79	8.56
18	49.5	3 894	0.3	0.005	0.17	0.83	8.64
30	82.5	18 010	0.19	0.002	0.10	0.90	8.78
50	137.5	36 484	0.13	0.0006	0.06	0.94	8.87
∞	∞	∞	0	0	0	1	9

N.B. Calculations made for octahedral crystals. More exact calculations involving "unfinished" crystals, cubo-octahedrons, etc. do not differ appreciably.

shape of smallest particles themselves, the relation between shape and epitaxy, shape and the glueing layer between metals and carriers (oxide, hydroxide, silicate layers, etc.) and similar problems, is still highly desirable. It is interesting to notice once more that there are already data available showing that the geometry of the smallest particles follows indeed, the indicated rules[27-29].

Particles smaller than a certain limit (arbitrary put e.g. at 4.0 nm) expose the low coordination sites in excessive amounts such that their (possibly) different catalytic behaviour should be observable (see table 1, ref.30).

Small particles of some transition metals adsorb hydrogen to a higher ratio: H_{tot}/M_{metal} than unity, as shown by Kip et al.[31]. It means that the "valency" of some surface atoms of small metal particles of metals such as Rh or Ir is different from that of atoms constituting macroscopic flat planes. This is a very interesting finding but unfortunately it also casts some doubt on the determination of the total metal surface area (and from that, by using models - the particle size -) determination with supported metals[26]. This discovery[21] demonstrates clearly how important it is to determine the particle size by other than adsorption methods, e.g. by Electron Microscopy or by EXFAS.

LOW COORDINATED SITES

Discussion on the role of low coordinated sites and on electronic structure effects induced by diminishing the particle size (this affects also atoms on the planes) was not started yesterday.

Already in the fifties, vivid discussions took place between the school of Roginskij[32] and Kobozev[33] on one side and the school of

Boreskov[34] on the other side; while the first mentioned scientists claimed a decisive role of the special (low coordinated) sites in all catalytic reactions, Boreskov stated that the steady state activity of metals in any reaction is a function of the electronic structure of metal atoms and varies with the metal particle size only marginally (within one order of magnitude in rates). Important progress in these matters was achieved by Boudart[26,35], who recognized that there are two types of reactions:

I. structure insensitive, or better to say now, particle size insensitive, and
II. particle size sensitive reactions.

Simple hydrogenation reactions are an example of the first group, hydrocarbon reactions or ammonia synthesis (here the dissociative adsorption of N_2 is the sensitive step) of the second one[36].

It has been shown that the smallest metal particles have an energy spectrum which is rather different from the energy spectrum of bulk metals[37-39]. What is not known with the desired certainty is, whether this is also reflected by a substantially different catalytic behaviour of a metal atom in small and large particle surfaces.

There have been many speculations in the literature about the possible effects of the so called "electron deficiency" of small particles. It is in principle possible (although not yet proven) that when a metal particle is placed above an electron acceptor centre, such as Al in SiO_2 structure, an electron is transferred from a metal particle to the support, making the particle positively charged. However, some people go so far that they expect a "catalytic transmutation" (according to them) by electron transfer. A particle of Pd should behave more like Rh, that of Pt more like Ir, Au should became active as Pt, etc.[40,41]. This matter has been discussed elsewhere[42,43] and thus only very briefly here: a charge in a metal particle is very limited, it is localized and screened and it should not substantially influence the reactivity of all other metal surface atoms. Catalytic "alchymie" is not possible.

SIDE EFFECTS OF THE PARTICLE SIZE VARIATIONS

Small metal particles behave differently from large ones in many reactions, but this does not always mean a size sensitivity of the main reaction. This fact has been correctly recognized already by Katzer[44], who also supplied the first examples showing that differences in the behaviour towards side reactions simulate a particle size sensitivity of a catalytic reaction followed (in his case, oxidation of ammonia and/or ethylene).

Lankhorst et al.[45] showed, that the skeletal reactions of hexane [see scheme (3) above] are characterized by an activity and a selectivity which does not vary too much in the critical range of particle sizes (1.8 – 7.0 nm), when the particles are kept clean. However, when catalysts with different particle sizes were subjected to selfpoisoning (deposition of carbonaceous non-reactive layers formed from hexane), the activity (remaining activity) of small particles was much higher than that of the large ones. Also selectivity changes can be induced by that effect. Obviously, it is the behaviour towards a side reaction which induces a particle size sensitivity of the catalysts in hexane skeletal reactions.

Since the role of side reactions (in the last mentioned case, the selfpoisoning by carbonaceous layers) varies in its importance with various metals, and it can combine with the properties of a metal, one can expect all types of correlations of the catalytic activity with the particle size sympathetic or antipathic ones or functions with a maximum[16]. When going from one metal to another, or from one reaction to another it is the task of science to establish, which is the intrinsic (not induced by side effects) correlation with a given reaction.

MULTIPLE BOND FORMATION

Reactions of hydrocarbons are initiated usually by dissociation of the first C-H bond. However, when thereafter a C-C bond has to be split, probably (at least) two H atoms have to be removed, to enable the splitting to run with a low activation energy:

$$
\begin{array}{c}
\mid\ \mid \\
-C-C- \\
\mid\ \mid \\
*\ \ *
\end{array}
\ \rightarrow\
\begin{array}{c}
/ \\
-C \\
\parallel \\
*
\end{array}
\ +\
\begin{array}{c}
\backslash \\
C- \\
\parallel \\
*
\end{array}
\qquad\qquad (4)
$$

There are good reasons to believe that both isomerization and dehydrocyclization can profit from the external multiple bond (external formation means "outside the molecule") formation, although in the latter cases, the internal multiple bond formation might be of greater importance[16,46].

In any case hydrogenolytic cracking is most likely most dependent of all on the formation of metal carbon multiple bonds. In compliance with this expectation Kemball[47] and Van Broekhoven[48] established a very clear correlation between the propensity of metals to form multiple bonds (a) and the reactivity of metals in cracking (b)

(a) Ru < Ni < Co < Ir < Rh << Pt < Pd
(b) Ru ≃ Ni ≃ Co < Ir < Rh < Pt, Pd

Let us mention that the propensity of a metal to form metal-carbon multiple bonds can be, for example, determined by monitoring an exchange reaction of a hydrocarbon with deuterium. CH_4 and cyclopentane have already been used to this end[48].

Van Broekhoven decided to check whether the formation of multiple bonds is particle size sensitive. He used methane/D_2 exchange for which the ratio of CD_4/CH_3D in the initial product distribution can be taken as a measure of the multiple bond formation. The results are shown in fig.5. It is evident that small particles of various metals form multiple bonds less readily, than large ones.

In compliance with the low activity in cracking, the small particles form less readily carbonaceous layers and the curved surface is more difficult to be covered by carbonaceous layers than the flat planes. This explains the finding by Lankhorst et al.[45].

Let us mention in passing that the smallest particles of various metals have been prepared by filling the cages of zeolites by metal precursor solution, drying (oxidation of ligands) and reduction. An EM photo of such materials is shown in fig.6.

Fig. 5. Propensity of metals (here characterized by the CD_4/CDH_3 ratio measured at standard conditions) to form metal-to-carbon multiple bonds, as a function of the particle size. Pt, Ir and Ni (from ref.49, with permission).

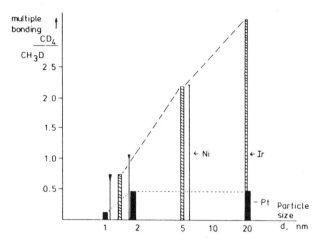

Fig. 6. Pt in zeolites; HREM by H.W. Zandbergen, Leiden University, (1987). The regular zeolite lattice has been partially removed by electron bombardment, in order to visualize better the Pt particles.

CONCLUSIONS

Practically used catalysts consist often of small particles. It is important to study the structure, shape, thermal behaviour, etc. of these particles. Surface structure, the local site geometries, are important for catalysis, as demonstrated by several examples above. In the studies mentioned, EM and EXFAS are very important tools.

REFERENCES

1. E.W. Müller, Ergeb. Exact. Naturwiss., 27:290 (1953);
   R. Gomer, Adv. Catal., 7:93 (1955);
   E.W. Müller and T.T. Tsong, Field on Microscopy, Elsevier, N.Y. 1969.
2. B.E. Nieuwenhuys, Diss. Thesis, Leiden University, 1974.
3. R.V. Culver and F.C. Tompkins, Adv. Catal., 11:67 (1959);
   C. Herring and M.H. Nichols, Rev. Mod. Phys., 21:185 (1949).
4. V.V. Gorodetskii and B.E. Nieuwenhuys, Surf. Sci., 105:229 (1981).
5. B.E. Nieuwenhuys and W.M.H. Sachtler, Surf. Sci., 34:317 (1973);
   Colloid Interface Sci., 58:65 (1977).
6. V.V. Gorodetskii, B.E. Nieuwenhuys, W.M.H. Sachtler and Y.K. Boreskov,
   Surf. Sci., 108:225 (1981).
7. B.E. Nieuwenhuys, Surf. Sci., 126:307 (1983).
8. H.A.C.M. Hendrickx and B.E. Nieuwenhuys, Surf. Sci., 175:85 (1986).
9. G.A. Somorjai, "Chemistry in two dimensions, Surfaces", Cornell Univ.
   Press, Ithaca, 1981.
10. G.A. Somorjai, Catal. Rev., 18:173 (1978); Adv. Catal., 26:1 (1977).
11. R. Lang, R.W. Joyner and G.A. Somorjai, Surf. Sci., 30:454 (1972).
12. S.M. Davis, F. Zaera and G.A. Somorjai, J. Catal., 85:206 (1984);
    J. Am. Chem. Soc., 104:7453 (1982).
13. E.H. Van Broekhoven and V. Ponec, Progress in Surf. Sci., 19 (4)
    (1985).
14. F.G. Gault, Adv. Catal., 30:1 (1981).
15. V. Ponec, Catal. Rev., 11(1):1 (1975).
16. V. Ponec, Adv. Catal., 32:149 (1983).
17. J.H. Sinfelt, Accounts Chem. Res., 10:15 (1977).
18. J.K.A. Clarke, Chem. Rev., 75:291 (1975).
19. R. Burch, Accounts Chem. Res., 15:24 (1982).
20. R.C. Yeates and G.A. Somorjai, J. Catal., 103:208 (1987).
21. T. Engel and G. Ertl, Adv. Catal., 29:2 (1979);
    M. Boudart, A. Aldag, J.E. Benson, N.A. Dongharty and C.G. Harkins,
    J. Catal., 6:92 (1966);
    D.R. Kahn, E.E. Peterson and G.A. Somorjai, J. Catal., 34:294 (1974).
22. R.L. Burwell, H.H. Kung and R.J. Pellet, Proc. 6th Int. Congr. on
    Catal., London, 1976; Chem Soc. London (1977), Vol.1, p.108.
23. A. Gavezotti, M. Simonnetta, M.A. Van Hove and G.A. Somorjai, Surf.
    Sci., 122:292 (1982).
24. A.M. Bradshaw and F.M. Hoffmann, Surf. Sci., 72:513(1978).
25. A.M. Bradshaw and F.M. Hoffmann, Surf. Sci., 52:449 (1975).
26. M. Boudart, J. Mol. Catal. Review Issue, 1986, p.29.
27. J.J. Burton, Catal. Rev. Sci. Eng., G-209 (1974).
28. J.G. Allpress and J.V. Sanders, Surf. Sci., 7:1 (1967).
29. P. Gallezot, A. Bienenstock and M. Boudart, Nouveau J. Chim., 2:263
    (1978);
    P. Gallezot, Catal. Rev. Sci. Eng., 20:121 (1979).
30. O.M. Poltorak, V.S. Borozin and A.N. Mitrofanova, Proc. 4th Int.
    Congr. on Catal., Moscow, 1968; Akademiai Kiado, Budapest (1971),
    p.276.
31. B.J. Kip, F.B.M. Duivenvoorden, D.C. Koningsberger and R. Prins, J.
    Am. Chem. Soc., 108:5633 (1986).
32. S.Z. Roginskij, Problemy Kin. Katal., 7:72 (1949) (in Russian);
    S.Z. Roginskij, in "Heterogeneous Catalysis in Chem. Industry"
    (in Russian), Goschimizdat 1955.
33. N.I. Kobozev, Zhur. Fiz. Kchim., 13 (1939) (in Russian); Usp. Kchim.
    No.5, 546 (1956) (in Russian);
    A.A. Balandin, Adv. Catal., 19:1 (1969).
34. G.K. Boreskov, in "Preparation of Catalysts", B. Delmon, P.A. Jacobs
    and G. Poncelet eds, Elsevier, 1976, p.223 (and refs therein).
35. M. Boudart, Proc. 6th Int. Congr. on Catal., London, 1976, p.2.

36. G.A. Somorjai, Proc. 8th Int. Congr. on Catal., Berlin, 1984, Vol.I, p.113.
37. P.P. Messmen, Surf. Sci., 106:225 (1981) and refs therein.
38. R.C. Baetzold, Surf. Sci., 106:243 (1981) and refs therein.
39. M. Cini, J. Catal., 37:187 (1975).
40. R.A. Dalla Betta and M. Boudart in "Catalysis", Proc. 5th Int. Congr. on Catal., Miami Beach, 1972, (ed. J.W. Hightower), North-Holland, Amsterdam, 1971, Vol.2, p.1329.
41. G.C. Bond, Gold Bulletin, 6(4):102 (1973).
42. V. Ponec, in "Metal. Support and Metal-Additive Effects in Catalysis", B. Imelik et al. eds, Elsevier, 1982, p.63.
43. V. Ponec, Proc. 9th Int. Vacuum Congr. and 5th Int. Conf. Solid Surf., Madrid, 1973, p.121.
44. J.R. Katzer, J. Catal., 32:166 (1974);
    J.J. Ostermaier, J.R. Katzer and M.H. Manoque, J. Catal., 41:277 (1976).
45. P.P. Lankhorst, H.C. de Jongste and V. Ponec, in "Catalyst Deactivation",
    B. Delmon and G.F. Froment eds, Elsevier, 1980.
46. O.E. Finlayson, J.K.A. Clarke and J.J. Roney, J. Chem. Soc. Faraday Trans. I, 80:191 (1984).
47. C. Kemball, Cat. Rev., 5:33 (1971).
48. E.H. van Broekhoven and V. Ponec, J. Mol. Catal., 25:109 (1984).
49. E.H. van Broekhoven and V. Ponec, Surf. Sci., 162:731 (1985).

313

SUBJECT INDEX

Aberration, 31 et seq., 60, 198,
        204, 207, 228
Adsorbate, 287
    desorption, 286 et seq.
    scattering, 212
Adsorption, complexes, 302
    sites, 306
AES, 89 et seq.
Atomic Force Microscope, 255 et
        seq.
Attenuation length, see Mean free
        path
Auger
    angular distribution, 229
    damage processes, 27
    decay, 291, 292
    excitation process, 51, 294
    imaging and spectroscopy, 26, 108,
        225 et seq.
    and oxide studies, 52
    and REM imaging, 133
    scanning microscopy, 17, 108
    signal and surface
        cleanliness, 70
    spectroscopy, 89 et seq.
    yield, 225

Backscattering, 6, 209-212, 226, 229
    and chemical effects, 6, 204
    and 7x7 structure, 67
Bloch waves, 6, 141

Carbonaceous layers, 309
Catalysis, definitions 301, 302
Catalyst
    Ag, 12, 25
    exhaust, 91
    MoS_2, 12
    particle shapes, 11
    Pt on C (STEM images), 26
    Ru, 15
    selectivity, 302
    spinel, 49
    structure (in)sensitivity, 306

Channeling
    and orientation contrast, 6, 16,
        204, 205
    and 7x7 structure, 67
    and surface resonance, 136, 146
Chemical shift
    in EELS, 16, 94, 153
    in XPS, 93, 94
Coincidence techniques, 99
Column approximation, 7
    and REM contrast, 141
Contrast transfer, see Aberration
Convergent beam diffraction
        in REM, 131
Corrosion, 90

Damage
    electron stimulated
        desorption, 52, 285 et
        seq., 296
    ionization, 26-28
    in CdS, 51
    in REM, 152
    in STM, 253
Density of states,
    local from STM, 242, 250, 253
    from XPS, 97
Depth
    escape, 163, 195, 277, 279
    penetration in RHEED, 23, 147,
        162
    profiling, 102, 104
Desorption, 204, 205, 229, 285 et
        seq.
Dielectric excitations, 19, 101
Diffraction
    contrast, 3, 136, 215, 219, 221
    RHEED, 159 et seq.
Diffusion to surface
        steps, 70, 149, 190
Dipole excitations, 22
Dislocations
    misfit, 1, 3
    subsurface, 151

Domains, 217, 219, 220
Dynamic recording, see In-situ
    observations
Dynamical diffraction, 6, 34, 143

EELS
  core loss spectroscopy (CLS), 94
  dielectric excitations, 19, 101
  spatial resolution, 15
  valence losses, 287
Electron emission, 201 et seq.,
    221 et seq.
Epitaxy, 1, 5
  Au on MgO, 152
  chemical beam, 160
  in-situ observations, 71, 185
  molecular beam, 160
  in multilayers, 83
ESCA, 89 et seq.
Escape depth, see Depth
ESD, 204, 205, 229, 285 et seq.
EXAFS (EXELFS, EELFS), 112, 153, 294

Facetting, 12, 47, 70, 219
Field emission, 195, 302
Forbidden reflections, 4, 17, 44, 66
Foreshortening of REM images, 134
Fresnel
  diffraction, 45
  fringe, 38
  imaging, 84

Graphite, 151
  STM images, 250

High resolution electron
    microscopy, 31 et seq.,
    55 et seq.
  of interfaces, 77 et seq.
Hydrocarbons
  skeletal reactions of, 305, 309

Icosahedral materials, 12, 47, 51,
    80
Image potential, 291
  for high energy electrons, 22
  in STM, 245
Image simulation, 36, 45
In-situ observations, 50-52, 71,
    148, 185 et seq., 216 et
    seq.
Inner potential, 174, 209, 212
Interfaces, 77 et seq.
  characterization, 81, 83
  compositional, 64, 82
  Co/Si, 72
  displacements, 62, 82
  $Ge_{0.5}Si_{0.5}/Si$, 64, 65
  industrial applications, 90
  Ni/NiO, 39

Interfaces (continued)
  $NiSi_2/Si$, 73
  $Si/CoSi_2/Si$, 58
  Si/GaAs, 57
  Si/SiC, 39
  Si/SiO_2, 56
  spectroscopy, 22, 23
  strain relaxation, 62, 83

Kikuchi lines in RHEED, 175
Kinematical theory, 3, 33

LEED, 7, 67, 171, 196, 207 et seq.,
    229,
  damage effects in, 292
Low energy electron microscopy
    (LEEM), 196 et seq.
    resolution in, 199 et seq.

Magnetic effects in spin
    polarization, 267 et
    seq.
Magnetic probing depth, 277, 279
Magnetic resolution, 268
Mean free path
  inelastic, 101, 162, 178, 212,
    226, 228,
  thermal diffuse scattering
    (phonon), 14, 162
  valence excitation, 21, 162
Microanalysis, quantitative, 104 et
    seq.
  in STEM, 14
Minimun Detectable Concentration,
    106
Microdiffraction, 17
Multilayer characterization, 83
Multilayers, see Interfaces
Multislice method, 7, 34, 59-62
  in REM, 143
Multiple bonds, 310

Near edge structure, in EELS, 97

Oxide, 2, 5, 23, 48, 52, 294

Particle
  Auger excitation, 51
  icosahedral, 12, 47, 80
  profile imaging of, 43 et seq.
  size, 307
  size sensitivity, 306, 307
  structural rearrangements, 51
  surface facetting, 47
Phonon scattering, 168
Photoemission
  microscopy, 201-203
  spectroscopy, 67, 71
Plan view images, 57

Plasmon excitation, see Mean free path, inelastic
Profile image, 40, 43 et seq.

Reciprocal lattice, 2, 164, 165
Reconstruction
  of Au, 50
  of GaAs, 46
  and reflection imaging, 136, 137
  of Si (113), 49
  of Si (100), 219
  of Si 7x7, 46, 66, 67, 69, 148, 220
  and transmission imaging, 5
Reflection High Energy Electron Diffraction, see RHEED
Reflection imaging, 44, 45, 127 et seq.
Refraction
  at surfaces, 139
  in RHEED, 172
Resolution (depth), see Depth
Resolution, lateral
  in EELS, 15
  in EXAFS, 112
  in LEEM, 199-201
  in SAM, 108, 225
  in STM, 246-250
  in XPS, 111
RHEED, 159 et seq.
  intensity calculations, 178
  oscillations, 185 et seq.
  refraction, 172
  steps and facets, 171
  streaks, 171

SAM, 108
Scanning tunneling microscopy, 69, 155, 235 et seq.
Secondary electron
  emission, 201 et seq.
  imaging, 25, 153, 200 et seq., 216, 231
  spin polarization, 267 et seq.
  Stoner excitation produced by, 274
Signal to noise ratio
  in Auger imaging, 227
  in HREM images, 46, 66
  in LEEM, 214
Slice method, see Multislice method
Specimen preparation
  for HREM, 78
Spectroscopy
  Auger, see Auger
  electron, 93 et seq.

Spectroscopy (continued)
  electron (continued)
    core loss, 94
    correlation studies, 97
    inelastic tunneling, 243
    in REM, 22, 131, 153
    in STEM, 15
    in STM, 243
    photoemission, see Photoemission
  X-ray
    in STEM, 15, 19
Spherical aberration, 32
Spin-flip, 221
Spin polarization, 267 et seq.
Spinodal decomposition, 79
Stacking fault, 67, 151
STEM imaging, 13, 131
Steps
  contrast effects, 7, 135, 139, 215-221
  effect in RHEED, 187
  influence on nucleation, 148, 219
  profile imaging, 43 et seq., 68
  reflection imaging, 130, 132, 135, 148, 150
  relaxation effects, 150
  transmission imaging, 4, 66, 68
Strain
  contrast at surfaces, 141
  relaxation at interfaces, 62
Streaks in RHEED, 170
Surface
  cleaning procedures, 70, 130, 152, 161
  exciton, 292
  plasmon, 21
  reconstruction, see Reconstruction
  relaxation, 40, 46, 64
  resonance, 136, 139, 146, 175, 212

Tunneling, see Scanning Tunneling Microscopy

Vacuum level and practice, 130, 132, 159 et seq.
Valence excitation, 288

Weak beam dark field, 3, 4, 45
Work function, 97, 196, 201, 203, 204, 238

XPS, 93

Z contrast imaging, 13

Participants included in the photograph

1. E. Bauer, FRG
2. D. Menzel, FRG
3. M.G. Walls, UK
4. N. Thangaraj, India
5. D.J. Smith, USA
6. S.K. Roy, India
7. J.P. Morniroli, France
8. A. Howie, UK
9. A. Casaleggi, Italy
10. J. Lynch, France

11. U. Valdrè, Italy
12. R.I. Fonseca, Brazil
13. E. Bielanska, Poland
14. R. Braglia, Italy
15. R.C. Spinella, Italy
16. M. Vittori, Italy
17. S. Miotkowska, Poland
18. A. Santoni, Italy
19. A. Rivacoba, Spain
20. G. Pozzi, Italy

21. S. De Bondt, Belgium
22. R. Sangiorgi, Italy
23. J. Cazaux, France
24. J. Kirschner, FRG
25. M. Essig, FRG
26. E. Ruedl, Italy
27. F. Phillipp, FRG
28. P.J. Dobson, UK
29. R. Baudoing, France
30. A.-J. Hoeven, The Netherlands

31. A.J. Kinneging, The Netherlands
32. J.N. Ness, UK
33. N.D.R. Goddard, UK
34. R.O. Toivanen, Finland
35. S. Fox, UK
36. A.L. Bleloch, UK
37. S. Johansson, Sweden
38. F.J. Mayer, FRG
39. L. Vazquez, Spain
40. F. Ericson, Sweden

41. P.L. Hansen, Denmark
42. J. Holender, Poland
43. W. Jablonski, Australia
44. D. Ajò, Italy
45. P. Mengucci, Italy
46. M.L. Muolo, Italy
47. I. Salerno, Italy
48. M.E. De Paolini, Italy
49. V.M. Castaño, Mexico
50. M. Vollmer, FRG

51. W. Carillo-Cabrera, USA
52. J.M. Gibson, USA
53. Mrs S. Stobbs, UK
54. W. Telieps, FRG
55. N. Garcia, Spain
56. M.W. Stobbs, UK
57. T. Epicier, France
58. J.M. Cowley, USA
59. R. Engin, Turkey
60. K. Sickafus, USA

61. A. Brokman, Israel
62. V. Marcu, Israel
63. C.A. Walsh, UK
64. M. Pitaval, France
65. G. Bratina, Yugoslavia
66. P. Ferret, France
67. M. Bentzon, Denmark
68. G. Dražič, Yugoslavia
69. S. Tarantino, Italy
70. L. Meda, Italy

71. B. Tøtdal, Norway
72. G. Ceccone, Italy
73. J.F. Bohonek, Switzerland
74. L.H. Chan, USA
75. L. Populin, Italy
76. A. Claverie, France
77. A.J. Bleeker, The Netherlands
78. F.D. Tichelaar, The Netherlands
79. M. Mauronihi, Greece
80. K. Hirose, Japan

81. U.A. Rolander, Sweden
82. L.G. Mattsson, Sweden